Current Aspects of Radiopharmaceutical Chemistry

Current Aspects of Radiopharmaceutical Chemistry

Special Issue Editor

Peter Brust

MDPI • Basel • Beijing • Wuhan • Barcelona • Belgrade

MDPI

Special Issue Editor
Peter Brust
Helmholtz-Zentrum Dresden-Rossendorf, Research Site Leipzig
Germany

Editorial Office
MDPI
St. Alban-Anlage 66
Basel, Switzerland

This is a reprint of articles from the Special Issue published online in the open access journal *Molecules* (ISSN 1420-3049) from 2017 to 2018 (available at: http://www.mdpi.com/journal/molecules/special issues/Radiopharmaceutical Chemistry)

For citation purposes, cite each article independently as indicated on the article page online and as indicated below:

LastName, A.A.; LastName, B.B.; LastName, C.C. Article Title. *Journal Name* **Year**, *Article Number*, Page Range.

ISBN 978-3-03897-162-7 (Pbk)
ISBN 978-3-03897-163-4 (PDF)

Cover image courtesy of Helmholtz-Zentrum Dresden-Rossendorf.

Contents

About the Special Issue Editor

Peter Brust, Prof., Dr., is a biologist. He received his M.S. in Immunology in 1981 and his Ph.D. in Neuroscience from Leipzig University in 1986. He worked as a postdoctoral fellow at Montreal Neurological Institute and Johns Hopkins University, Baltimore, from 1990 to 1991. He joined the Research Center Rossendorf (now known as Helmholtz-Zentrum Dresden-Rossendorf, HZDR) in 1992 and headed the Department of Biochemistry. Since 2002, he has been working in Leipzig, first at the Institute of Interdisciplinary Isotope Research and, after an operational transfer in 2010, again at the HZDR, where he leads the Department of Neuroradiopharmaceuticals. His main research interest is in radiotracer development for brain imaging with positron emission tomography, including brain tumor imaging (glioblastoma, brain metastases), imaging of blood-brain barrier transport of radiopharmaceuticals, and neuroimaging of the cholinergic system, second-messenger systems, and neuromodulatory processes. He has about 250 peer-reviewed publications and owns numerous patents.

Preface to "Current Aspects of Radiopharmaceutical Chemistry"

Positron emission tomography (PET) and single-photon emission computed tomography (SPECT) are in vivo molecular imaging techniques which are widely used in nuclear medicine for the diagnosis and treatment follow-up of many major diseases. They use target-specific molecules as probes, which are labeled with radionuclides of short half-lives, synthesized prior to the imaging studies. These probes are called radiopharmaceuticals. Their design and development require a rather interdisciplinary process involving many different disciplines of natural sciences and medicine. In addition to their diagnostic and therapeutic applications in the field of nuclear medicine, radiopharmaceuticals are powerful tools for in vivo pharmacology during the process of pre-clinical drug development to identify new drug targets, investigate the pathophysiology of diseases, discover potential drug candidates, and evaluate the pharmacokinetics and pharmacodynamics of drugs in vivo. Furthermore, they allow molecular imaging studies in various small-animal models of disease, including genetically engineered animals. The current collection of articles provides unique examples covering all major aspects in the field. The first half, radiopharmacy, is more chemistry-related, while the second half, radiopharmacology, deals with the preclinical development of radiopharmaceuticals.

The largest proportion of positron-emitting radionuclides is commonly produced in particle accelerators, usually cyclotrons. More recently, generator systems, e.g., the ^{68}Ge/^{68}Ga generator, have shown great potential as a source of positron-emitting radionuclides for PET. Gallium-68, which has a relatively short half-life (68 min), is particularly suitable for the labeling of peptides that show rapid target tissue accumulation and clearance. PET investigation of monoclonal antibodies, which represent one of the fastest growing therapeutic groups, requires radionuclides with much longer half-lives. The decay half-life of zirconium-89 (3.3 d) matches the circulation half-lives of antibodies (usually in the order of days); therefore, it emerged as a suitable PET radionuclide for labeling. The review of Bhatt et al. focuses on recent advances in zirconium-89 chelation chemistry.

Another major use of radiometals is in radionuclide therapy of cancer. Radium-223, an alpha-emitting radionuclide, has been approved for the treatment of bone metastasis in metastatic castration-resistant prostate cancer. In vivo generators are thought to combine the long half-life of a parent radionuclide with the high decay energy of the daughter to achieve high-dose targeted radiotherapy. On the other hand, they suffer from the nuclear recoil effect, causing at least a partial release of daughter radioactive nuclei from the targeting molecule or a delivery vehicle. In such cases, an unwanted radioactive burden is spread over the body, and its elimination is limited. The overview of recent developments in this field by Kozempel et al. discusses some pitfalls of this technology mainly related to the nuclear recoil effect.

The radioactive metal technetium-99m is regarded as the workhorse of nuclear medicine because of its ideal imaging properties as a pure gamma emitter and its constant availability as generator nuclide with a half-life of 6 h. For its clinical use in SPECT, thorough quality assurance is important. The impact of different conditions on the radiopharmaceuticals quality has to be verified before administration to humans. The article of Uccelli et al. deals with a minor, previously neglected detail in quality assurance: the influence of the storage temperature on the 99mTc-radiopharmaceuticals.

The short-lived radionuclides carbon-11 and fluorine-18 have the broadest applicability in PET, in particular for the labeling of small molecules which are biologically active on enzymes, receptors, transporters, and other proteins. The development of robust methods for the incorporation of the radionuclides is of high importance. The article of Roslin et al. deals with the urea functional group which is present in many small molecules and is thus an attractive target for the incorporation of carbon-11 in the form of [^{11}C]carbon monoxide.

Labeling with carbon-11 offers the advantage of allowing the radiolabeling of many conventional drugs without influencing their biological activity. Accordingly, this is the method of choice for exploratory studies, while labeling with fluorine-18 is preferred for its broader applicability in clinical routine. The latter requires additional structure–activity studies to evaluate the effect of the fluorine substitution on the biological activity of the labeled molecule. The review of Tago and Toyohara summarizes the results of ^{11}C- and ^{18}F-labeling and the biological evaluation of the PET ligands for imaging histone deacetylases, enzymes which are involved in epigenetic phenomena. The article of Schröder et al. deals with the development of ^{18}F-labelled PET ligands for molecular imaging of the cyclic nucleotide phosphodiesterase 2A, a key enzyme in the cellular metabolism of the second messengers cAMP and cGMP.

The molecular imaging of inflammation with PET and/or SPECT is a useful non-invasive tool to early detect pathophysiological changes in affected human tissues and is regarded to be of particular importance for prognostic purposes, therapy decision-making, and therapy follow-up. For over 25 years, the translocator protein 18 kDa (TSPO), formerly called peripheral benzodiazepine receptor, has been studied as a biomarker of reactive gliosis and inflammation associated with a variety of neuropathological conditions. The article of Vignal et al. describes an optimized radiosynthesis of the TSPO ligand [^{18}F]FEPPA and its evaluation as a PET radiotracer in a mouse model of brain inflammation to facilitate the use of this radiotracer in humans. It is followed by a review of Janssen et al. which describes alternative biological targets that have gained interest for PET imaging of microglial activation over recent years, such as the cannabinoid receptor type 2, cyclooxygenase-2, the P2X7 receptor, and reactive oxygen species.

An important aspect in PET radiotracer development is the characterization of metabolism and metabolites. The investigation of radiotracer metabolism in vivo needs special consideration, especially for neuroimaging. Because of the exceptionally great functional diversity of the brain compared to other organs, there is a need to precisely differentiate between various brain regions with regard to specific radiotracer binding and target density. Therefore, it has to be ensured that the PET image is derived from the radiotracer only and not blurred by the presence of radiolabeled, blood–brain barrier-penetrating metabolites. Consequently, the potential presence of radiometabolites in the brain needs to be investigated and ideally excluded. The article of Ludwig et al. is focused on the LC–MS/MS-aided identification of radiometabolites of (+)-[^{18}F]flubatine, a radiopharmaceutical which has successfully been used to identify deficits in cholinergic transmission in patients with Alzheimer's disease.

Since the development of 6-[^{18}F]fluoro-L-DOPA ([^{18}F]FDOPA) by Günther Firnau in 1984, Parkinson's disease has been a focus of molecular imaging with PET. The degeneration of dopaminergic neurons, which can be monitored with [^{18}F]FDOPA, is accompanied by a complex network of molecular changes in the parkinsonian brain, which involves various enzymes, receptors, transporters, and structural proteins, among them dopamine and sigma-1 receptors. The study of Mann et al. investigates the influence of the cholinergic system on the D_2/D_3 receptor availability in the hemiparkinsonian rat brain measured with [^{18}F]fallypride.

[^{18}F]FDG is the classical example of a universal PET radiopharmaceutical covering imaging tasks in neurology, oncology, cardiology, and other fields. The article by Kranz et al. deals with two PET radiotracers, namely, R- and S-[^{18}F]fluspidine, which originally were developed for use in neurology but later turned out to be useful also for tumor imaging. They bind with high selectivity to the sigma-1 receptor, which acts as a molecular chaperone and is involved in various neurodegenerative disorders and overexpressed in a variety of tumors. The study provides strong evidence that both enantiomers are suitable for the imaging of glioblastoma.

The folate receptor is another target which is highly expressed in many tumor types. Accordingly, it has been intensively investigated for the development of therapeutics and diagnostic agents, including those containing radionuclides. In this regard, lutetium-177-labelled radiopharmaceuticals have the potential for combined therapeutic application and imaging with SPECT. Folic acid-derived radiotracers usually show a high kidney accumulation, which is a major drawback for their therapeutic use. The study of Müller et al. combines the radiofolate [^{177}Lu]cm13 with the antifolate pemetrexed to increase the tumor-to-kidney ratio.

In summary, I regard this Special Issue as an interesting collection of papers with some hidden treasures for radiochemists

Peter Brust
Special Issue Editor

molecules

MDPI

Review

Recent Advances in Zirconium-89 Chelator Development

Nikunj B. Bhatt †, Darpan N. Pandya † and Thaddeus J. Wadas *

Department of Cancer Biology, Wake Forest University Health Sciences, Winston-Salem, NC 27157, USA;
nbhatt@wakehealth.edu (N.B.B.); dapandya@wakehealth.edu (D.N.P.)
* Correspondence: twadas@wakehealth.edu; Tel.: +01-336-716-5696
† These authors contributed equally to this work.

Received: 11 February 2018; Accepted: 9 March 2018; Published: 12 March 2018

Abstract: The interest in zirconium-89 (^{89}Zr) as a positron-emitting radionuclide has grown considerably over the last decade due to its standardized production, long half-life of 78.2 h, favorable decay characteristics for positron emission tomography (PET) imaging and its successful use in a variety of clinical and preclinical applications. However, to be utilized effectively in PET applications it must be stably bound to a targeting ligand, and the most successfully used ^{89}Zr chelator is desferrioxamine B (DFO), which is commercially available as the iron chelator Desferal®. Despite the prevalence of DFO in ^{89}Zr-immuno-PET applications, the development of new ligands for this radiometal is an active area of research. This review focuses on recent advances in zirconium-89 chelation chemistry and will highlight the rapidly expanding ligand classes that are under investigation as DFO alternatives.

Keywords: zirconium-89; chelator; positron emission tomography

1. Introduction

Over the last four decades, molecular imaging has had a transformative effect on the way research is conducted in academia, industry and on how medical care is managed in the clinic [1–8]. Of the modalities available to preclinical researchers and clinicians, the popularity of the nuclear medicine technique positron emission tomography (PET) has surged since it provides physiological data relating to disease pathophysiology, receptor expression levels, enzyme activity and cellular metabolism non-invasively and quantitatively [9–12]. PET imaging relies upon the unique decay characteristics of PET radionuclides, which decay by positron emission, and are chemically attached to ligands designed to probe biochemical phenomena in vivo [13,14]. As the radionuclide decays, it ejects a positron from its nucleus, which after travelling a short distance, undergoes a process called annihilation with an electron to release two 511 keV γ rays 180° apart. These coincident gamma rays have sufficient energy to escape the organism and can be detected by the PET scanner. Computer-based algorithms then convert the signal data into an image that reveals the distribution of the radiotracer within the organism. Historically, PET isotopes such as ^{18}F, ^{15}O, ^{13}N, ^{11}C and ^{68}Ga; which have relatively short half-lives, were developed for use with small molecules or peptides that demonstrated rapid target tissue accumulation and clearance, and facilitated the imaging of physiological processes within the first 24 h of radiopharmaceutical injection [15]. However, researchers engaged in the development of monoclonal antibodies, which represent one of the fastest growing therapeutic groups, were unable to take full advantage of PET as a molecular imaging technique. The aforementioned radionuclides had half-lives incompatible with the biological half-life of an antibody, and made imaging their biodistribution days after injection extremely difficult. While several PET radionuclides such as ^{64}Cu, ^{86}Y and ^{124}I have been used in the development of mAb-based radiopharmaceuticals, they possess undesirable physical, chemical or radioactive properties that have minimized their use [15–17]. For example, ^{64}Cu and ^{86}Y have half-lives, which are incompatible with the slow pharmacokinetics displayed by an antibody in vivo. Furthermore, dehalogenation of ^{124}I-radiolabeled antibodies in vivo coupled with the low resolution images they produce have left the molecular imaging community with little enthusiasm to apply this PET radionuclide for the diagnostic imaging of disease. However, the introduction of zirconium-89 (^{89}Zr) more than three decades ago has reinvigorated this rapidly expanding area of research known as immuno-PET [18,19]. Its impact on antibody and nanoparticle development, clinical trials and precision medicine strategies has been reviewed extensively [14–16,20–46].

2. Zirconium Chemistry and the Production of Zirconium-89

Zirconium, a second row transition metal, was first isolated by Berzelius in 1824 [47], and since that time numerous inorganic and organometallic complexes of Zr have been described with zircon ($ZrSiO_4$), being its most widely recognized inorganic form [48–51]. Zirconium can exist in several oxidation states including Zr(II), Zr(III) and Zr(IV), which is its preferred oxidation state [48]. Zirconium (II) complexes are known, but they typically require p-donor ligands to enhance stability even under inert atmosphere conditions, and even fewer reports describing the Zirconium (III) oxidation state exist. A significant portion of knowledge regarding this element's reactivity has been extrapolated from hafnium (Hf) chemistry since their atomic and ionic properties yield similar chemistries with a variety of ligands, and much of what is known about zirconium coordination chemistry has been discovered in the context of solid-state material or catalysis development [52,53]. While research in these areas has provided numerous societal benefits including heat and corrosion resistant coatings; fracture resistant ceramics; and the development of catalysts that play a role in the petroleum, plastics, and pharmaceutical industries, it has been difficult to translate this knowledge into the research fields of radiochemistry and molecular imaging. The requirements of zirconium complexes in the latter arenas are completely different from the former branches of scientific inquiry. For example, typical catalytic applications require a non-aqueous environment and a zirconium complex with labile ligands [54–62], but for molecular imaging applications, zirconium complexes must be extremely hydrophilic and inert to ligand substitution or loss [14]. Further complicating the exploration of zirconium radioisotopes in molecular imaging is its complex aqueous chemistry [14,16,63–65]. Currently, experimental evidence indicates that due to its high charge and small radius, hydrated Zr(IV) exists as multiple monomeric and polynuclear μ-oxy- and μ-hydroxy-bridged species in solution at low pH. The nature and abundance of these species can change depending upon pH, while an increasing solution pH favors the formation and precipitation of zirconium hydroxide species. This has made the accurate determination of stability constants with various chelating ligands very difficult.

While several isotopes of Zr including ^{86}Zr ($t_{1/2}$: 17 h, γ 100%, E_γ = 241 keV), ^{88}Zr ($t_{1/2}$: 85 d, γ 100%, E_γ = 390 keV), and ^{89}Zr ($t_{1/2}$: 78.4 h, β+ 22.8%, E β+$_{max}$ = 901 keV; 901 keV, EC 77%, E_γ = 909 keV) can be produced on a cyclotron [66,67], ^{89}Zr has received the most attention for radiopharmaceutical development because of its favorable nuclear decay properties that make it useful in the labeling of antibodies for immuno-PET applications (Figure 1) [68–70]. The availability of carrier-free ^{89}Zr as either zirconium-89 oxalate ([^{89}Zr]Zr(ox)$_2$) or zirconium-89 chloride ([^{89}Zr]Zr Cl$_4$) is essential to the development of effective immuno-PET agents. Link et al. were the first to produce ^{89}Zr by a (p,n) reaction by bombarding ^{89}Y foil with 13 MeV protons [18]. After irradiation, ^{89}Zr was purified by a double extraction protocol followed by anion exchange and elution with oxalic acid to afford ^{89}Zr (as [^{89}Zr]Zr(ox)$_2$) in an 80% yield and with a purity greater than 99%. Although incremental improvements were made in the production and purification of ^{89}Zr soon after that [71,72] a major advance in ^{89}Zr production was reported by Meijs and coworkers, who were able to produce ^{89}Zr using the (p,n) reaction and 14 MeV protons produced on a Philips AVF cyclotron [73]. After oxidation of the target material, other metal impurities were removed by anion exchange chromatography using a hydroxamate-modified resin, which was chosen because of this coordinating unit's ability to form complexes with ^{89}Zr(IV) under highly acidic conditions. This allowed the ^{89}Zr to be retained within the column while the other metal impurities were removed under low pH conditions. The purified ^{89}Zr was then eluted in 95% yield using 1 M oxalic acid, which was removed by sublimation under vacuum. Using this method the authors were able to prepare highly pure [^{89}Zr]Zr(ox)$_2$ for subsequent radiochemical applications, which were later incorporated into comprehensive procedures for preparing ^{89}Zr-labeled antibodies [74]. Later, Holland et al. demonstrated how to maximize recovery of isotopically pure ^{89}Zr with an achievable molar activity of more than 1000 Ci/nmol by examining ^{89}Zr production as a function of cyclotron irradiation time, and purification of the target material as a function of the concentration-dependent loading efficiencies of the hydroxamate resin [75]. Additionally, the authors described improved processes for making [^{89}Zr]ZrCl$_4$. These findings were instrumental in automating [^{89}Zr]Zr(ox)$_2$ production and seized upon by other research groups, who have endeavored to increase the availability of high molar activity [^{89}Zr]Zr(ox)$_2$ and [^{89}Zr]ZrCl$_4$ [67,76–82].

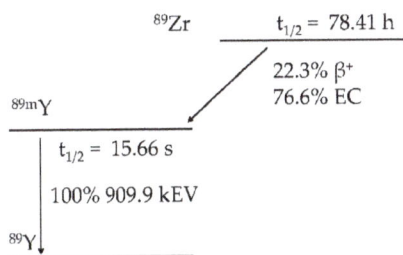

Figure 1. Zirconium-89 decay scheme. Zirconium-89 decays by positron emission and electron capture to metastable yttrium-89. Metastable yttrium-89 decays by gamma emission to stable yttrium-89.

3. The Rationale for New Zirconium-89 Chelation Strategies

Medical researchers have always found inspiration in nature when developing new treatments to combat disease. In a similar manner, chemists have developed ligands for ^{89}Zr chelation, which have been inspired by siderophores or the chelating agents produced by bacteria and fungi to sequester metal ions from the environment [83–85]. The desferrioxamines are a class of iron (III) binding-siderophores that are synthesized from the amino acids lysine and ornithine and contain a tris-hydroxamate coordination motif [84–86]. Given Zr's preference for hard, anionic donor groups, and its ability to form complexes with mono-hydroxamates, it was reasonable to assume these types of iron-binding ligands would be valuable in ^{89}Zr radiochemistry. This rationale led Meijs et al. to perform the first evaluation of desferrioxamine B (DFO; **1**) as a ^{89}Zr chelator, which was observed to be highly stable in human serum (Figure 2) [87]. Since that time, many derivatives have been prepared to facilitate bioconjugation to antibodies using the strategies depicted in Figure 3 [88,89]. Initially, the derivative, *N*-(*S*-acetyl)mercatopacetyldesferrioxamine B (SATA-DFO; **2**) was prepared for mAb coupling using a strategy that involved reacting SATA-DFO with maleimide-modified lysine side chains on the mAb surface to yield a thioether linkage between the DFO chelator and targeting mAb [90]. However, due to instability at physiological pH, this method was abandoned. Later, reacting the activated 2,3,5,6-tetrafluorphenol ester-modified DFO (**3**) with the primary amine side chains of solvent accessible lysine residues located on the mAb surface, Verel et al. were able to conjugate **3** to the U36 mAb through a succinamide linkage (**4**) [74]. Using this conjugate the authors then prepared [^{89}Zr]Zr-DFO-*N*-SUC-U36 mAb, and evaluated it in a murine model bearing xenografts derived from the HNX-OE, human head and neck carcinoma cell line. Tumor-to-non-target background contrast improved over the time course of the study with tumors being easily visualized at 72 h post-injection. Acute biodistribution studies demonstrated that radioactivity retention in tissue was consistent with a ^{89}Zr-labeled mAb. Despite the promising results obtained, the cumbersome preparation strategy, which involved chelation of Fe(III) and its EDTA-mediated removal from DFO before ^{89}Zr radiochemistry could be performed, was also abandoned due to its complexity.

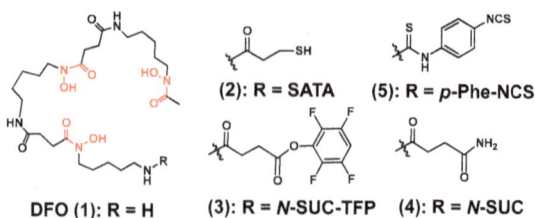

Figure 2. DFO-based bifunctional chelators for ^{89}Zr. The coordinating units are depicted in red font.

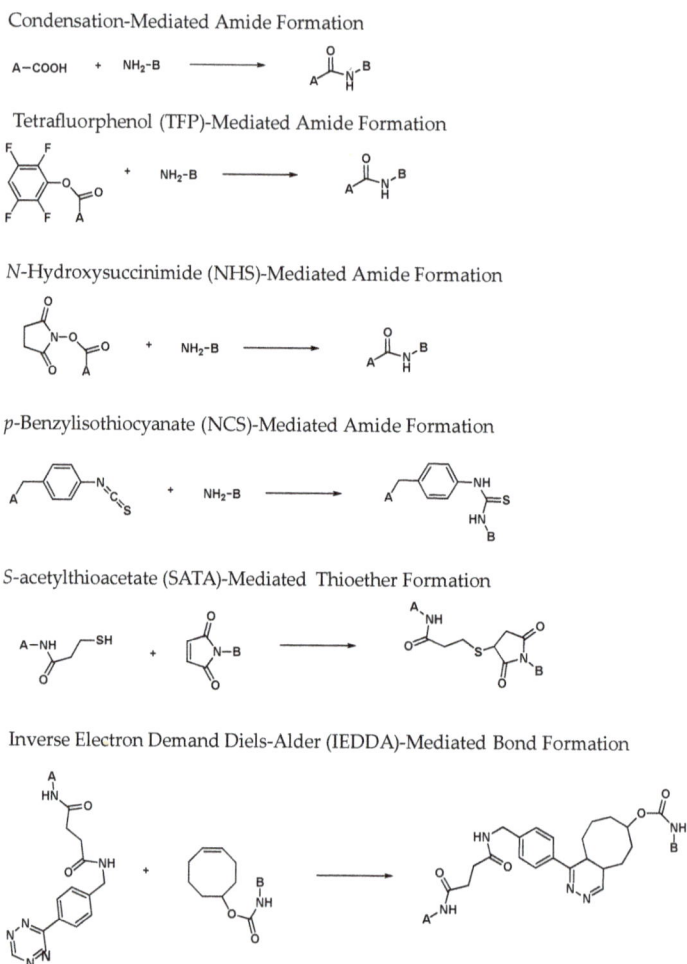

Condensation-Mediated Amide Formation

Tetrafluorphenol (TFP)-Mediated Amide Formation

N-Hydroxysuccinimide (NHS)-Mediated Amide Formation

p-Benzylisothiocyanate (NCS)-Mediated Amide Formation

S-acetylthioacetate (SATA)-Mediated Thioether Formation

Inverse Electron Demand Diels-Alder (IEDDA)-Mediated Bond Formation

Figure 3. Selected bioconjugation reactions used to link ^{89}Zr-bifunctional chelators A with targeting ligands B that are described in this text. For clarity, leaving groups and reaction conditions are not shown.

Several years later, Perk et al. described the new bifunctional chelator *p*-isothiocyanato-benzyl-desferrioxamine B (DFO-*p*-Phe-NCS; **5**) as superior to the TFP-*N*-SUC-DFO and SATA-DFO BFC analogs [91]. The underlying conjugation strategy relied upon the stable formation of a thiourea linkage between the antibody and the chelator, and the one-step coupling process was complete within 60 min when the ligand and the mAb were reacted at 37 °C under highly basic conditions. To demonstrate utility, the authors prepared DFO-*p*-Phe-NCS-U36 mAb with an achievable chelator-to-mAb ratio of 1.5. They then compared the radiochemistry of this conjugate with that of DFO-*N*-SUC-U36, which was prepared using the seven step, TFP method. Radiochemical studies demonstrated comparable radiochemical yields were achieved for both conjugates allowing the authors to conclude the thiourea bond did not interfere with the radiochemistry of the conjugate. Despite facile radiochemistry, the authors did note that the NCS-derived conjugates were less stable in solution. Although no radiolysis experiments were conducted, this instability was attributed to in situ radiolysis, which could be mitigated by formulating the ^{89}Zr-radiopharmaceutical in serum. In an effort to further compare conjugation strategies, the biodistribution of [^{89}Zr]Zr-DFO-*p*-Phe-NCS-U36 and the [^{89}Zr]Zr-DFO-*N*-SUC-U36 were compared in mice bearing FaDu human xenografts that were derived from the human pharynx squamous carcinoma cell line. In vivo results were similar for both radiopharmaceuticals indicating that the different conjugation strategies did not alter the biodistribution in this murine model, and these results were corroborated by small animal PET/CT imaging. After 72 h, the subcutaneous xenografts were clearly visible with excellent image contrast. To further demonstrate the applicability of this approach, the authors also prepared [^{89}Zr]Zr-DFO-*p*-Phe-NCS-rituximab and evaluated it in a nude mouse model bearing tumors that were derived from the A431 human squamous carcinoma cell line. Results in this model were similar to those obtained using the U36 mAb and FaDu animal model. Since these initial reports, this strategy has been universally adopted for preclinical research and clinical trials because of its advantages, which include facile reaction chemistry and its adaptability to good manufacturing compliant processes (cGMP).

Despite extensive use of DFO-*p*-Phe-NCS in ^{89}Zr-Immuno-PET applications, questions regarding [^{89}Zr]Zr-DFO instability during the extended circulation of the radiolabeled mAb in vivo have appeared in the literature [14,16,27,92,93]. Current scientific consensus suggests that the unsaturated coordination sphere of [^{89}Zr]Zr-DFO in combination with perturbation by endogenous serum proteins during the extended circulation of the mAb-based radiopharmaceutical is responsible for this observed instability, ^{89}Zr transchelation, and eventual deposition into the phosphate-rich hydroxylapatite matrix found in bone [94–96]. Unfortunately, the presumed instability of the [^{89}Zr]Zr-DFO complex has complicated the preclinical evaluation of therapeutic antibodies and also may complicate the interpretation of clinical trial results designed to improve clinical care [97]. These reports of instability fueled a desire within the research community to understand the requirements needed to form a stable ^{89}Zr complex and generate new ligands that enhance the stability of the resulting ^{89}Zr complex [98]. The remainder of this review will discuss recent progress in ^{89}Zr chelator research and highlight the major coordinating units being incorporated into their design [99,100].

3.1. Zirconium-89 Chelators Containing Hydroxamate Coordinating Units

In addition to the desferrioxamines, additional siderophores have stimulated the creativity of molecular imaging scientists. For example, Zhai et al. examined fusarine C (FSC; **6**), which was previously evaluated as a ^{68}Ga chelator, and its triacetylated analog TFAC (**7**) as ^{89}Zr chelators [101]. They are depicted in Figure 4. The design benefits of these ligands include the three hydroaxamtae groups for ^{89}Zr coordination, the cyclic structure to improve stability and three primary amine groups, which are amenable to a variety of bioconjugation strategies and also offer the possibility of multivalent targeting. Initially the authors studied TFAC radiochemistry and observed excellent complexation kinetics. Within 90 min the hydrophilic complex, [^{89}Zr]Zr-TFAC could be prepared from [^{89}Zr]Zr(ox)$_2$ with a molar activity of 25 GBq/μmol. Interestingly, this research group also examined the preparation of ^{Nat}Zr-TFAC using ^{Nat}ZrCl$_4$. Analysis of their results led the research team to support the initial claims of Holland et al., who stated that [^{89}Zr]ZrCl$_4$ might be superior to [^{89}Zr]Zr(ox)$_2$ [75]. In vitro, [^{89}Zr]Zr-TFAC demonstrated greater stability against EDTA challenge compared to [89Zr]Zr-DFO. Additionally, biodistribution and small animal PET/CT studies of [89Zr]Zr-TFAC revealed rapid blood clearance with predominate renal excretion and minimal bone uptake suggesting that the [89Zr]Zr-TFAC complex was stable over the short time course of the study. In additional studies the authors prepared [89Zr]Zr-FSC-RGD and [89Zr]Zr-DFO-RGD. They then evaluated each radiopharmaceutical using receptor binding studies and the M21 ($\alpha_v\beta_3^+$) and M21L ($\alpha_v\beta_3^-$) human melanoma cells to determine if the new chelator had any effect on $\alpha_v\beta_3$ binding in vitro. Results of these studies demonstrated that the [89Zr]Zr-FSC complex did not disrupt RGD-$\alpha_v\beta_3$ binding in vitro, and this finding was corroborated using biodistribution and small animal PET/CT studies in nude mice bearing contralateral human melanoma M21 and M21L tumors. These studies revealed excellent retention of radioactivity in integrin positive tumors, and a complete biodistribution profile that was consistent with RGD-based radiopharmaceuticals [102]. In a recent publication, Summers, et al. extended the evaluation of the FSC ligand by conjugating it to the anti-EGFR affibody, ZEGFR:23377 using a maleimide-based bioconjugation strategy to produce FSC-ZEGFR:23377 [103]. This bioconjugate was radiolabeled in a facile manner using [89Zr]Zr(ox)$_2$ and radiochemical methods typically used to prepare [89Zr]Zr–DFO-mAbs. Binding studies and biodistribution studies demonstrated that antigen reactivity was retained in vitro and in vivo, but at 24 h post-injection, the radioactivity level in the bone tissues of mice receiving the radiolabeled affibody was comparable to levels in the bones of mice receiving the [89Zr]Zr-DFO-bioconjugate. Clearly, significant progress has been made exploring FSC and its 89Zr radiochemistry, but additional studies to examine radioactivity levels in bone tissue at much later time points will be necessary to fully appreciate its potential as a 89Zr-chealtor.

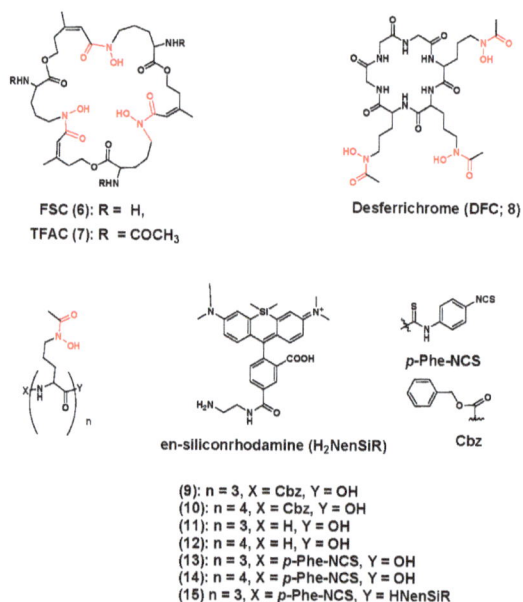

FSC (6): R = H,
TFAC (7): R = COCH$_3$

Desferrichrome (DFC; 8)

en-siliconrhodamine (H$_2$NenSiR)

Cbz

p-Phe-NCS

(9): n = 3, X = Cbz, Y = OH
(10): n = 4, X = Cbz, Y = OH
(11): n = 3, X = H, Y = OH
(12): n = 4, X = H, Y = OH
(13): n = 3, X = p-Phe-NCS, Y = OH
(14): n = 4, X = p-Phe-NCS, Y = OH
(15) n = 3, X = p-Phe-NCS, Y = HNenSiR

Figure 4. Hydroxamate-containing Zirconium-89 chelators inspired by the siderophores fusarine C and desferrichrome. The coordinating units are depicted in red font.

The Boros group recently described the desferrichrome (DFC; **18**)-inspired 89Zr chelators **9–15** (Figure 4) [104]. Desferrichrome (DFC; **18**) is an ornithine-derived hexapeptidyl siderophore secreted by bacteria and fungi, but to ensure accessibility of the tris-hydroxamate coordinating groups during chelation, the naturally occurring ligand was reverse-engineered to be acyclic, and modified with the near infrared (NIR) dye, silicon rhodamine (SiR). Attaching the NIR dye allowed the researchers to monitor coordination kinetics during metal complexation, to identify the Zr-DFC complex during any subsequent purification steps and provide a multi-modal imaging platform to describe tissue residualization after in vivo injection. Radiochemical studies demonstrated that these ligands could be radiolabeled quantitatively at room temperature; while EDTA challenge studies, revealed that ^{89}Zr[Zr]-**12** and [^{89}Zr]Zr-DFO demonstrated comparable resistance to transchelation. Although biodistribution studies involving the radiometal chelates were not reported, **14**, which was an NCS-modified version of **12**, was conjugated to trastuzumab and radiolabeled with ^{89}Zr in order to compare its stability to DFO when incorporated into a mAb-based radiopharmaceutical. Biodistribution studies conducted in normal C57Bl6 mice revealed accelerated blood clearance compared to [^{89}Zr]Zr-DFO-trastuzumab, but animals injected with [^{89}Zr]Zr-**14**-trastuzumab retained significantly more radioactivity in liver tissue, which may preclude the imaging of tumors within the abdominal cavity. Finally, radioactivity levels in bone tissue of mice receiving either [^{89}Zr]Zr-**14**- or the [^{89}Zr]Zr-DFO-trastuzumab were similar. Current experiments are underway to examine the NIR properties of **15** to determine if it can be applied in a multi-modal strategy that enables preclinical antibody development.

Siebold and coworkers described the rational design and solid phase synthesis of CTH36 (**16**) as a ligand for ^{89}Zr chelation (Figure 5) [105]. To maximize the potential of this new ligand its rational design was predicated on extensive computational studies and several important design characteristics including (1) the inclusion of four hydroxamate coordinating units; (2) a macrocyclic structure to take advantage of the macrocyclic effect; (3) rotational symmetry to limit isomers; (4) hydrophilic character; and (5) an optimal cavity size to provide a balance between ring strain and entropic effects. They also developed Tz-CTH36 (**17**) and conjugated it to a transcyclooctene-modified c(RGDfK) analog using and inverse electron demand Diels-Alder coupling strategy so that they could compare the radiochemistry and in vitro properties of this conjugate with [^{89}Zr]Zr-DFO-c(RGDfK). Interestingly, both conjugates underwent facile radiolabeling, but [^{89}Zr]Zr-CTH36-c(RGDfK) was more resistant to EDTA challenge. The authors postulated that the higher kinetic stability of [^{89}Zr]Zr-CTH36-c(RGDfK) was a consequence of the rationally designed ligand and its ability to complex the ^{89}Zr^{4+} ion in an octadentate manner. However, the authors did not report any in vivo data regarding the biodistribution of [^{89}Zr]Zr-CTH36-c(RGDfK) or any data regarding the preparation and evaluation of a [^{89}Zr]Zr-CTH36-mAb conjugate. Completion of these remaining studies will be crucial in validating this rational design approach and determining if **16** and its BFC version **17** will be useful in ^{89}Zr-immuno-PET applications.

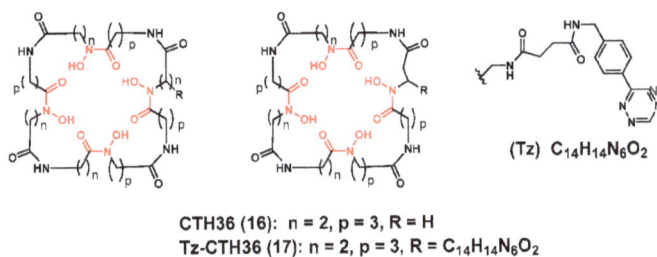

CTH36 (**16**): n = 2, p = 3, R = H
Tz-CTH36 (**17**): n = 2, p = 3, R = C$_{14}$H$_{14}$N$_6$O$_2$

Figure 5. Macrocycle-based hydroxamate chelators for Zirconium-89. The coordinating units are depicted in red font.

Another attempt to develop ligands that completely satisfy the ^{89}Zr-coordination sphere was communicated by the Smith group, who evaluated the hybrid molecule DFO-1-hydroxy-2-pyridone (**18**), which was originally designed for plutonium (IV) sequestration (Figure 6) [106]. Although nuclear magnetic resonance (NMR) analysis of ^{Nat}Zr-**18** was not performed due its low solubility, the authors deduced a 1:1 metal-ligand complex using high resolution mass spectrometry. Radiochemical experiments demonstrated that the ligand could be quantitatively radiolabeled with excellent molar activity using comparable experimental conditions needed to prepare [^{89}Zr]Zr-DFO, but unlike the latter, the former radiometal complex was inert to EDTA and serum challenge. Small animal PET/CT and acute biodistribution studies were also conducted on normal mice injected with either [^{89}Zr]Zr-**18** or [^{89}Zr]Zr-DFO. In contrast to the clearance profile of [^{89}Zr]Zr-DFO, which underwent renal excretion exclusively, [^{89}Zr]Zr-**18** demonstrated a bimodal excretion pattern. At early time points post-injection, rapid renal clearance was observed, but as the experimental time course progressed, hepatobiliary clearance predominated. The authors hypothesized that differences in hydrophilicity were responsible for the dichotomous excretion pattern of the former radiometal chelate. Moreover, radioactivity levels in the bone tissue of mice injected with [^{89}Zr]Zr-**18** were significantly lower when compared to the radioactivity levels in the bone tissue of mice receiving [^{89}Zr]Zr-DFO. Although current results demonstrate promise for this hybrid ligand, the BFC version must be evaluated before any conclusions can be made regarding its utility in ^{89}Zr-immuno-PET applications.

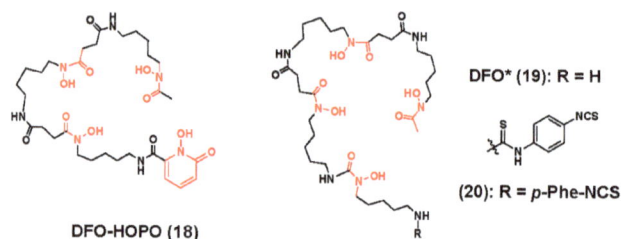

Figure 6. Octa-coordinate chelators for ^{89}Zr inspired by DFO. The coordinating units are depicted in red font.

Taking cues from Guerard et al. [98], Patra and coworkers, reported the DFO analog, DFO* (**19**), which was easily synthesized from the DFO-mesylate salt and the protected hydroxamic acid precursor (Figure 6) [107]. While derivatization to introduce functional groups for bioconjugation into **19** was facile, preparing and characterizing the ^{Nat}Zr-**19** complex was challenging due to its poor solubility. Nevertheless, the authors were able to deduce a 1:1 metal-to-ligand binding motif using high resolution mass spectrometry and NRM analysis, but also noted the presence of structural isomers, which the authors attributed to the numerous coordination modes that the acyclic ligand can adopt during complexation of the Zr^{4+} ion. To understand how the addition of the fourth hydroxamate coordinating unit influenced the radiochemistry of this ligand, the authors prepared [^{89}Zr]Zr-**19**-[NIe14]BBS(7–14) and [^{89}Zr]Zr-DFO-[NIe14]BBS(7–14) with achievable molar activities of 5–6 GBq-µmol^{-1}. Although no uncomplexed ^{89}Zr was observed in the reaction mixture, radio-high performance liquid chromatography (HPLC) did detect the presence of isomers, which was consistent with the NMR solution data of the non-radioactive complex. When challenged with excess DFO, the [^{89}Zr]Zr-**19** conjugate resisted transchelation. More importantly, LogD$_{7.4}$ and in vitro binding studies using gastrin-releasing peptide receptor positive cell lines suggested that the addition of the fourth hydroxamate coordinating unit had minimal influence on the physical properties of the radiopharmaceuticals in vitro. Several years later, Vugts et al. extended this work by synthesizing **20**, which was prepared by reacting **19** with *p*-phenylenediisothiocyante, and then conjugating it to trastuzumab to yield a bioconjugate with an observed chelator-to-mAb ratio of 0.6–0.9 [108]. Additionally, comparative radiochemistry studies revealed that [^{89}Zr]Zr-**DFO***-*p*-Phe-NCS-trastuzumab and [^{89}Zr]Zr-DFO-trastuzumab could be radiolabeled in high radiochemical yield within 60 min, but only [^{89}Zr]Zr-DFO*-*p*-Phe-NCS-trastuzumab was more stable in a variety of storage media, and demonstrated a more robust immunoreactivity when challenged with the HER2/neu antigen. The authors attributed the superior in vitro stability of the DFO*-based radiopharmaceutical to the coordinatively saturated environment that the octa-coordinate ligand, **20** provides the ^{89}Zr^{4+} ion. Biodistribution studies in nude mice bearing HER2/neu positive tumors derived from the human, N87 human gastric cancer cell line were also very descriptive. Although both radiopharmaceuticals had very similar blood clearance profiles, animals injected with the [^{89}Zr]Zr-**DFO***-*p*-Phe-NCS-trastuzumab demonstrated significantly lower radioactivity levels in liver, spleen, and bone tissue, which was corroborated by small animal PET/CT studies in the same xenograft model. The development of **20** represents a significant achievement in ^{89}Zr chelator design, and since recent efforts to improve the water solubility of this ligand have been successful the radiopharmaceutical community anxiously awaits its evaluation in a clinical setting [109–111].

Additional hydroxamate-based ligands **21–24** were reported by Boros and coworkers [112]. These ligands are depicted in Figure 7. Creatively, the authors used the macrocycles cyclen and cyclam as molecular scaffolds, and alkylated the secondary amines within each macrocycle to yield chelators with three or four hydroxamte-functionalized pendant arms in relatively good overall yields. Further computational studies to optimize the length of the pendant arm generated a cyclam-based ligand (**25**) that underwent facile radiolabeling to produce an [^{89}Zr]Zr-complex with improved stability over the cyclen-based chelators. EDTA challenge studies revealed [^{89}Zr]Zr-**25** was 95% intact after 72 h, while LogD$_{7.4}$ studies revealed a comparable hydrophilicity with ^{89}Zr-DFO. Furthermore, biodistribution and small animal PET/CT studies in normal mice confirmed the in vitro results with rapid renal clearance of the administered radioactivity after only 30 min. Biodistribution data obtained 24 h post-injection revealed a biodistribution pattern similar to ^{89}Zr-DFO, but to further assess **25** as a ^{89}Zr chelator, the authors compared the in vivo pharmacokinetics of [^{89}Zr]Zr-**25**-*N*-SUC-trastuzumab, which was prepared using a TFP-mediated approach, and [^{89}Zr]Zr-DFO-trastuzumab using small animal PET/CT in a nude mouse model bearing bilateral BT474 (HER2+) and BT20 (HER2−) human breast cancer xenografts. As expected, efficient and specific targeting of the HER2/neu receptor was observed with HER2+ tumors easily visualized as image contrast improved over the experimental time course. However, image analysis and post-PET biodistribution results revealed that the radioactivity levels in the bones of animals injected with [^{89}Zr]Zr-**25**-*N*-SUC-trastuzumab were 9-fold higher than that observed in the bones of mice receiving [^{89}Zr]Zr-DFO-trastuzumab. This suggests that further optimization of the ligand structure will be necessary to improve ^{89}Zr-chealte stability before these ligands are made available to the research community.

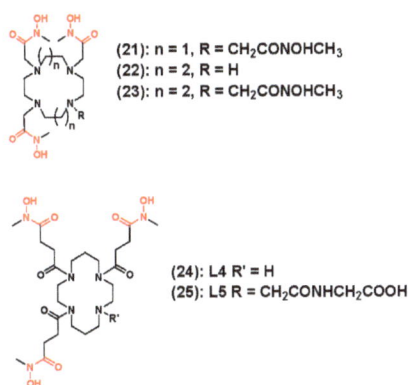

(21): n = 1, R = CH$_2$CONOHCH$_3$
(22): n = 2, R = H
(23): n = 2, R = CH$_2$CONOHCH$_3$

(24): L4 R' = H
(25): L5 R = CH$_2$CONHCH$_2$COOH

Figure 7. Cyclen-and cyclam-based ^{89}Zr chelators containing hydroxamate pendant arms. The coordinating units are depicted in red font.

3.2. Zirconium-89 Chelators Containing Hydroxyisopthalamide, Terepthalamide and Hydroxypyridinone Coordinating Units

Over the last three decades, the prolific work by the Raymond group and others has led to the development of a vast library of rationally designed, pre-organized ligands containing multi-dentate catecholate, hydroxypyridinoate, hydroxyisopthalamide and terepthalamide coordinating units [86]. Although originally designed as decorporation agents for numerous radioactive ions such as plutonium (IV), uranium (IV) and thorium (IV), the radiochemistry community has recently rediscovered these ligands and made numerous attempts to adapt them for zirconium (IV) radiochemistry.

Bhatt et al. reported a pair of ligands containing hydroxyisopthalamide coordinating units, which were believed to bind ^{89}Zr in an octadentate manner through a combination of phenolic and carbonyl oxygen atoms (Figure 8) [113]. Ligand **26** was developed as a rigid trimacrocycle comprised of 24 and 30 member rings whereas ligand **27** formed a more flexible bimacrocycle comprised of 24 and 27 member rings. The influence of these structural differences on in vitro and in vivo behavior was readily apparent. While both ligands could be quantitatively radiolabeled with ^{89}Zr to yield high molar activity radiometal complexes, the more rigid **26** required forcing conditions. However, [^{89}Zr]Zr-**26** was more resistant to DTPA and serum challenge, and demonstrated greater in vivo stability. For example, mice injected with [^{89}Zr]Zr-**26** retained less activity in their liver, kidney and bone tissue than did those animals injected with [^{89}Zr]Zr-**27**, but it still did not demonstrate pharmacokinetic properties that surpassed those of [^{89}Zr]Zr-DFO.

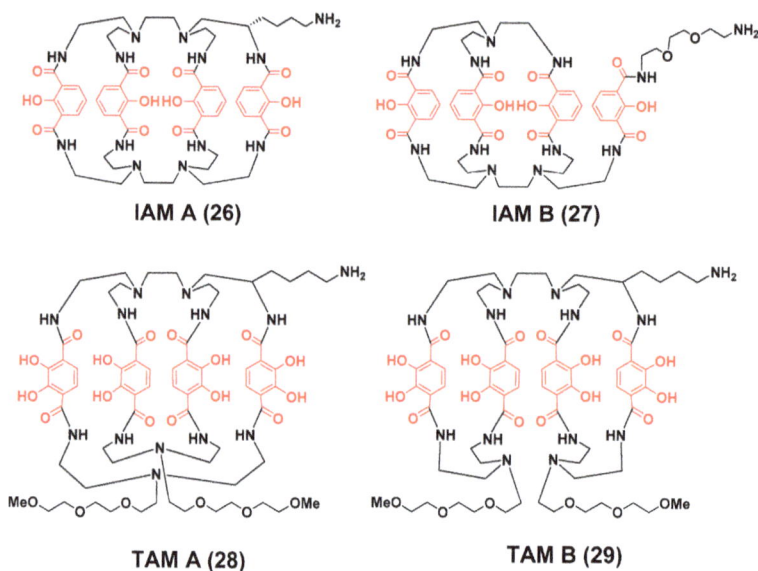

IAM A (26) **IAM B (27)**

TAM A (28) **TAM B (29)**

Figure 8. Bifunctional chelators for ^{89}Zr containing hydroxyisopthalamide and terepthalamide coordinating units. The coordinating units are depicted in red font.

Pandya et al. investigated the use of the 2,3-dihydroxyterepthalamide (TAM) coordinating units in ^{89}Zr-BFC design [114]. These functional groups are highly acidic and exist as di-anions at neutral pH. The authors believed their incorporation into a macrocycle with an appropriate cavity size would yield an octadentate ligand with high avidity for the ^{89}Zr^{4+} cation [114]. Moreover, Raymond and coworkers previously demonstrated how this coordination motif could be used to engineer the solution properties of the resultant chelating ligand [86]. Using high dilution conditions to avoid side reactions, the authors were able to isolate two distinct regioisomers. Regioisomer **28** was identified to have a more rigid structure containing two 29 atom macrocycles, while regioisomer **29** was identified as a more flexible "clam-shell" like system composed of two-26 atom ring systems. Despite the differences in rigidity, both ligands were quantitatively radiolabeled under the same conditions needed to prepare [^{89}Zr]Zr-DFO, and neither radiometal complex demonstrated transchelation when challenged with excess DTPA or human serum. Interestingly, these complexes were more resistant to transmetallation than the IAM analogs and illustrate the greater affinity of the TAM coordination motif for the Zr^{4+} cation. Biodistribution studies conducted in normal mice revealed that the more rigid complex [^{89}Zr]Zr-**28** was more stable in vivo than [^{89}Zr]Zr-**29** and underwent more rapid clearance from all tissues over the experimental time course. The authors asserted that since **29** was a less rigid bi-macrocycle, the resulting ^{89}Zr complex was more susceptible to perturbation by endogenous serum proteins and resulted in greater transchelation of the ^{89}Zr^{4+} ion. The in vivo behavior of [^{89}Zr]Zr-**28** was also compared with [^{89}Zr]Zr-DFO. Although both radiometal complexes demonstrated a similar clearance pattern from the blood pool, [^{89}Zr]Zr-**28** had significantly elevated levels of radioactivity in liver and kidney tissue, which the authors attributed to multifactorial processes including aggregation and changes in the molecular structure of [^{89}Zr]Zr-**28** due to changes in intracellular pH within these tissues. Finally, levels of radioactivity observed in bone tissue of mice receiving [^{89}Zr]Zr-**28** were comparable to that of mice injected with [^{89}Zr]Zr-DFO. Typically, low incorporation of radioactivity in bone would stimulate further research into this ligand class, but the excessive retention of radioactivity in kidney tissue will most likely prohibit their further development as ^{89}Zr chelators.

The Blower group reported CP256 (**30**) [115], which is based upon three 1,6,-dimethyl-3-hydroxy-pyridin-4-one groups, and is depicted in Figure 9. Although a single molecule crystal structure was not reported, high resolution mass spectrometry analysis in conjunction with ^{1}H and ^{13}C-NMR studies revealed that ^{Nat}Zr-**30** formed with a 1:1 ligand-to-metal stoichiometric ratio. However a high degree of fluxionality was observed within the NMR spectra and suggested to the authors that multiple structural isomers existed in solution. Initial radiochemistry studies indicated that **30** could be quantitatively radiolabeled at a ligand concentration of 10 mM, but this became more difficult when the concentration of the latter approached the nanomolar range. Biodistribution studies in normal mice injected with [^{89}Zr]Zr-**30** revealed a renal excretion pattern similar to that of [^{89}Zr]Zr-DFO suggesting that [^{89}Zr]Zr-**30** was stable on this rapid time scale. However, stability differences became more pronounced when the research team compared [^{89}Zr]Zr-**31**-trastuzumab, which contained the bifunctional chelating version of **30**, and [^{89}Zr]Zr-DFO-trastuzumab. At early biodistribution time points, similar levels of radioactivity were observed in the blood pool, spleen, liver, kidney and bone tissues regardless of the injected radiopharmaceutical. However, as the experimental study progressed, animals injected with [^{89}Zr]Zr-DFO-trastuzumab seemed to retain less activity within tissue. For example, by the end of the study, the radioactivity associated with the bone tissue of animals receiving [^{89}Zr]Zr-**31**-trastuzmab was 6-fold higher than that of animals receiving [^{89}Zr]Zr-DFO-trastuzumab. Moreover, this accumulation of radioactivity in bone was clearly visible in small animal PET images, and forced the authors to conclude that **31** would not be a reliable ligand for the stable chelation of ^{89}Zr where prolonged stability was required in vivo.

In an attempt to overcome the limitations of ligands **30** and **31**, Buchwalder and coworkers rationally designed 1,3,-propanediamine-N,N,N′,N′-tetrakis[(2-(aminoethy)-3-hydroxy-1-6-dimethyl-4 (1*H*)-pyridinone)acetamide] (THPN, **32**), which is based upon four 3-hydroxy-4-pyridinone (3,4-HOPO) coordinating groups, in an attempt to fully satisfy the octadentate coordination sphere preferred by the ^{89}Zr^{4+} ion [116]. Although traditional synthetic routes to this ligand proved laborious, a breakthrough was achieved when the research team adopted a microwave-assisted approach, which reduced the reaction time from six days to six hours and substantially increased the product yield. Computational studies and high resolution mass spectrometry of ^{Nat}Zr-**32** revealed a 1:1 metal-to-ligand stoichiometry, while radiochemistry studies revealed radiolabeling kinetics that were superior to DFO in the micromolar range. LogP studies indicated that [^{89}Zr]Zr-**32** was a highly hydrophilic complex and demonstrated a similar hydrophilicity to other [^{89}Zr]Zr-HOPO and [^{89}Zr]Zr-TAM complexes.

Moreover, while serum stability and EDTA challenge studies at physiological pH revealed comparable stability between both radiotracers, studies under low pH conditions revealed that [^{89}Zr]Zr-**32** was more resistant to demetallation suggesting it would remain stable in vivo. Finally, in comparative biodistribution and small animal PET/CT studies, the authors injected [^{89}Zr]Zr-**32** or [^{89}Zr]Zr-DFO into normal mice to evaluate the clearance of radioactivity from normal tissues. Image analysis revealed minimal amounts of radioactivity in tissue with a majority of the injected activity being excreted renally. A 24 h post-PET biodistribution analysis revealed a similar pharmacokinetic profile as [^{89}Zr]Zr-DFO, and comparable radioactivity levels were observed in the bone tissues of both animal cohorts. Although the authors speculate that **32** could be a reasonable alternative to DFO, biodistribution and small animal PET imaging data describing the stability of this ligand as part of mAb-based radiopharmaceutical must be completed, and the radiopharmaceutical community anxiously awaits these results.

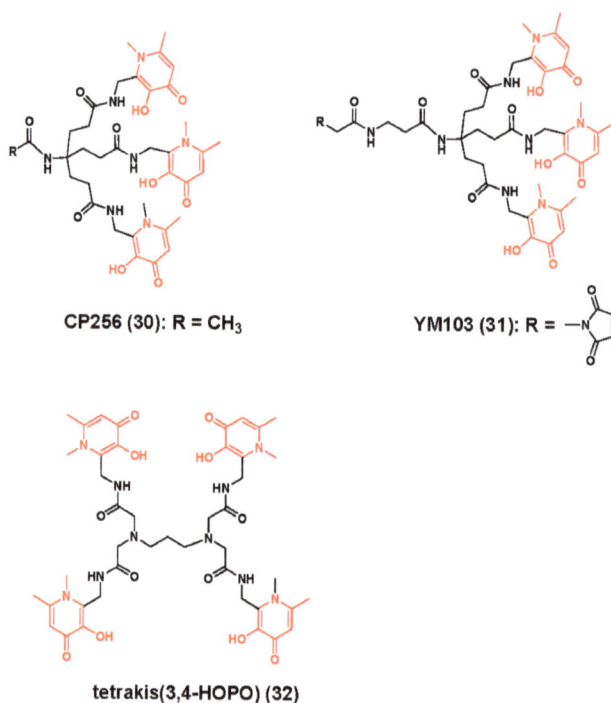

CP256 (30): R = CH$_3$

YM103 (31): R =

tetrakis(3,4-HOPO) (32)

Figure 9. Zirconium-89 chelators containing 3,4-hydroxypyridinone coordinating units. The coordinating units are depicted in red font.

A recent academic-industrial collaboration between Genentech, Inc., Lumiphore, Inc. and The Wake Forest School of Medicine resulted in the development of BPDETLysH22-2,3-HOPO (**33**), which contained four 3,2-HOPO coordinating units (Figure 10) [93]. The novel "clam shell" structure was envisioned to offer rapid ^{89}Zr^{4+} complexation kinetics that would be similar to those of DFO, and improved radiometal complex stability because of the pre-organized macrocyclic design. Similar to ligands reported by Bhatt and Pandya [113,114], the synthesis of **33** involved the condensation of tetraamine and activated diacid intermediates under high dilution conditions to avoid oligomeric and polymeric by-products. While a 1:1 meta-to-ligand stoichiometric ratio was observed by high resolution mass spectrometry, NMR and HPLC analysis revealed the presence of structural isomers after ^{Nat}Zr^{4+} ion coordination. Initial radiochemistry experiments using the conditions described for the preparation of [⁸⁹Zr]Zr-DFO revealed that this ligand could be quantitatively radiolabeled with excellent molar activity in less than 30 min, and the resulting hydrophilic complex was more resistant to DTPA challenge than [⁸⁹Zr]Zr-DFO. Biodistribution studies in normal mice were conducted with [⁸⁹Zr]Zr-33 and [⁸⁹Zr]Zr-DFO to compare the clearance kinetics of each tracer, and revealed efficient blood clearance of both radiotracers over the 72 h experimental time course. However elevated levels of radioactivity were observed in the kidney, liver and bone tissues of mice receiving [⁸⁹Zr]Zr-33. Next, the authors prepared the BFC version 34 and conjugated it to trastuzumab. They then compared [⁸⁹Zr]Zr-2,3-HOPO-*p*-Phe-NCS-trastuzumab and [⁸⁹Zr]Zr-DFO-trastuzumab in a HER2/neu mouse model of ovarian carcinoma using small animal PET/CT imaging. Time activity curves for tumor, blood, liver and bone were obtained over the 144 h experimental time course. As expected, radioactivity in the blood pool was comparable for each cohort regardless of the injected immuno-PET agent, and in vivo targeting of the HER2/neu tumor was efficient suggesting that the chelator did not alter trastuzumab specificity in vivo. However, animals receiving [⁸⁹Zr]Zr-2,3-HOPO-*p*-Phe-NCS-trastuzumab did retain more radioactivity in liver and bone tissues when compared to animals injected with [⁸⁹Zr]Zr-DFO-trastuzumab forcing the authors to conclude that further optimization of the macrocyclic scaffold would be necessary to improve radiometal complex stability and make these ligands viable DFO alternatives.

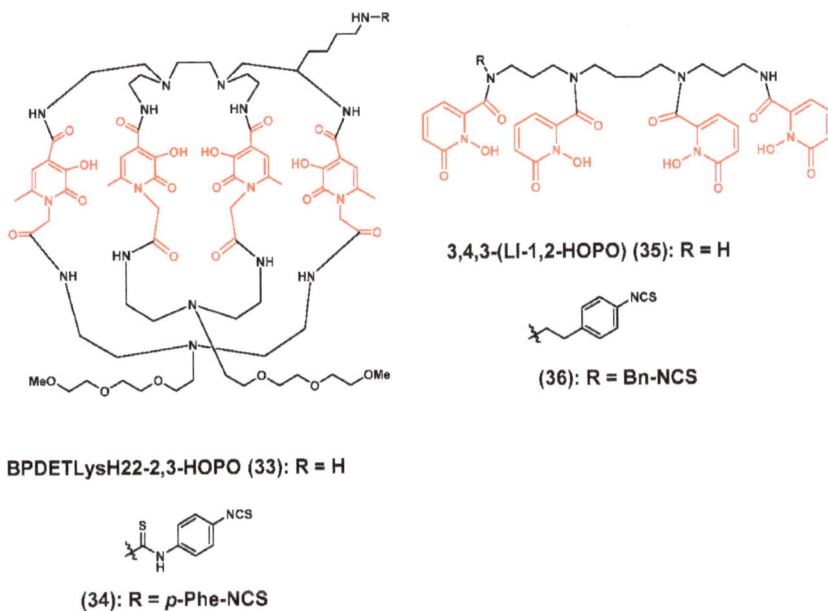

3,4,3-(LI-1,2-HOPO) (35): R = H

(36): R = Bn-NCS

BPDETLysH22-2,3-HOPO (33): R = H

(34): R = *p*-Phe-NCS

Figure 10. Zirconium-89 chelators containing 2,3-hydroxypyridinone and 1,2-hydroxypyridinone coordinating units. The coordinating units are depicted in red font.

To date, the most successful application of the hydroxypyridinone coordinating unit in ⁸⁹Zr chelator design was published by Deri and collaborators who described the acyclic 3,4,3-(Li-1,2-HOPO) ligand 35 [117], which comprised a linear spermine backbone appended with four 1,2-hydroxypyridinone coordinating units. The acyclic nature of the ligand, the low pKa values of the coordinating units and an ideal cavity size were believed to be important design characteristics that would facilitate ⁸⁹Zr complexation and result in an ultra-stable radiometal complex. Similar to other ligands containing hydroxypyridinone coordinating units, ⁸⁹Zr radiolabeling kinetics were rapid in the presence of excess ligand, but during these studies, the authors noted the presence of two distinct species, whose formation was concentration-dependent. Experimental evidence led the research team to hypothesize that the radiometal complex initially formed reflected a dimeric species involving two ligands binding one $^{89}Zr^{4+}$ ion. However, over time, this dimeric complex interconverted to a second species, which the authors attributed to [^{89}Zr]Zr-**35** containing the expected 1:1 metal-to-ligand stoichiometry. After isolating the later species, the authors evaluated its kinetic inertness, its resistance to transchelation and its selectivity for $^{89}Zr^{4+}$ ion by challenging [^{89}Zr]Zr-**35** with human serum, excess EDTA and biologically relevant metal ions, respectively. In human serum both [^{89}Zr]Zr-**35** and [^{89}Zr]Zr-DFO were more than 98% intact after 7 days at 37 °C, but when challenged with excess EDTA or biologically relevant metal cations, [^{89}Zr]Zr-**35** was observed to be more resistant to demetallation. Biodistribution and small animal PET/CT studies revealed that in contrast to animals injected with [^{89}Zr]Zr-DFO, animals injected with [^{89}Zr]Zr-**35** displayed a biomodal excretion pattern with renal excretion occurring at early time points and hepatobiliary clearance occurring at later time points. Biodistribution studies indicated significantly more radioactivity in the bone tissue of mice injected with [^{89}Zr]Zr-**35** than in the bones of mice receiving [^{89}Zr]Zr-DFO, but the authors attributed these observations to differences in the perfusion and clearance kinetics exhibited by the two radiometal complexes rather than transchelation of the ^{89}Zr into the phosphate rich bone matrix.

In a subsequent publication, Deri and coworkers reported the bifunctional chelator analog of **35**, SCN-Bn-HOPO (**36**) [117]. Synthesis of this molecule was non-trivial as it required the incorporation of the *p*-benzylisothiocyanate pendant arm, which was used for antibody conjugation, into the symmetrical ligand. Although deprotection steps to achieve the final BFC proved challenging, it was eventually conjugated to trastuzumab yielding a bioconjugate with a 3:1 chealtor:mAb ratio. Satisfyingly, the ^{89}Zr chelation kinetics of the BFC when attached to trastuzumab were similar to the initial ligand; to that observed with DFO-trastuzumab, and both conjugates were quantitatively radiolabeled with a specific activity of 74 MBq/mg in less than an hour. More importantly, both radiopharmaceuticals demonstrated immunoreactivities greater than 85% suggesting that the presence of the [^{89}Zr]Zr-HOPO complex on the antibody surface did not affect in vitro stability or HER2/neu receptor affinity. The in vivo performance of both radiopharmaceuticals was compared using small animal PET/CT imaging and acute biodistribution studies in a murine model of HER2/neu positive breast cancer. Both agents demonstrated efficient tumor accumulation with improving image contrast as they cleared from non-target tissues. However, mice injected with the HOPO-based radiopharmaceutical demonstrated reduced accumulation of radioactivity in their skeleton over the experimental time course of this study. Biodistribution data corroborated the imaging results and revealed that the radioactivity level in the skeleton of mice receiving [^{89}Zr]Zr-1,2-HOPO-Bn-SCN-trastuzumab was approximately 8-fold lower than for mice receiving the [^{89}Zr]Zr-DFO conjugate. Accordingly, the former radiopharmaceutical afforded a better tumor-to-bone ratio, which may be an important criterion in reducing false positive rates associated with ^{89}Zr-immuno-PET-based bone metastasis detection strategies [97,118]. Additionally, the improved contrast may enable more accurate biodistribution and dosimetry in advance of targeted systemic radiotherapy, and the successful use of this ligand in immuno-PET applications stands out as another excellent success story in the preclinical development of ^{89}Zr chelators.

3.3. ^{89}Zr Chelators Containing Tetraazamacrocycles

Polyaminocarboxylate ligands, which include acyclic ligands such as ethylenediamine-tetraacetic acid (EDTA), diethylenetriaminepentaacetic acid (DTPA) or tetraazamacrocycles such as 1,4,7,10-tetraazacyclododecane-1,4,7,10-tetraacetic acid (DOTA, **37**) represent one ligand class that has been at the forefront of radiopharmaceutical development for nearly half a century (Figure 11) [14,15]. While useful for stably chelating a variety of radiometals, reports describing them as ^{89}Zr chelators are absent in the literature despite a strong preference of Zr(IV) for polyanionic hard donor ligands and the very impressive stability constants of some of the resulting Zr-complexes. Recently, Pandya et al. reported their initial observations on $^{Nat/89}Zr$-tetraazamacrocycle complexes by preparing Zr-**37**,

Zr-**38** and Zr-**39** and characterizing each complex using high resolution mass spectrometry, and ^{1}H- and ^{13}C-NMR [119]. Fortunately, the molecular structure of Zr-**37** was elucidated by single crystal x-ray diffraction analysis. It provided irrefutable proof of an octadentate coordination environment where all four macrocycle nitrogen atoms and acetate pendant arms participated in Zr^{4+} ion coordination. Furthermore, the compressed, square anti-prismatic geometry and low-symmetry saddle-like ligand conformation displayed by Zr-**37** was consistent with the limited number of reports describing Zr-catalysts containing analogous ligands with a Zr(IV) metal center that lacks crystal-field stabilization [120–128].

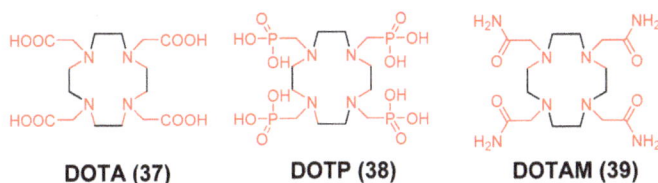

Figure 11. Zirconium-89 chelators containing tetraazamacrocycles. The coordinating units are depicted in red font.

After completing synthesis and characterization of the reference complexes, attempts to prepare the radioactive analogs using standard procedures published for the preparation of [[89]Zr]Zr-DFO resulted in low radiochemical yields. However, the authors seized upon earlier work by Holland et al., who described the use of [[89]Zr]ZrCl$_4$ as potential alternative to [[89]Zr]Zr(ox)$_2$ in [89]Zr-immuno-PET synthesis [75]. Using this radioactive precursor, they were able to quantitatively prepare [[89]Zr]Zr-**37**, [[89]Zr]Zr-**38**, and [[89]Zr]Zr-**39** with specific activities that were similar to other [89]Zr-complexes published in the literature. To explain differences in ligand reactivity, the authors rationalized the [Nat/89]Zr species present in solution dictated complex formation [63]. Zirconium oxalate is a highly stable complex, even under highly acidic conditions and at very low molar concentrations. Accordingly, when [[89]Zr]Zr(ox)$_2$ is reacted with the tetraazamacrocycle, the oxalate anion's ability to form a stable complex in aqueous media effectively competes with the macrocycle, resulting in a low radiochemical yield of the [[89]Zr]Zr-tetraazamacrocycle complex. Conversely, ZrCl$_4$, is highly charged, oxophilic and readily undergoes aquation in solution to form multiple μ-hydroxo-and μ-oxo-bridged species. Thus, in the absence of a competing oxalate ligand, [[89]Zr]Zr-tetraazamacrocycle complex formation was favored when [[89]Zr]ZrCl$_4$ was reacted with the macro-cycle in solution [63].

In vitro, [[89]Zr]Zr-tetrazamacrocycle complexes demonstrated remarkable stability with [[89]Zr]Zr-**37** being the most inert to serum, EDTA or biologically relevant metal ion challenge. Acute biodistribution studies were also conducted to examine stability in vivo. Mice intravenously injected with [[89]Zr]Zr-**37** retained significantly less radioactivity in their tissues compared to mice injected with [[89]Zr]Zr-**38** or [[89]Zr]Zr-**39**. More importantly, biodistribution and small animal PET/CT studies demonstrated that mice injected with [[89]Zr]Zr-**37** retained less radioactivity in their tissues than did animals injected with [[89]Zr]Zr-DFO suggesting the former is superior to the latter in terms of in vivo stability. Although results describing **37** as part of an [89]Zr-immuno-PET agent are still unreported; if successful, use of tetraazamacrocycles in [89]Zr-immuno-PET may pave the way for truly theranostic approach to targeted systemic radiotherapy since it would be possible to radiolabel one FDA approved DOTA-mAb conjugate with [89]Zr and therapeutic radionuclides [70,129–142]. This strategy may increase dosimetric accuracy, reduce regulatory burden, and minimize costs associated with cGMP-compliant radiopharmaceutical development, so that they may be more readily integrated into personalized medicine strategies in the future.

Molecules **2018**, 23, 638

4. Conclusions

Zirconium-89 radiopharmaceutical research has progressed rapidly since this radionuclide was first produced more than three decades ago; great strides have been made to standardize its production, understand aqueous zirconium chemistry, and design ligands that stably chelate this important PET isotope. As a consequence, numerous siderophore-inspired ligands containing hydroxamate; hydroxyisopthalamide; terepthalamide, and hydroxypiridinoate coordinating units have been scrutinized and reported to effectively chelate ^{89}Zr. Thus far, only DFO* and HOPO derivatives have proven effective as ^{89}Zr-chelators when incorporated into an antibody-based radiopharmaceutical, while others chelators still await evaluation in this context. Interest in tetraazamacrocycles as ^{89}Zr chelators has also increased given the recent revelation that this ligand class can form ultra-stable ^{89}Zr complexes. Undoubtedly, the next few years will see this highly dynamic research area yield new insights and exciting breakthroughs in ^{89}Zr-immuno-PET radiopharmaceutical design that will have a transformative impact on how precision medicine strategies are implemented in the clinic.

Acknowledgments: This research was supported by Wake Forest University Health Sciences, Wake Forest Innovations, and the North Carolina Biotechnology Center (2016-BIG-6524). The authors acknowledge the editorial assistance of Karen Klein, MA, in the Wake Forest Clinical and Translational Science Institute (UL1 TR001420; PI: McClain).

Author Contributions: All authors wrote and edited the manuscript.

Conflicts of Interest: N.B.B., D.N.P. and T.J.W. have filed patents relating to work described in this text.

Abbreviations

BFC	bifunctional chelator
Bq	bequerel
Ci	curie
DFC	desferrichrome
DFO	desferrioxamine B
DOTA	1,4,7,10-tetraazacyclododecane-1,4,7,10-tetraacetic acid
DTPA	diethylenetriaminepentaacetic acid
EDTA	ethylenediaminetetraacetic acid
FSC	fusarine C
GBq	gigabequerel
HOPO	hydroxpyridinoate
IAM	hydroxyisopthalamide
kBq	kilobequerel
keV	kiloelectron volt
MBq	Megabequerel
mCi	milicurie
MeV	megaelectron volt
PET	positron emission tomography
TAM	terepthalamide
TFAC	triacetylfusarine C
THPN	1,3-propanediamine-*N,N,N′,N′*-tetrakis[(2-(aminoethy)-3-hydroxy-1-6-dimethyl-4(1*H*)-pyridinone)acetamide]
uCi	microcurie
α++	alpha particle
β−	beta minus particle
β+	positron
γ	gamma photon

References

1. Wang, Y.X.; Choi, Y.; Chen, Z.; Laurent, S.; Gibbs, S.L. Molecular imaging: From bench to clinic. *Biomed. Res. Int.* **2014**, *2014*, 357258. [CrossRef] [PubMed]
2. Ossenkoppele, R.; Prins, N.D.; Pijnenburg, Y.A.; Lemstra, A.W.; van der Flier, W.M.; Adriaanse, S.F.; Windhorst, A.D.; Handels, R.L.; Wolfs, C.A.; Aalten, P.; et al. Impact of molecular imaging on the diagnostic process in a memory clinic. *Alzheimers Dement.* **2013**, *9*, 414–421. [CrossRef] [PubMed]
3. Leuschner, F.; Nahrendorf, M. Molecular imaging of coronary atherosclerosis and myocardial infarction: Considerations for the bench and perspectives for the clinic. *Circ. Res.* **2011**, *108*, 593–606. [CrossRef] [PubMed]
4. Hruska, C.B.; Boughey, J.C.; Phillips, S.W.; Rhodes, D.J.; Wahner-Roedler, D.L.; Whaley, D.H.; Degnim, A.C.; O'Connor, M.K. Scientific Impact Recognition Award: Molecular breast imaging: A review of the Mayo Clinic experience. *Am. J. Surg.* **2008**, *196*, 470–476. [CrossRef] [PubMed]
5. Kipper, M.S. Steps in moving molecular imaging to the clinic. *J. Nucl. Med.* **2008**, *49*, 58N–60N. [PubMed]
6. Schafers, M. The future of molecular imaging in the clinic needs a clear strategy and a multidisciplinary effort. *Basic Res. Cardiol.* **2008**, *103*, 200–202. [CrossRef] [PubMed]
7. Glunde, K.; Bhujwalla, Z.M. Will magnetic resonance imaging (MRI)-based contrast agents for molecular receptor imaging make their way into the clinic? *J. Cell. Mol. Med.* **2008**, *12*, 187–188. [CrossRef] [PubMed]
8. Eckelman, W.C.; Rohatagi, S.; Krohn, K.A.; Vera, D.R. Are there lessons to be learned from drug development that will accelerate the use of molecular imaging probes in the clinic? *Nucl. Med. Biol.* **2005**, *32*, 657–662. [CrossRef] [PubMed]
9. Blankenberg, F.G. Molecular imaging with single photon emission computed tomography. How new tracers can be employed in the nuclear medicine clinic. *IEEE Eng. Med. Biol. Mag.* **2004**, *23*, 51–57. [CrossRef] [PubMed]
10. Bohnen, N.I.; Minoshima, S. FDG-PET and molecular brain imaging in the movement disorders clinic. *Neurology* **2012**, *79*, 1306–1307. [CrossRef] [PubMed]
11. Wang, X.; Feng, H.; Zhao, S.; Xu, J.; Wu, X.; Cui, J.; Zhang, Y.; Qin, Y.; Liu, Z.; Gao, T.; et al. SPECT and PET radiopharmaceuticals for molecular imaging of apoptosis: From bench to clinic. *Oncotarget* **2017**, *8*, 20476–20495. [CrossRef] [PubMed]
12. Bernard-Gauthier, V.; Collier, T.L.; Liang, S.H.; Vasdev, N. Discovery of PET radiopharmaceuticals at the academia-industry interface. *Drug Discov. Today Technol.* **2017**, *25*, 19–26. [CrossRef] [PubMed]
13. Phelps, M.E. *PET Molecular Imaging and Its Biological Implications*; Springer: New York, NY, USA, 2004.
14. Wadas, T.J.; Wong, E.H.; Weisman, G.R.; Anderson, C.J. Coordinating radiometals of copper, gallium, indium, yttrium, and zirconium for PET and SPECT imaging of disease. *Chem. Rev.* **2010**, *110*, 2858–2902. [CrossRef] [PubMed]
15. Anderson, C.J.; Welch, M.J. Radiometal-labeled agents (non-technetium) for diagnostic imaging. *Chem. Rev.* **1999**, *99*, 2219–2234. [CrossRef] [PubMed]
16. Deri, M.A.; Zeglis, B.M.; Francesconi, L.C.; Lewis, J.S. PET imaging with (8)(9)Zr: From radiochemistry to the clinic. *Nucl. Med. Biol.* **2013**, *40*, 3–14. [CrossRef] [PubMed]
17. Welch, M.J.; Redvanly, C.S. *Handbook of Radiopharmaceuticals*; Wiley: Chichester, UK, 2003; p. 847.
18. Link, J.M.; Krohn, K.A.; Eary, J.F.; Kishore, R.; Lewellen, T.K.; Johnson, M.W.; Badger, C.C.; Richter, K.Y.; Nelp, W.B. Zr-89 for antibody labeling and positron emission tomography. *J. Label. Compd. Radiopharm.* **1986**, *23*, 1297–1298.
19. Eary, J.F.; Link, J.M.; Kishore, R.; Johnson, M.W.; Badger, C.C.; Richter, K.Y.; Krohn, K.A.; Nelp, W.B. Production of Positron Emitting Zr89 for Antibody Imaging by Pet. *J. Nucl. Med.* **1986**, *27*, 983.
20. Ikotun, O.F.; Lapi, S.E. The rise of metal radionuclides in medical imaging: copper-64, zirconium-89 and yttrium-86. *Future Med. Chem.* **2011**, *3*, 599–621. [CrossRef] [PubMed]
21. Jauw, Y.W.; Menke-van der Houven van Oordt, C.W.; Hoekstra, O.S.; Hendrikse, N.H.; Vugts, D.J.; Zijlstra, J.M.; Huisman, M.C.; van Dongen, G.A. Immuno-Positron Emission Tomography with Zirconium-89-Labeled Monoclonal Antibodies in Oncology: What Can We Learn from Initial Clinical Trials? *Front. Pharmacol.* **2016**, *7*, 131–141. [CrossRef] [PubMed]

22. Van Dongen, G.A.; Huisman, M.C.; Boellaard, R.; Harry Hendrikse, N.; Windhorst, A.D.; Visser, G.W.; Molthoff, C.F.; Vugts, D.J. 89Zr-immuno-PET for imaging of long circulating drugs and disease targets: Why, how and when to be applied? *Q. J. Nucl. Med. Mol. Imaging* **2015**, *59*, 18–38. [PubMed]

23. van Dongen, G.A.; Vosjan, M.J. Immuno-positron emission tomography: Shedding light on clinical antibody therapy. *Cancer Biother. Radiopharm.* **2010**, *25*, 375–385. [CrossRef] [PubMed]

24. Vugts, D.J.; Visser, G.W.; van Dongen, G.A. 89Zr-PET radiochemistry in the development and application of therapeutic monoclonal antibodies and other biologicals. *Curr. Top. Med. Chem.* **2013**, *13*, 446–457. [CrossRef] [PubMed]

25. Price, E.W.; Orvig, C. Matching chelators to radiometals for radiopharmaceuticals. *Chem. Soc. Rev.* **2014**, *43*, 260–290. [CrossRef] [PubMed]

26. Price, T.W.; Greenman, J.; Stasiuk, G.J. Current advances in ligand design for inorganic positron emission tomography tracers (68)Ga, (64)Cu, (89)Zr and (44)Sc. *Dalton Trans.* **2016**, *45*, 15702–15724. [CrossRef] [PubMed]

27. Heskamp, S.; Raave, R.; Boerman, O.; Rijpkema, M.; Goncalves, V.; Denat, F. (89)Zr-Immuno-Positron Emission Tomography in Oncology: State-of-the-Art (89)Zr Radiochemistry. *Bioconj. Chem.* **2017**, *28*, 2211–2223. [CrossRef] [PubMed]

28. Boros, E.; Holland, J.P. Chemical aspects of metal ion chelation in the synthesis and application antibody-based radiotracers. *J. Label. Comp. Radiopharm.* **2017**. [CrossRef] [PubMed]

29. Chen, F.; Goel, S.; Valdovinos, H.F.; Luo, H.; Hernandez, R.; Barnhart, T.E.; Cai, W. In Vivo Integrity and Biological Fate of Chelator-Free Zirconium-89-Labeled Mesoporous Silica Nanoparticles. *ACS Nano* **2015**, *9*, 7950–7959. [CrossRef] [PubMed]

30. Cheng, L.; Kamkaew, A.; Shen, S.; Valdovinos, H.F.; Sun, H.; Hernandez, R.; Goel, S.; Liu, T.; Thompson, C.R.; Barnhart, T.E.; et al. Facile Preparation of Multifunctional WS2 /WOx Nanodots for Chelator-Free (89) Zr-Labeling and In Vivo PET Imaging. *Small* **2016**, *12*, 5750–5758. [CrossRef] [PubMed]

31. Cheng, L.; Shen, S.; Jiang, D.; Jin, Q.; Ellison, P.A.; Ehlerding, E.B.; Goel, S.; Song, G.; Huang, P.; Barnhart, T.E.; et al. Chelator-Free Labeling of Metal Oxide Nanostructures with Zirconium-89 for Positron Emission Tomography Imaging. *ACS Nano* **2017**, *11*, 12193–12201. [CrossRef] [PubMed]

32. Fairclough, M.; Ellis, B.; Boutin, H.; Jones, A.K.P.; McMahon, A.; Alzabin, S.; Gennari, A.; Prenant, C. Development of a method for the preparation of zirconium-89 radiolabelled chitosan nanoparticles as an application for leukocyte trafficking with positron emission tomography. *Appl. Radiat. Isot.* **2017**, *130*, 7–12. [CrossRef] [PubMed]

33. Goel, S.; Chen, F.; Luan, S.; Valdovinos, H.F.; Shi, S.; Graves, S.A.; Ai, F.; Barnhart, T.E.; Theuer, C.P.; Cai, W. Engineering Intrinsically Zirconium-89 Radiolabeled Self-Destructing Mesoporous Silica Nanostructures for In Vivo Biodistribution and Tumor Targeting Studies. *Adv. Sci.* **2016**, *3*, 1600122. [CrossRef] [PubMed]

34. Heneweer, C.; Holland, J.P.; Divilov, V.; Carlin, S.; Lewis, J.S. Magnitude of enhanced permeability and retention effect in tumors with different phenotypes: 89Zr-albumin as a model system. *J. Nucl. Med.* **2011**, *52*, 625–633. [CrossRef] [PubMed]

35. Kaittanis, C.; Shaffer, T.M.; Bolaender, A.; Appelbaum, Z.; Appelbaum, J.; Chiosis, G.; Grimm, J. Multifunctional MRI/PET Nanobeacons Derived from the in Situ Self-Assembly of Translational Polymers and Clinical Cargo through Coalescent Intermolecular Forces. *Nano Lett.* **2015**, *15*, 8032–8043. [CrossRef] [PubMed]

36. Karmani, L.; Bouchat, V.; Bouzin, C.; Leveque, P.; Labar, D.; Bol, A.; Deumer, G.; Marega, R.; Bonifazi, D.; Haufroid, V.; et al. (89)Zr-labeled anti-endoglin antibody-targeted gold nanoparticles for imaging cancer: Implications for future cancer therapy. *Nanomedicine* **2014**, *9*, 1923–1937. [CrossRef] [PubMed]

37. Karmani, L.; Labar, D.; Valembois, V.; Bouchat, V.; Nagaswaran, P.G.; Bol, A.; Gillart, J.; Leveque, P.; Bouzin, C.; Bonifazi, D.; et al. Antibody-functionalized nanoparticles for imaging cancer: Influence of conjugation to gold nanoparticles on the biodistribution of 89Zr-labeled cetuximab in mice. *Contrast Media Mol. Imaging* **2013**, *8*, 402–408. [CrossRef] [PubMed]

38. Li, N.; Yu, Z.; Pham, T.T.; Blower, P.J.; Yan, R. A generic (89)Zr labeling method to quantify the in vivo pharmacokinetics of liposomal nanoparticles with positron emission tomography. *Int. J. Nanomed.* **2017**, *12*, 3281–3294. [CrossRef] [PubMed]

39. Majmudar, M.D.; Yoo, J.; Keliher, E.J.; Truelove, J.J.; Iwamoto, Y.; Sena, B.; Dutta, P.; Borodovsky, A.; Fitzgerald, K.; Di Carli, M.F.; et al. Polymeric nanoparticle PET/MR imaging allows macrophage detection in atherosclerotic plaques. *Circ. Res.* **2013**, *112*, 755–761. [CrossRef] [PubMed]

40. Miller, L.; Winter, G.; Baur, B.; Witulla, B.; Solbach, C.; Reske, S.; Linden, M. Synthesis, characterization, and biodistribution of multiple 89Zr-labeled pore-expanded mesoporous silica nanoparticles for PET. *Nanoscale* **2014**, *6*, 4928–4935. [CrossRef] [PubMed]

41. Perez-Medina, C.; Abdel-Atti, D.; Tang, J.; Zhao, Y.; Fayad, Z.A.; Lewis, J.S.; Mulder, W.J.; Reiner, T. Nanoreporter PET predicts the efficacy of anti-cancer nanotherapy. *Nat. Commun.* **2016**, *7*, 11838–11840. [CrossRef] [PubMed]

42. Perez-Medina, C.; Abdel-Atti, D.; Zhang, Y.; Longo, V.A.; Irwin, C.P.; Binderup, T.; Ruiz-Cabello, J.; Fayad, Z.A.; Lewis, J.S.; Mulder, W.J.; et al. A modular labeling strategy for in vivo PET and near-infrared fluorescence imaging of nanoparticle tumor targeting. *J. Nucl. Med.* **2014**, *55*, 1706–1711. [CrossRef] [PubMed]

43. Perez-Medina, C.; Tang, J.; Abdel-Atti, D.; Hogstad, B.; Merad, M.; Fisher, E.A.; Fayad, Z.A.; Lewis, J.S.; Mulder, W.J.; Reiner, T. PET Imaging of Tumor-Associated Macrophages with 89Zr-Labeled High-Density Lipoprotein Nanoparticles. *J. Nucl. Med.* **2015**, *56*, 1272–1277. [CrossRef] [PubMed]

44. Polyak, A.; Naszalyi Nagy, L.; Mihaly, J.; Gorres, S.; Wittneben, A.; Leiter, I.; Bankstahl, J.P.; Sajti, L.; Kellermayer, M.; Zrinyi, M.; et al. Preparation and (68)Ga-radiolabeling of porous zirconia nanoparticle platform for PET/CT-imaging guided drug delivery. *J. Pharm. Biomed. Anal.* **2017**, *137*, 146–150. [CrossRef] [PubMed]

45. Shaffer, T.M.; Harmsen, S.; Khwaja, E.; Kircher, M.F.; Drain, C.M.; Grimm, J. Stable Radiolabeling of Sulfur-Functionalized Silica Nanoparticles with Copper-64. *Nano Lett.* **2016**, *16*, 5601–5604. [CrossRef] [PubMed]

46. Shi, S.; Fliss, B.C.; Gu, Z.; Zhu, Y.; Hong, H.; Valdovinos, H.F.; Hernandez, R.; Goel, S.; Luo, H.; Chen, F.; et al. Chelator-Free Labeling of Layered Double Hydroxide Nanoparticles for in Vivo PET Imaging. *Sci. Rep.* **2015**, *5*, 16930–16940. [CrossRef] [PubMed]

47. Lide, D.R. *CRC Handbook of Chemistry and Physics*, 99th ed.; CRC Press: New York, NY, USA, 2007; p. 2560.

48. Cotton, F.A.; Wilkinson, G.; Gaus, P.L. *Basic Inorganic Chemistry*, 3th ed.; John Wiley & Sons, Inc.: Singapore, 1995.

49. Blumenthal, W.B. Zirconium chemsitry in industry. *J. Chem. Educ.* **1962**, *39*, 604–610. [CrossRef]

50. Blumenthal, W.B. Properties of zirconium compounds likely to be of interest to the analytical chemist. *Talanta* **1968**, *15*, 877–882. [CrossRef]

51. Blumenthal, W.B. Zirconium in the ecology. *Am. Ind. Hyg. Assoc. J.* **1973**, *34*, 128–133. [CrossRef] [PubMed]

52. Page, E.M.; Wass, S.A. Zirconium and hafnium 1994. *Coord. Chem. Rev.* **1996**, *152*, 411–466. [CrossRef]

53. Hollink, E.; Stephan, D.W. Zirconium and hafnium. *ChemInform* **2004**, *35*, 106–160. [CrossRef]

54. Cano Sierra, J.; Huerlander, D.; Hill, M.; Kehr, G.; Erker, G.; Frohlich, R. Formation of dinuclear titanium and zirconium complexes by olefin metathesis—Catalytic preparation of organometallic catalyst systems. *Chemistry* **2003**, *9*, 3618–3622. [CrossRef] [PubMed]

55. Despagnet-Ayoub, E.; Henling, L.M.; Labinger, J.A.; Bercaw, J.E. Addition of a phosphine ligand switches an N-heterocyclic carbene-zirconium catalyst from oligomerization to polymerization of 1-hexene. *Dalton Trans.* **2013**, *42*, 15544–15547. [CrossRef] [PubMed]

56. Luconi, L.; Giambastiani, G.; Rossin, A.; Bianchini, C.; Lledos, A. Intramolecular sigma-bond metathesis/ protonolysis on zirconium(IV) and hafnium(IV) pyridylamido olefin polymerization catalyst precursors: Exploring unexpected reactivity paths. *Inorg. Chem.* **2010**, *49*, 6811–6813. [CrossRef] [PubMed]

57. Pacheco, J.J.; Davis, M.E. Synthesis of terephthalic acid via Diels-Alder reactions with ethylene and oxidized variants of 5-hydroxymethylfurfural. *Proc. Natl. Acad. Sci. USA* **2014**, *111*, 8363–8367. [CrossRef] [PubMed]

58. Zuckerman, R.L.; Krska, S.W.; Bergman, R.G. Zirconium-mediated metathesis of imines: A study of the scope, longevity, and mechanism of a complicated catalytic system. *J. Am. Chem. Soc.* **2000**, *122*, 751–761. [CrossRef] [PubMed]

59. Kehr, G.; Frohlich, R.; Wibbeling, B.; Erker, G. (N-pyrrolyl)B(C6F5)2—A new organometallic Lewis acid for the generation of group 4 metallocene cation complexes. *Chemistry* **2000**, *6*, 258–266. [CrossRef]

60. Millward, D.B.; Sammis, G.; Waymouth, R.M. Ring-opening reactions of oxabicyclic alkene compounds: Enantioselective hydride and ethyl additions catalyzed by group 4 metals. *J. Org. Chem.* **2000**, *65*, 3902–3909. [CrossRef] [PubMed]

61. Kobayashi, S.; Ishitani, H. Novel binuclear chiral zirconium catalysts used in enantioselective strecker reactions. *Chirality* **2000**, *12*, 540–543. [CrossRef]

62. Kobayashi, S.; Kusakabe, K.; Ishitani, H. Chiral catalyst optimization using both solid-phase and liquid-phase methods in asymmetric aza Diels-Alder reactions. *Org. Lett.* **2000**, *2*, 1225–1227. [CrossRef] [PubMed]

63. Illemassene, M.; Perrone, J. *Chemical Thermodynamics of Compounds and Complexes of U, Np, Pu, Am, Tc, Se, Ni and Zr with Selected Organic Liquids*; Elsevier Science: Amsterdam, The Netherlands, 2005.

64. Ekberg, C.; Kallvenius, G.; Albinsson, Y.; Brown, P.L. Studies on the hydrolytic behavior of zirconium(IV). *J. Solut. Chem.* **2004**, *33*, 47–79. [CrossRef]

65. Singhal, A.; Toth, L.M.; Lin, J.S.; Affholter, K. Zirconium(IV) tetramer/octamer hydrolysis equilibrium in aqueous hydrochloric acid solution. *J. Am. Chem. Soc.* **1996**, *118*, 11529–11534. [CrossRef]

66. Saha, G.B.; Porile, N.T.; Yaffe, L. (P,Xn) and (P,Pxn) Reactions of Yttrium-89 with 5-85-Mev Protons. *Phys. Rev.* **1966**, *144*, 962–982. [CrossRef]

67. Kasbollah, A.; Eu, P.; Cowell, S.; Deb, P. Review on production of 89Zr in a medical cyclotron for PET radiopharmaceuticals. *J. Nucl. Med. Technol.* **2013**, *41*, 35–41. [CrossRef] [PubMed]

68. Alzimami, K.S.; Ma, A.K. Effective dose to staff members in a positron emission tomography/CT facility using zirconium-89. *Br. J. Radiol.* **2013**, *86*, 20130318. [CrossRef] [PubMed]

69. Makris, N.E.; Boellaard, R.; Visser, E.P.; de Jong, J.R.; Vanderlinden, B.; Wierts, R.; van der Veen, B.J.; Greuter, H.J.; Vugts, D.J.; van Dongen, G.A.; et al. Multicenter harmonization of 89Zr PET/CT performance. *J. Nucl. Med.* **2014**, *55*, 264–267. [CrossRef] [PubMed]

70. Houghton, J.L.; Zeglis, B.M.; Abdel-Atti, D.; Sawada, R.; Scholz, W.W.; Lewis, J.S. Pretargeted Immuno-PET of Pancreatic Cancer: Overcoming Circulating Antigen and Internalized Antibody to Reduce Radiation Doses. *J. Nucl. Med.* **2016**, *57*, 453–459. [CrossRef] [PubMed]

71. DeJesus, O.T.; Nickles, R.J. Pretargeted immuno-PET of pancreatic cancer: Overcoming circulating antigen and internalized antibody to reduce radiation doses. *Appl. Radiat. Isot.* **1990**, *42*, 789–795. [CrossRef]

72. Zweit, J.; Downey, S.; Sharma, H.L. Production of No-Carrier-Added Zirconium-89 for Positron Emission Tomography. *Appl. Radiat. Isot.* **1991**, *42*, 199–215. [CrossRef]

73. Meijs, W.E.; Herscheid, J.D.M.; Haisma, H.J.; van Leuffen, P.J.; Mooy, R.; Pinedo, H.M. Production of Highly Pure No-Carrier Added Zr-89 for the Labeling of Antibodies with a Positron Emitter. *Appl. Radiat. Isot.* **1994**, *45*, 1143–1149. [CrossRef]

74. Verel, I.; Visser, G.W.; Boellaard, R.; Stigter-van Walsum, M.; Snow, G.B.; van Dongen, G.A. 89Zr immuno-PET: Comprehensive procedures for the production of 89Zr-labeled monoclonal antibodies. *J. Nucl. Med.* **2003**, *44*, 1271–1281. [PubMed]

75. Holland, J.P.; Sheh, Y.; Lewis, J.S. Standardized methods for the production of high specific-activity zirconium-89. *Nucl. Med. Biol.* **2009**, *36*, 729–739. [CrossRef] [PubMed]

76. Infantino, A.; Cicoria, G.; Pancaldi, D.; Ciarmatori, A.; Boschi, S.; Fanti, S.; Marengo, M.; Mostacci, D. Prediction of (89)Zr production using the Monte Carlo code FLUKA. *Appl. Radiat. Isot.* **2011**, *69*, 1134–1137. [CrossRef] [PubMed]

77. Lin, M.; Mukhopadhyay, U.; Waligorski, G.J.; Balatoni, J.A.; Gonzalez-Lepera, C. Semi-automated production of (8)(9)Zr-oxalate/(8)(9)Zr-chloride and the potential of (8)(9)Zr-chloride in radiopharmaceutical compounding. *Appl. Radiat. Isot.* **2016**, *107*, 317–322. [CrossRef] [PubMed]

78. Queern, S.L.; Aweda, T.A.; Massicano, A.V.F.; Clanton, N.A.; El Sayed, R.; Sader, J.A.; Zyuzin, A.; Lapi, S.E. Production of Zr-89 using sputtered yttrium coin targets (89)Zr using sputtered yttrium coin targets. *Nucl. Med. Biol.* **2017**, *50*, 11–16. [CrossRef] [PubMed]

79. Sharifian, M.; Sadeghi, M.; Alirezapour, B.; Yarmohammadi, M.; Ardaneh, K. Modeling and experimental data of zirconium-89 production yield. *Appl. Radiat. Isot.* **2017**, *130*, 206–210. [CrossRef] [PubMed]

80. Siikanen, J.; Tran, T.A.; Olsson, T.G.; Strand, S.E.; Sandell, A. A solid target system with remote handling of irradiated targets for PET cyclotrons. *Appl. Radiat. Isot.* **2014**, *94*, 294–301. [CrossRef] [PubMed]

81. Wooten, A.L.; Madrid, E.; Schweitzer, G.D.; Lawrence, L.A.; Mebrahtu, E.; Lewis, B.C.; Lapi, S.E. Routine production of ^{89}Zr using an automated module. *Appl. Sci.* **2013**, *3*, 593–613. [CrossRef]

82. O'Hara, M.J.; Murray, N.J.; Carter, J.C.; Kellogg, C.M.; Link, J.M. Hydroxamate column-based purification of zirconium-89 ((89)Zr) using an automated fluidic platform. *Appl. Radiat. Isot.* **2018**, *132*, 85–94. [CrossRef] [PubMed]

83. Petrik, M.; Zhai, C.; Novy, Z.; Urbanek, L.; Haas, H.; Decristoforo, C. In Vitro and In Vivo Comparison of Selected Ga-68 and Zr-89 Labelled Siderophores. *Mol. Imaging Biol.* **2016**, *18*, 344–352. [CrossRef] [PubMed]

84. Saha, M.; Sarkar, S.; Sarkar, B.; Sharma, B.K.; Bhattacharjee, S.; Tribedi, P. Microbial siderophores and their potential applications: A review. *Environ. Sci. Pollut. Res. Int.* **2016**, *23*, 3984–3999. [CrossRef] [PubMed]

85. Saha, R.; Saha, N.; Donofrio, R.S.; Bestervelt, L.L. Microbial siderophores: A mini review. *J. Basic Microbiol.* **2013**, *53*, 303–317. [CrossRef] [PubMed]

86. Gorden, A.E.; Xu, J.; Raymond, K.N.; Durbin, P. Rational design of sequestering agents for plutonium and other actinides. *Chem. Rev.* **2003**, *103*, 4207–4282. [CrossRef] [PubMed]

87. Meijs, W.E.; Herscheid, J.D.M.; Haisma, H.J.; Pinedo, H.M. Evaluation of desferal as a bifunctional chelating agent for labeling antibodies with Zr-89. *Appl. Radiat. Isot.* **1992**, *43*, 1443–1447. [CrossRef]

88. Rudd, S.E.; Roselt, P.; Cullinane, C.; Hicks, R.J.; Donnelly, P.S. A desferrioxamine B squaramide ester for the incorporation of zirconium-89 into antibodies. *Chem. Commun.* **2016**, *52*, 11889–11892. [CrossRef] [PubMed]

89. Hermanoson, G.T. *Bioconjugate Techniques*, 2th ed.; Elsevier: Amsterdam, The Netherlands, 2008.

90. Meijs, W.E.; Haisma, H.J.; Van der Schors, R.; Wijbrandts, R.; Van den Oever, K.; Klok, R.P.; Pinedo, H.M.; Herscheid, J.D. A facile method for the labeling of proteins with zirconium isotopes. *Nucl. Med. Biol.* **1996**, *23*, 439–448. [CrossRef]

91. Perk, L.R.; Vosjan, M.J.; Visser, G.W.; Budde, M.; Jurek, P.; Kiefer, G.E.; van Dongen, G.A. *p*-Isothiocyanatobenzyl-desferrioxamine: A new bifunctional chelate for facile radiolabeling of monoclonal antibodies with zirconium-89 for immuno-PET imaging. *Eur. J. Nucl. Med. Mol. Imaging* **2010**, *37*, 250–259. [CrossRef] [PubMed]

92. Holland, J.P.; Divilov, V.; Bander, N.H.; Smith-Jones, P.M.; Larson, S.M.; Lewis, J.S. 89Zr-DFO-J591 for immunoPET of prostate-specific membrane antigen expression in vivo. *J. Nucl. Med.* **2010**, *51*, 1293–1300. [CrossRef] [PubMed]

93. Tinianow, J.N.; Pandya, D.N.; Pailloux, S.L.; Ogasawara, A.; Vanderbilt, A.N.; Gill, H.S.; Williams, S.P.; Wadas, T.J.; Magda, D.; Marik, J. Evaluation of a 3-hydroxypyridin-2-one (2,3-HOPO) Based Macrocyclic Chelator for (89)Zr(4+) and Its Use for ImmunoPET Imaging of HER2 Positive Model of Ovarian Carcinoma in Mice. *Theranostics* **2016**, *6*, 511–521. [CrossRef] [PubMed]

94. Pandit-Taskar, N.; O'Donoghue, J.A.; Beylergil, V.; Lyashchenko, S.; Ruan, S.; Solomon, S.B.; Durack, J.C.; Carrasquillo, J.A.; Lefkowitz, R.A.; Gonen, M.; et al. (8)(9)Zr-huJ591 immuno-PET imaging in patients with advanced metastatic prostate cancer. *Eur. J. Nucl. Med. Mol. Imaging* **2014**, *41*, 2093–2105. [CrossRef] [PubMed]

95. Holland, J.P.; Vasdev, N. Charting the mechanism and reactivity of zirconium oxalate with hydroxamate ligands using density functional theory: Implications in new chelate design. *Dalton Trans.* **2014**, *43*, 9872–9884. [CrossRef] [PubMed]

96. Abou, D.S.; Ku, T.; Smith-Jones, P.M. In vivo biodistribution and accumulation of 89Zr in mice. *Nucl. Med. Biol.* **2011**, *38*, 675–681. [CrossRef] [PubMed]

97. Ulaner, G.A.; Hyman, D.M.; Ross, D.S.; Corben, A.; Chandarlapaty, S.; Goldfarb, S.; McArthur, H.; Erinjeri, J.P.; Solomon, S.B.; Kolb, H.; et al. Detection of HER2-Positive Metastases in Patients with HER2-Negative Primary Breast Cancer Using 89Zr-Trastuzumab PET/CT. *J. Nucl. Med.* **2016**, *57*, 1523–1528. [CrossRef] [PubMed]

98. Guerard, F.; Lee, Y.S.; Tripier, R.; Szajek, L.P.; Deschamps, J.R.; Brechbiel, M.W. Investigation of Zr(IV) and 89Zr(IV) complexation with hydroxamates: Progress towards designing a better chelator than desferrioxamine B for immuno-PET imaging. *Chem. Commun.* **2013**, *49*, 1002–1004. [CrossRef] [PubMed]

99. Bailey, G.A.; Price, E.W.; Zeglis, B.M.; Ferreira, C.L.; Boros, E.; Lacasse, M.J.; Patrick, B.O.; Lewis, J.S.; Adam, M.J.; Orvig, C. H(2)azapa: A versatile acyclic multifunctional chelator for (67)Ga, (64)Cu, (111)In, and (177)Lu. *Inorg. Chem.* **2012**, *51*, 12575–12589. [CrossRef] [PubMed]

100. Price, E.W.; Zeglis, B.M.; Lewis, J.S.; Adam, M.J.; Orvig, C. H6phospa-trastuzumab: Bifunctional methylenephosphonate-based chelator with 89Zr, 111In and 177Lu. *Dalton Trans.* **2014**, *43*, 119–131. [CrossRef] [PubMed]

101. Zhai, C.; Summer, D.; Rangger, C.; Franssen, G.M.; Laverman, P.; Haas, H.; Petrik, M.; Haubner, R.; Decristoforo, C. Novel Bifunctional Cyclic Chelator for (89)Zr Labeling-Radiolabeling and Targeting Properties of RGD Conjugates. *Mol. Pharm.* **2015**, *12*, 2142–2150. [CrossRef] [PubMed]

102. Jacobson, O.; Zhu, L.; Niu, G.; Weiss, I.D.; Szajek, L.P.; Ma, Y.; Sun, X.; Yan, Y.; Kiesewetter, D.O.; Liu, S.; et al. MicroPET imaging of integrin alphavbeta3 expressing tumors using 89Zr-RGD peptides. *Mol. Imaging Biol.* **2011**, *13*, 1224–1233. [CrossRef] [PubMed]

103. Summer, D.; Garousi, J.; Oroujeni, M.; Mitran, B.; Andersson, K.G.; Vorobyeva, A.; Lofblom, J.; Orlova, A.; Tolmachev, V.; Decristoforo, C. Cyclic versus Noncyclic Chelating Scaffold for (89)Zr-Labeled ZEGFR:2377 Affibody Bioconjugates Targeting Epidermal Growth Factor Receptor Overexpression. *Mol. Pharm.* **2018**, *15*, 175–185. [CrossRef] [PubMed]

104. Adams, C.J.; Wilson, J.J.; Boros, E. Multifunctional Desferrichrome Analogues as Versatile (89)Zr(IV) Chelators for ImmunoPET Probe Development. *Mol. Pharm.* **2017**, *14*, 2831–2842. [CrossRef] [PubMed]

105. Seibold, U.; Wangler, B.; Wangler, C. Rational Design, Development, and Stability Assessment of a Macrocyclic Four-Hydroxamate-Bearing Bifunctional Chelating Agent for (89) Zr. *ChemMedChem* **2017**, *12*, 1555–1571. [CrossRef] [PubMed]

106. Allott, L.; Da Pieve, C.; Meyers, J.; Spinks, T.; Ciobota, D.M.; Kramer-Marek, G.; Smith, G. Evaluation of DFO-HOPO as an octadentate chelator for zirconium-89. *Chem. Commun.* **2017**, *53*, 8529–8532. [CrossRef] [PubMed]

107. Patra, M.; Bauman, A.; Mari, C.; Fischer, C.A.; Blacque, O.; Haussinger, D.; Gasser, G.; Mindt, T.L. An octadentate bifunctional chelating agent for the development of stable zirconium-89 based molecular imaging probes. *Chem. Commun.* **2014**, *50*, 11523–11525. [CrossRef] [PubMed]

108. Vugts, D.J.; Klaver, C.; Sewing, C.; Poot, A.J.; Adamzek, K.; Huegli, S.; Mari, C.; Visser, G.W.; Valverde, I.E.; Gasser, G.; et al. Comparison of the octadentate bifunctional chelator DFO*-pPhe-NCS and the clinically used hexadentate bifunctional chelator DFO-pPhe-NCS for (89)Zr-immuno-PET. *Eur. J. Nucl. Med. Mol. Imaging* **2017**, *44*, 286–295. [CrossRef] [PubMed]

109. Briand, M.; Aulsebrook, M.L.; Mindt, T.J.; Gasser, G. A solid phase-assisted approach for the facile synthesis of a highly water-soluble zirconium-89 chelator for radiopharmaceutical development. *Dalton Trans.* **2017**, *46*, 16387–16389. [CrossRef] [PubMed]

110. Richardson-Sanchez, T.; Tieu, W.; Gotsbacher, M.P.; Telfer, T.J.; Codd, R. Exploiting the biosynthetic machinery of Streptomyces pilosus to engineer a water-soluble zirconium(IV) chelator. *Org. Biomol. Chem.* **2017**, *15*, 5719–5730. [CrossRef] [PubMed]

111. Gotsbacher, M.P.; Telfer, T.J.; Witting, P.K.; Double, K.L.; Finkelstein, D.I.; Codd, R. Analogues of desferrioxamine B designed to attenuate iron-mediated neurodegeneration: Synthesis, characterisation and activity in the MPTP-mouse model of Parkinson's disease. *Metallomics* **2017**, *9*, 852–864. [CrossRef] [PubMed]

112. Boros, E.; Holland, J.P.; Kenton, N.; Rotile, N.; Caravan, P. Macrocycle-Based Hydroxamate Ligands for Complexation and Immunoconjugation of (89)Zirconium for Positron Emission Tomography (PET) Imaging. *Chempluschem* **2016**, *81*, 274–281. [CrossRef] [PubMed]

113. Bhatt, N.B.; Pandya, D.N.; Xu, J.; Tatum, D.; Magda, D.; Wadas, T.J. Evaluation of macrocyclic hydroxyisophthalamide ligands as chelators for zirconium-89. *PLoS ONE* **2017**, *12*, e0178767–e0178777. [CrossRef] [PubMed]

114. Pandya, D.N.; Pailloux, S.; Tatum, D.; Magda, D.; Wadas, T.J. Di-macrocyclic terephthalamide ligands as chelators for the PET radionuclide zirconium-89. *Chem. Commun.* **2015**, *51*, 2301–2303. [CrossRef] [PubMed]

115. Ma, M.T.; Meszaros, L.K.; Paterson, B.M.; Berry, D.J.; Cooper, M.S.; Ma, Y.; Hider, R.C.; Blower, P.J. Tripodal tris(hydroxypyridinone) ligands for immunoconjugate PET imaging with (89)Zr(4+): Comparison with desferrioxamine-B. *Dalton Trans.* **2015**, *44*, 4884–4900. [CrossRef] [PubMed]

116. Buchwalder, C.; Rodriguez-Rodriguez, C.; Schaffer, P.; Karagiozov, S.K.; Saatchi, K.; Hafeli, U.O. A new tetrapodal 3-hydroxy-4-pyridinone ligand for complexation of (89)zirconium for positron emission tomography (PET) imaging. *Dalton Trans.* **2017**, *46*, 9654–9663. [CrossRef] [PubMed]

117. Deri, M.A.; Ponnala, S.; Zeglis, B.M.; Pohl, G.; Dannenberg, J.J.; Lewis, J.S.; Francesconi, L.C. Alternative chelator for (8)(9)Zr radiopharmaceuticals: Radiolabeling and evaluation of 3,4,3-(LI-1,2-HOPO). *J. Med. Chem.* **2014**, *57*, 4849–4860. [CrossRef] [PubMed]

118. Ulaner, G.A.; Goldman, D.A.; Gonen, M.; Pham, H.; Castillo, R.; Lyashchenko, S.K.; Lewis, J.S.; Dang, C. Initial Results of a Prospective Clinical Trial of 18F-Fluciclovine PET/CT in Newly Diagnosed Invasive Ductal and Invasive Lobular Breast Cancers. *J. Nucl. Med.* **2016**, *57*, 1350–1356. [CrossRef] [PubMed]

119. Pandya, D.N.; Bhatt, N.; Yuan, H.; Day, C.S.; Ehrmann, B.M.; Wright, M.; Bierbach, U.; Wadas, T.J. Zirconium tetraazamacrocycle complexes display extraordinary stability and provide a new strategy for zirconium-89-based radiopharmaceutical development. *Chem. Sci.* **2017**, *8*, 2309–2314. [CrossRef] [PubMed]

120. Alves, L.G.; Hild, F.; Munha, R.F.; Veiros, L.F.; Dagorne, S.; Martins, A.M. Synthesis and structural characterization of novel cyclam-based zirconium complexes and their use in the controlled ROP of rac-lactide: Access to cyclam-functionalized polylactide materials. *Dalton Trans.* **2012**, *41*, 14288–14298. [CrossRef] [PubMed]

121. Alves, L.G.; Madeira, F.; Munha, R.F.; Barroso, S.; Veiros, L.F.; Martins, A.M. Reactions of heteroallenes with cyclam-based Zr(IV) complexes. *Dalton Trans.* **2015**, *44*, 1441–1455. [CrossRef] [PubMed]

122. Rogers, A.; Solari, E.; Floriani, C.; Chiesi-Villa, A.; Rizzoli, C. New directions in amido-transition metal chemistry: The preparation and reaction of mixed amino-amido macrocyclic ligands. *J. Chem Soc. Dalton Trans.* **1997**, 2385–2386. [CrossRef]

123. Munha, R.F.; Veiros, L.F.; Duarte, M.T.; Fryzuk, M.D.; Martins, A.M. Synthesis and structural studies of amido, hydrazido and imido zirconium(IV) complexes incorporating a diamido/diamine cyclam-based ligand. *Dalton Trans.* **2009**, 7494–7508. [CrossRef] [PubMed]

124. Angelis, S.; Solari, E.; Floriani, C.; Chiesi-Villa, A.; Rizzoli, C. Mono- and bis(dibenzotetramethyltetraaza [14]annulene) complexes of Group IV metals including the structure of the lithium derivative of the macrocyclic ligand. *Inorg. Chem.* **1992**, *31*, 2520–2527. [CrossRef]

125. Jewula, P.; Berthet, J.-C.; Chambron, Y.; Rousselin, P.T.; Meyer, M. Synthesis and Structural Study of Tetravalent (Zr4+, Hf4+, Ce4+, Th4+, U4+) Metal Complexes with Cyclic Hydroxamic Acids. *Eur. J. Inorg. Chem.* **2015**, 1529–1541. [CrossRef]

126. Li, A.; Ma, H.; Huang, J. Highly Thermally Stable Eight-Coordinate Dichloride Zirconium Complexes Supported by Tridentate [ONN] Ligands: Syntheses, Characterization, and Ethylene Polymerization Behavior. *Organometallics* **2013**, *32*, 7460–7469. [CrossRef]

127. Solari, E.; Maltese, C.; Franceschi, F.; Floriani, C.; Chiesi-Villa, A.; Rizzoli, C. Geometrical isomerism and redox behaviour in zirconium-Schiff base complexes: The formation of C-C bonds functioning as two-electron reservoirs. *J. Chem. Soc. Dalton Trans.* **1997**, 2903–2910. [CrossRef]

128. Kato, C.; Shinohara, A.; Hayashi, K.; Nomiya, K. Syntheses and X-ray crystal structures of zirconium(IV) and hafnium(IV) complexes containing monovacant Wells-Dawson and Keggin polyoxotungstates. *Inorg. Chem.* **2006**, *45*, 8108–8119. [CrossRef] [PubMed]

129. Cives, M.; Strosberg, J. Radionuclide Therapy for Neuroendocrine Tumors. *Curr. Oncol. Rep.* **2017**, *19*, 9–15. [CrossRef] [PubMed]

130. Pandya, D.N.; Hantgan, R.; Budzevich, M.M.; Kock, N.D.; Morse, D.L.; Batista, I.; Mintz, A.; Li, K.C.; Wadas, T.J. Preliminary Therapy Evaluation of (225)Ac-DOTA-c(RGDyK) Demonstrates that Cerenkov Radiation Derived from (225)Ac Daughter Decay Can Be Detected by Optical Imaging for In Vivo Tumor Visualization. *Theranostics* **2016**, *6*, 698–709. [CrossRef] [PubMed]

131. Virgolini, I.; Traub, T.; Novotny, C.; Leimer, M.; Fuger, B.; Li, S.R.; Patri, P.; Pangerl, T.; Angelberger, P.; Raderer, M.; et al. Experience with indium-111 and yttrium-90-labeled somatostatin analogs. *Curr. Pharm. Des.* **2002**, *8*, 1781–1807. [CrossRef] [PubMed]

132. Virgolini, I.; Britton, K.; Buscombe, J.; Moncayo, R.; Paganelli, G.; Riva, P. In- and Y-DOTA-lanreotide: Results and implications of the MAURITIUS trial. *Semin. Nucl. Med.* **2002**, *32*, 148–155. [CrossRef] [PubMed]

133. Fiebiger, W.C.; Scheithauer, W.; Traub, T.; Kurtaran, A.; Gedlicka, C.; Kornek, G.V.; Virgolini, I.; Raderer, M. Absence of therapeutic efficacy of the somatostatin analogue lanreotide in advanced primary hepatic cholangiocellular cancer and adenocarcinoma of the gallbladder despite in vivo somatostatin-receptor expression. *Scand. J. Gastroenterol.* **2002**, *37*, 222–225. [CrossRef] [PubMed]

134. Zeglis, B.M.; Davis, C.B.; Abdel-Atti, D.; Carlin, S.D.; Chen, A.; Aggeler, R.; Agnew, B.J.; Lewis, J.S. Chemoenzymatic strategy for the synthesis of site-specifically labeled immunoconjugates for multimodal PET and optical imaging. *Bioconj. Chem.* **2014**, *25*, 2123–2128. [CrossRef] [PubMed]

135. Zeglis, B.M.; Emmetiere, F.; Pillarsetty, N.; Weissleder, R.; Lewis, J.S.; Reiner, T. Building Blocks for the Construction of Bioorthogonally Reactive Peptides via Solid-Phase Peptide Synthesis. *ChemistryOpen* **2014**, *3*, 48–53. [CrossRef] [PubMed]

136. Zeglis, B.M.; Lewis, J.S. The bioconjugation and radiosynthesis of 89Zr-DFO-labeled antibodies. *J. Vis. Exp.* **2015**, *96*, 354–366.

137. Laforest, R.; Lapi, S.E.; Oyama, R.; Bose, R.; Tabchy, A.; Marquez-Nostra, B.V.; Burkemper, J.; Wright, B.D.; Frye, J.; Frye, S.; et al. [(89)Zr]Trastuzumab: Evaluation of Radiation Dosimetry, Safety, and Optimal Imaging Parameters in Women with HER2-Positive Breast Cancer. *Mol. Imaging Biol.* **2016**, *18*, 952–959. [CrossRef] [PubMed]

138. Makris, N.E.; Boellaard, R.; van Lingen, A.; Lammertsma, A.A.; van Dongen, G.A.; Verheul, H.M.; Menke, C.W.; Huisman, M.C. PET/CT-derived whole-body and bone marrow dosimetry of 89Zr-cetuximab. *J. Nucl. Med.* **2015**, *56*, 249–254. [CrossRef] [PubMed]

139. Makris, N.E.; van Velden, F.H.; Huisman, M.C.; Menke, C.W.; Lammertsma, A.A.; Boellaard, R. Validation of simplified dosimetry approaches in (8)(9)Zr-PET/CT: The use of manual versus semi-automatic delineation methods to estimate organ absorbed doses. *Med. Phys.* **2014**, *41*, 102503–102533. [CrossRef] [PubMed]

140. Natarajan, A.; Gambhir, S.S. Radiation Dosimetry Study of [(89)Zr]rituximab Tracer for Clinical Translation of B cell NHL Imaging using Positron Emission Tomography. *Mol. Imaging Biol.* **2015**, *17*, 539–547. [CrossRef] [PubMed]

141. Perk, L.R.; Visser, O.J.; Stigter-van Walsum, M.; Vosjan, M.J.; Visser, G.W.; Zijlstra, J.M.; Huijgens, P.C.; van Dongen, G.A. Preparation and evaluation of (89)Zr-Zevalin for monitoring of (90)Y-Zevalin biodistribution with positron emission tomography. *Eur. J. Nucl. Med. Mol. Imaging* **2006**, *33*, 1337–1345. [CrossRef] [PubMed]

142. Rizvi, S.N.; Visser, O.J.; Vosjan, M.J.; van Lingen, A.; Hoekstra, O.S.; Zijlstra, J.M.; Huijgens, P.C.; van Dongen, G.A.; Lubberink, M. Biodistribution, radiation dosimetry and scouting of 90Y-ibritumomab tiuxetan therapy in patients with relapsed B-cell non-Hodgkin's lymphoma using 89Zr-ibritumomab tiuxetan and PET. *Eur. J. Nucl. Med. Mol. Imaging* **2012**, *39*, 512–520. [CrossRef] [PubMed]

Sample Availability: Samples of the compounds are not available from the authors.

molecules

MDPI

Review

Progress in Targeted Alpha-Particle Therapy. What We Learned about Recoils Release from *In Vivo* Generators

Ján Kozempel [1,*], Olga Mokhodoeva [2] and Martin Vlk [1]

1   Czech Technical University in Prague, Faculty of Nuclear Sciences and Physical Engineering, Prague CZ-11519, Czech Republic; martin.vlk@fjfi.cvut.cz
2   Vernadsky Institute of Geochemistry and Analytical Chemistry, Moscow 119991, Russia; olga.mokhodoeva@mail.ru
*   Correspondence: jan.kozempel@fjfi.cvut.cz; Tel.: +420-224-358-253

Received: 17 January 2018; Accepted: 28 February 2018; Published: 5 March 2018

Abstract: This review summarizes recent progress and developments as well as the most important pitfalls in targeted alpha-particle therapy, covering single alpha-particle emitters as well as *in vivo* alpha-particle generators. It discusses the production of radionuclides like ^{211}At, ^{223}Ra, ^{225}Ac/^{213}Bi, labelling and delivery employing various targeting vectors (small molecules, chelators for alpha-emitting nuclides and their biomolecular targets as well as nanocarriers), general radiopharmaceutical issues, preclinical studies, and clinical trials including the possibilities of therapy prognosis and follow-up imaging. Special attention is given to the nuclear recoil effect and its impacts on the possible use of alpha emitters for cancer treatment, proper dose estimation, and labelling chemistry. The most recent and important achievements in the development of alpha emitters carrying vectors for preclinical and clinical use are highlighted along with an outlook for future developments.

Keywords: targeted alpha therapy; nuclear recoil; *in vivo* generators; radium; ^{223}Ra; actinium; astatine; bismuth; alpha particle; decay

1. Introduction

Targeted alpha-particle therapy (TAT) is the most rapidly developing field in nuclear medicine and radiopharmacy. The basic advantage of TAT over commonly used β^- emitting radionuclides therapy lies in the irradiation of fewer cancer cells, micrometastases or tumors by an emission of a single alpha particle or by a cascade of heavy alpha particles from close vicinity. The 2^+ charged α particles with high linear-energy transfer (LET) lose the maximum of their energy close to the Bragg peak at the end of their track. The range in tissues is about 50–100 μm depending on the alpha-particle energy. The energy deposition then occurs in a very small tissue volume and with high relative biological effectiveness (RBE) [1]. This is fully true for single α particle decays. However, so called *in vivo* generators [2] that provide, typically, four α decays, depending on the selected radionuclide system, suffer from the nuclear recoil effect, causing at least partial release of daughter radioactive nuclei from the targeting molecule or a delivery vehicle. In such cases, an unwanted radioactive burden is spread over the body and its elimination is limited [3].

Even though recent developments brought significant clinical results [4,5] and novel insights into the problem of the nuclear recoil effect were gained [6–9], neither a detailed analysis nor exhaustive discussion has been undertaken to solve this problem. Furthermore, proper targeting and dosimetry on a subcellular level has become crucial, and advantageous use of theranostic isotopes or theranostic isotope pairs is becoming very important in therapy prognosis [10].

The nuclear recoil effect causes the release of radioactive daughter nuclei from the original radiopharmaceutical preparations. It may lead to unwanted irradiation of healthy tissues that may cause severe radiotoxic effects like organ dysfunction (e.g., kidneys), secondary tumorigenesis, etc. [11]. The released activity and radioactive daughter nuclei fraction as well as their metabolic fate, therefore, need to be estimated and carefully evaluated. Additionally, the key *in vivo* parameters of the radiopharmaceuticals for TAT like e.g., biological half-life, carrier *in vivo* stability, uptake in the reticulo-endothelial system (RES), plasma clearance, elimination routes, etc. may play an important role.

Dosimetric studies should separately evaluate in detail the contributions of a radiolabelled targeted vector, its labelled metabolites, liberated mother nuclide as well as daughter recoils. The evaluation should be performed either experimentally or using mathematical models. Various techniques were used for *ex vivo* evaluation of activity distribution in tissue samples. They include, e.g., an alpha camera [12] or a timepix detector [13] to assess the distribution in sub-organ or cellular levels. Also the possibility of the Cherenkov radiation luminescence imaging technique for α emitters employing the co-emitted β^- radiations [14] was reported.

Several different approaches were developed regarding the carriers for TAT. Small molecules, particularly those labelled with single α emitters, brought the advantage of fast kinetics even though their *in vivo* stability was not always good. Additional approaches to mitigate radiotoxic effects were studied, e.g., to protect kidneys [15]. Immunoactive molecules like antibodies, antibody fragments, nanobodies or receptor-specific peptides represent another group of highly selective targeting vectors [16].

A relatively novel concept of at least partially recoil-resistant carriers for TAT was developed. It employs nanoconstructs composed of various nanoparticulate materials [6,8,17] that allow further surface chemistry, including antibody targeting. However, the major disadvantageous property of large molecular vectors, e.g., of TiO_2 nanoparticles (NPs) [18], is their typical uptake in RES and slower *in vivo* kinetics, e.g., when using antibody without surface detergent modulation [19].

This review tries to cover all aspects of TAT from the research and development of production of alpha emitters and labelling techniques to the preclinical and clinical research and applications of the developed radiopharmaceuticals. In order to estimate the potential risks and benefits of TAT, we survey important features of different stages of radiopharmaceutical preparation and the directions of required investigation and development.

2. Production of Alpha Emitters

Production of alpha-particle emitters includes, in general, practically all methods for preparation of radionuclide sources—irradiation with charged particles in accelerators, neutron irradiation in a nuclear reactor, separation from long-lived natural radionuclides and various combinations thereof [20–35]. The great advantage of nuclides decaying in a series over single alpha particle emitters is not only in the higher energy deposition in target tissue but, thanks to the good nuclear characteristics, also the possibility of construction of a radionuclide generator. Selected characteristics and the main production methods for the most common alpha emitters used in various phases of research and use in nuclear medicine are summarized in Table 1. The challenges encountered in the production of alpha emitters were discussed in a recent review [21]; however, a wider clinical spread of alpha emitters depends more on the end-users' confidence and better understanding of the TAT concept that should help in overcoming the sometimes negative historical experience (e.g., with ^{226}Ra).

Table 1. Summary and properties of the most relevant alpha particle emitters suitable for nuclear medicine applications.

Radionuclide System *	Half-Life	$E_{\alpha max}/E_{chain}$ [MeV]	Production	Status	References
[149]Tb	4.12 h	3.97	[152]Gd(p, 4n)[149]Tb	Research	[22,23]
[211]At	7.2 h	5.87	[209]Bi(α, 3n)[211]At	Clinical trials	[24,25]
[229]Th/ [225]Ac// [213]Bi	7340 years 10 days 46 min	5.83/27.62	[229]Th decay [226]Ra(p, 2n)[225]Ac [232]Th(p, x)[225]Ac	Clinical trials	[26–28]
[227]Ac/ [227]Th/ [223]Ra	27 years 18 days 11 days	5.87/26.70	[227]Ac/[227]Th/[223]Ra decay	Clinical praxis	[29–32]
[228]Th/ [224]Ra/ [212]Pb	1.9 years 3.7 days 10.6 h	5.69/27.54	[228]Th/[224]Ra decay	Formerly in c.p., Research	[33–35]

* Note that in the case of chain decaying nuclides, not all members of the decay chain are included. Further characteristics of the mainly used chain nuclide are provided in bold.

3. General Radiopharmaceutical Issues

Direct and indirect radiolabeling methods are available for single alpha-particle emitters. Since the nuclear recoil effect does not affect the spread of radioactive burden originating from the recoiling radioactive daughters, particularly the [211]At, a halogen that uses chemistry similar to iodine is very attractive. Furthermore, the radiometals like [149]Tb and several latter decay series members like [213]Bi appear to be very promising. Radionuclides decaying by a series of several α decays release radioactive daughter nuclei from the radiopharmaceutical preparations due to the nuclear recoil effect. This effect complicates the labelling strategies and successful dose targeting. In its presence, both the pharmacokinetic properties of the radiopharmaceuticals and the strategies for elimination of the released radioactive burden need to be optimized.

3.1. Nuclear Recoil Effect and the Release of Daughter Nuclei

Due to the momentum conservation law, part of the decay energy is transferred to a daughter nucleus. An approximate value of this energy can be calculated by the mathematical relation:

$$E_r = \frac{m_\alpha}{M_r} Q \tag{1}$$

where E_r is the recoil energy, m_α the rest mass of an α particle, M_r mass of the recoil and Q is the decay energy. The energy distribution ratio between the alpha particle and the recoiling atom is typically 98% to 2%. The amount of energy that the recoil atom reaches is some 100 keV and that is not negligible. Such energy is sufficient to break some 10,000 chemical bonds (assuming 10 eV/one bond). An example of such 109 keV recoil is the [219]Rn with the range of some 88 nm in a water-like environment (e.g., cells or extracellular matrix). The comparison of LET and ion ranges of α particles and [219]Rn recoils originating from [223]Ra decay is shown in Figure 1 and the ranges of [219]Rn recoils in various materials are summarized in Table 2. Simulations were performed using SRIM code [36]. These factors have a direct impact on radiopharmaceutical stability and purity, as well as on dosimetry and daughter recoils' distribution in tissues, especially when so called *in vivo* generators are employed. In some cases the radioactive recoils are removed from the radiopharmaceutical preparations and their final formulations before use [37,38].

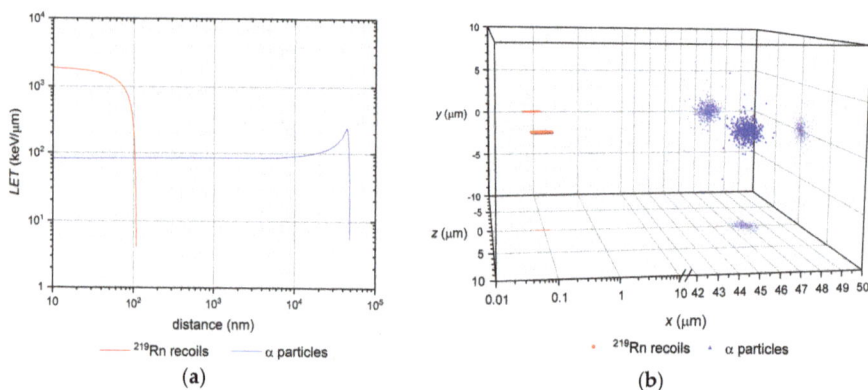

Figure 1. (**a**) Log/log plot of linear-energy transfer (LET) of α particles and ^{219}Rn ions vs. their path (distance) in water up to the rest; (**b**) Semi-log 3D plot of final at rest positions of α particles and ^{219}Rn ions with their *xy*, *xz* and *yz* plane projections. The recoil, in fact, travels in opposite direction to the emitted α particle (common decay-event origin at *x,y,z* = 0,0,0).

Table 2. Ranges of 109 keV ^{219}Rn ions in selected materials.

Material	Range (nm)
Au	11
ZrO_2 ICRU-712	26
Al_2O_3 ICRU-106	27
TiO_2 ICRU-652	28
SiO_2 ICRU-245	46
adult cortical bone	53
human blood	85
prostate tissue	87
water	89
nitrogen gas	76,000

To mitigate the consequences of the nuclear recoil effect in the body, we propose three methods based on the corresponding theorems:

Theorem 1. *Recoils spread mitigation by time—the spread of daughter radioactive ions takes time, so their spread in the organism would also depend on their half-life.*

Proof of Theorem 1. The blood flow measured in terms of red blood cells velocity in capillaries ranges between about 1–3 mm/s [39]. Taking into account this value as a reference for passive transport of radiopharmaceuticals in extracellular matrix or in a capillary blood stream, one may compare this displacement time and the half-life of the corresponding released daughter nuclide. While only one half of ^{219}Rn atoms (T = 3.96 s) decay roughly in 4 s, the number of atoms of further decay series member ^{215}Po (T = 1.78 ms) decreases to $^1/_{1000}$ of its initial amount in 17.8 ms, and it thus has practically no time to escape or to be translocated. Thus, the selection of nuclides with favorable decay properties determines this approach.

Theorem 2. *Recoils spread mitigation by nanoconstruct size/material—daughter-recoiling nuclide consumes some of its energy while getting through the nanoconstruct.*

Proof of Theorem 2. Depending on the nanoconstruct design the stopping power of various materials affects the recoils range. The material and size of the nanoconstruct thus determine the energy loss of recoils in nanoconstruct material. Not only the atomic structure but also the molecular structure and chemical-bond environment affect the stopping ability of the nanoconstruct as a whole [36]. Furthermore, the recoil ion range in nanoconstruct material is limited and its energy is significantly decreased. In general, the stopping power increases with various parameters like atomic weight, electronic density, bond structure, etc. The advantage of spherical nanoconstruct geometry in terms of the stopping efficiency of the nanoconstructs is obvious, and the mother nuclide should be preferably placed in the nanoconstruct core. On the other hand, in case of surface-bonded radionuclide, the probability of daughter recoil ion back-implantation into the nanoconstruct is about 50%.

Theorem 3. *Recoils spread mitigation by the nanoconstructs number/depot—even though the recoil ion may escape a nanoconstruct, the probability of its back-implantation or its implantation into surrounding nanoconstruct units is relatively high.*

Proof of Theorem 3. In cases when time, nanoconstruct material and size are not sufficient to degrade the recoils energy completely, the released ions may be trapped by a depot of surrounding nanoconstructs or even as mentioned in Theorem 2, by the nanoconstruct itself. This proof is also supported by the fact that both surface and intrinsic labelling strategies yielded quite similar data on *in vitro* stabilities results in terms of total released activity [17]. This method is, however, limited to topical applications of radiopharmaceuticals based on larger nanoconstruct aggregates or agglomerates.

3.2. Labelling Chemistry

A fundamental concept of small molecule labelling, e.g., the antibody fragments, peptides and also surface-modified nanocarriers, is based on chelators conjugated throughout a spacer with the vector or the nanocarriers themselves. Spacers are aliphatic or aromatic moieties (C4–C10 or longer) able to establish chemical bonds (e.g., amides, esters, etc.) via nucleophilic substitution, amide formation using carboimides (e.g., dicyclohexylcarbodiimide, diisopropylcarbodiimide) or the Schotten–Baumann reaction of acylhalogenides with amines. The "click reactions" of azides with moieties containing triple bonds play the most important role, e.g., the Huisgen's 1,3-dipolar addition at elevated temperatures resulting in 1,5- or 1,4-isomers, or Cu(I) catalyzed azide-alkine cycloaddition (preferably resulting in 1,4-product). Cycloaddition reactions help to establish a bond between the spacers and targeted moieties very quickly and efficiently.

Excellent chelators of trivalent metals are the azamacrocyclic ligands based on DOTA, NOTA or TETA analogues (e.g., carboxylic or phosphonic)—see Figure 2. Most of them are commercially available with various spacer lengths and as protected (e.g., *t*-butyl or benzyl) or unprotected derivatives.

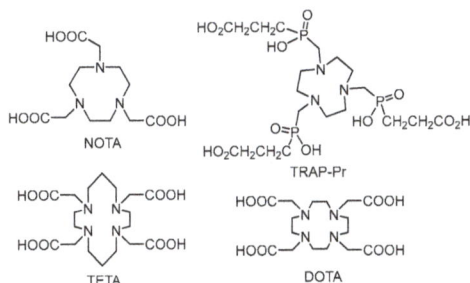

Figure 2. Chemical formulas of cyclic chelators.

These chelators provide very fast trivalent ions complexation kinetics (e.g., Ga, Lu, Tb, Ac, Bi, etc.) depending on pH and temperature. Most of them are used with coordinated stable metals (e.g., Gd) as contrast agents in magnetic resonance imaging (MRI) and they are very often employed as chelators for diagnostic positron emission tomography (PET) radionuclides (e.g., ^{68}Ga) as well as beta decaying therapeutical nuclides (e.g., ^{177}Lu) [40–44]. During the past few years macrocyclic ligands were also used in TAT as chelators suitable for ^{225}Ac or ^{213}Bi [44,45]. Thus, DOTA/NOTA like bifunctional chelators are fulfilling the theranostic concept according to which one chelator may be employed for multimodal diagnostic purposes or as α/β^- therapeutic agents. Concerning the α emitters, it is interesting that even though the energy released during α decay exceeds several hundred times the Me–C, Me–O or Me–N bond energy (Me—radiometal) and the recoils are released from the carrier, *in vivo* experiments indicate that the use of such delivery systems is also feasible [4,46].

As already mentioned, labelling procedures proceed quite rapidly, taking dozens of minutes at laboratory or elevated temperatures (up to 95 °C) at pH = 1–5 depending on the central atom and also ligand structure. Several studies indicate that coordination of trivalent gallium by TRAP-Pr at pH = 1–3 and room temperature is more efficient than NOTA, DOTA, TETA analogues under similar conditions. Optimal labelling protocol was established within 10–30 min for ^{68}Ga at pH = 3–4 (acetate or citrate buffer) at elevated temperatures (90–95 °C). It was also observed that the presence of trace metal impurities like Zn^{2+}, Cu^{2+}, Fe^{3+}, Al^{3+}, Ti^{4+} or Sn^{4+} does not significantly decrease the radiochemical yield while gallium labelling proceeds [47,48]. This ligand is, thus, promising also for other radiometals like ^{213}Bi, ^{225}Ac. However, under certain conditions macrocyclic ligands form mostly in-cage structures. Depending on the reaction conditions and basicity of the ligands, less thermodynamically stable out-of-cage structures may occur usually when the reaction has been performed at lower temperatures. Employing microwave irradiation may also significantly help to ensure faster formation of in-cage complexes. Experimental ^{225}Ac-DOTA-PSMA-617 was synthesized in a microwave reactor at pH = 9 (*TRIS* buffer) within 5 min with radiochemical purity over 98% and specific activity 0.17 MBq/nmol. Similar protocols were employed when synthesized ^{213}Bi-DOTATOC and ^{213}Bi-Substance P were synthesized, hexadentate DOTA-peptide conjugate being used [49,50]. Both ^{213}Bi and ^{212}Bi are considered for the purpose. A ^{212}Pb-TCMC-trastuzumab conjugate was studied on patients with HER-2 receptor carcinoma and its toxicity, pharmacokinetics and dosimetry were investigated. However, the use of DOTA analogues as chelators of ^{225}Ac or ^{213}Bi did not solve the toxicity of daughter recoils. A very interesting alternative to the presented α emitters is the ^{149}Tb, currently studied in a preclinical immunotherapy. Terbium-149 was separated from isobaric and other impurities including stable zinc by extraction with α-hydroxyisobutiric acid solution (pH = 4) and was directly added to DOTANOC (incubation: 15 min at 95 °C). Subsequent high-performance liquid chromatography (HPLC) confirmed an over 98% purity and high specific activity (5 MBq/nmol) of ^{149}Tb-DOTANOC. A similar approach was used for ^{149}Tb-DOTA-folate (incubation: 10 min at 95 °C) [51,52]. Labelling of monoclonal antibody CD20 rituximab with ^{149}Tb in a mixture of ammonium acetate, ascorbic acid and phosphate-buffered saline (PBS) buffer (pH = 5.5) and 10 min incubation at room temperature resulted in 99% yield and specific activity of 1.11 GBq/mg. Conjugate ^{149}Tb-rituximab was prepared using cyclohexane diethylene triamine pentaacetic acid (CHX-A''-DTPA) [53]. This pentaacetic acid analogue is a very interesting ring-opening chelator used in several studies with ^{213}Bi-HuM195 on patients with human myeloid leukemia. The TCMC and CHX-A''-DTPA chelators are shown in Figure 3.

While ^{225}Ac, ^{213}Bi, ^{149}Tb and other radionuclides may be easily coordinated using macrocyclic or DTPA chelating agents, efficient chelator for ^{223}Ra, which is currently used in palliative treatment of bone metastasis of prostate cancer, is still not available. Thus, direct sorption of ^{223}Ra onto surface or intrinsic labelling of nanocariers, e.g., nanohydroxyapatites, LaPO$_4$, SPIONs and others was investigated [6,8,17]. Due to the problematic chemistry of ^{211}At several studies were focused on the possibility of trapping astatine into a nanoconstruct (e.g., gold or silver nanoparticles (NPs), ^{211}AtCl@US-tubes, TiO$_2$), attached to targeting vector via a linker [54–57]. Synthesized nanoconstructs might be stabilised with polyethyleneoxide or polyethylene glycol (PEG). Retention of the α emitter is also significant in liposomes, where about 81% ^{225}Ac retention was observed but the recoil retention was not evaluated [58]. Whereas both the labelling of nanoconstructs or liposomes and stabilization of recoils are quite efficient in comparison with small molecules, the stability of their dispersions (e.g., the hydrodynamic diameter) may significantly vary depending on used material.

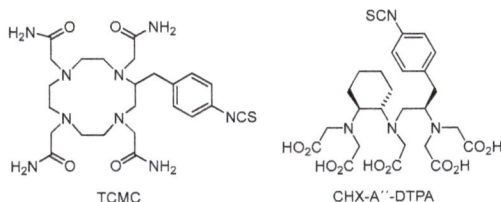

Figure 3. Chemical formulas of TCMC and CHXA''-DTPA chelators.

3.3. Targeting and Clearance

Investigation of how to deliver short-range, high LET radiation to target sites is of key importance. Short α particle range in soft tissues favors their use in the therapy of small lesions, metastases or system-spread diseases like some kinds of leukemia. Depending on the biochemical properties of the radiopharmaceuticals, three targeting strategies could be defined:

1. "self-targeting" based on physiological affinity of the radioisotope to a given tissue; thus radium tends to accumulate in bones or pertechnetate, astatine or iodide in the thyroid;
2. "passive targeting" or "blood circulation and extravasation" is based on accumulation of nanoparticles in the areas around the tumors with leaky vasculature; commonly referred to as the enhanced permeation and retention (EPR) effect [59];
3. "active induced targeting" based on specific ligand-receptor interactions between labelled small molecules, peptides, mAbs and their fragments and target cells; externally activated exposure is also possible (temperature, magnetic field or other activators) [60].

Taking into account the half-lives of the therapeutic nuclides and the recoiling daughters, their circulation time, biodistribution and clearance play a critical role. Matching radionuclide half-lives and pharmacokinetic profiles of the vehicle systems remains a significant criterion [61]. Radionuclides with half-lives long enough to allow differential tumor accumulation and possibly cellular internalization of radiolabeled molecules have some advantages in therapeutic application, but their toxicity for non-targeted sites should be minimized. The features of recoils' distribution in the body was discussed by de Kruijff et al. [62]. Pharmacokinetics of the injected radiopharmaceutical could be a function of both time and tumor size. As an example, the data of a preclinical study with ^{213}Bi-DOTATATE in animals bearing small and large tumors (50 and 200 mm^3) using two tumor models: H69 (human small cell lung carcinoma) and CA20948 (rat pancreatic tumor) are demonstrated in Figure 4 [63].

Different approaches have been explored to inhibit the accumulation of both parent and daughter radionuclides in critical organs or acceleration of their clearance: co-injection of lysine with ^{213}Bi-labelled conjugate can reduce kidney uptake of ^{213}Bi [64], bismuth citrate pre-treatment blocks renal retention of ^{213}Bi [65], and oral administration of BaSO$_4$ known as a coprecipitating agent of radium reduces the ^{223}Ra accumulation in the large intestine [66]. In some cases only locoregional therapy (not intravenous injection) is suited because of the large size or high hydrophilicity of the delivery agent, e.g., encapsulated liposomes or multi-layered nanoconstructs [67]. Imaging methods with the potential for *in vivo* evaluation of the pharmacokinetics of the radionuclides, such as single-photon emission computed tomography (SPECT)/PET/CT imaging are of great importance for assessing the outcome of the therapy.

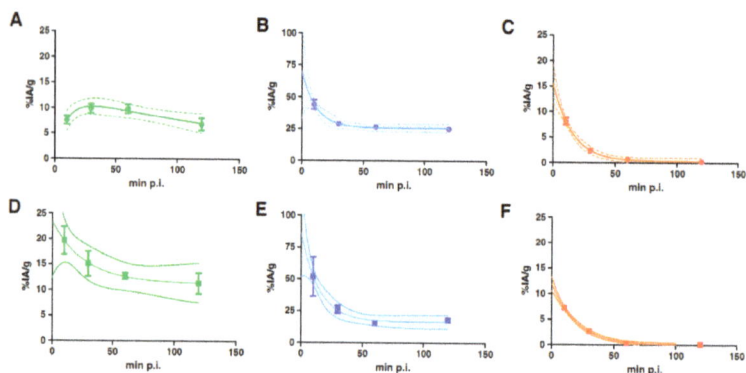

Figure 4. Selected pharmacokinetics of ^{213}Bi-DOTATATE in H69 (**A–C**) and CA20948 (**D–F**) tumor-bearing animals: uptake in tumors (**A,D**) and kidney (**B,E**), and radioactivity in blood (**C,F**) [63].

3.4. Dosimetry

The absorbed dose is defined as an energy delivered to a unit of mass (see Equation (2)).

$$D = \frac{E_x \text{ [J]}}{m_{irr.} \text{ [kg]}} \text{ [Gy]} \tag{2}$$

where the dose, D is defined as a ratio of the energy E_x deposited by the radiation passage to the matter in a unit of mass m_{irr}. This definition is however quite general and does not reflect the specific situation when α emitters and chain decays are used in TAT. This requires precise and accurate dose estimation on all levels, starting from whole body biodistribution down to subcellular level. The example of ^{223}Ra decay that produces one α particle and recoiling ^{219}Rn ion gives a clear picture of such situation. Let us assume that the cell density equals 1 g/cm^3, the mass m_{irr} taken into dose calculation is expressed as the mass of a sphere with the diameter of the ^{219}Rn recoil path, and the energy E_x equals the recoil total energy deposition (109.5 keV). In the case of α particle, a sphere with the diameter of 20 μm (single cell dimension) and only partial energy deposition calculated on the basis of LET is considered. Thus the absorbed dose D delivered by the ^{219}Rn recoil corresponds to 40 kGy in such small volume (total deposited energy of 109.5 keV) while for the α particle it amounts only to 70 mGy over its single-cell path (though the total energy deposited by an alpha particle is 1.83 MeV). To compare the dose in the same mass (or volume), e.g., of one cell, the ratio of the doses delivered by a single α particle and ^{219}Rn recoil turns then to 70 mGy to 4 mGy, respectively. Thus the implications for radionuclide targeting on the subcellular level (e.g., internalization into the nucleus or destruction of cell organelles) play an important role and the contribution of recoil ions should not be neglected. In general, the dosimetry should be evaluated separately in the following levels.

3.4.1. Body Level

In vivo whole body scans with α emitters may provide very helpful and quite detailed information on the pharmacokinetic and pharmacodynamic properties of radiopharmaceuticals [68,69]. Organ intake values, renal clearance or fecal excretion may be evaluated in this way and the recoil release could be possibly visualized by employing the multiple energetic windows data analysis.

3.4.2. Organ and Sub-Organ Levels

The *ex vivo* sample measurements in animal models and also the *in vivo* imaging can provide overall information on the biodistribution and organ uptake of radiopharmaceuticals [12,70]. Sub-organ distribution may also be visualized and more detailed information on target organ uptake compartments may be gained. Such information is again very important for the estimation of tumor therapy prognosis since some tumors do not express their specific antigens or do not accumulate the targeting vectors in their whole volume.

3.4.3. Cellular and Subcellular Level

Dosimetry on a cellular level should clarify the cell-death mechanisms induced by radiation and damage of cellular compartments including DNA damage. Direct (e.g., DNA double strand breaks) and indirect damage mechanisms (e.g., reactive oxygen species generation) should be considered and further analysis is needed, taking into account also the recoil effects. The standard condition of the radionuclide internalization in the cell need not be necessary. The dose distribution on a subcellular level differs significantly for α particles and for the recoil ion—see Section 3.1. The studies published so far did not evaluate the complete decay and the energy distribution in decay products even though microautoradiographic techniques, in a combination with immuno-staining methods, are available [71]. Single α particle-induced damage visualized in real time was also reported [72] and the stochastic simulation of ^{223}Ra α particle irradiation effects on subcellular level was recently performed [73]; however, recoils were not taken into account. The cell-to-cell fluctuations in dose deposition ranged up to about 40%. Interesting results were reported in [74]. In a simplified cellular model, the average number of hits by α particles resulting in a 90% probability of killing exactly one cell was estimated to range from 3.5 to 17.6. However, a better understanding of α particles and the damage induced by the hot recoil atoms is needed to achieve precise proper dose estimation.

Contrary to the efforts of trapping the recoils, an innovative approach that is actually based on the controlled release of recoiling atoms with radioactive nuclei was developed. A novel concept of diffusing α emitter radiation therapy (DaRT) was proposed as a new form of brachytherapy. To treat solid tumors, the method uses α particles employing implantable ^{224}Ra-loaded wire sources that continually release short-lived α particle emitting recoils that spread over a few millimeters inside the tumor [75]. Immunogenic cell death seems to significantly influence the overall effect of the therapy.

4. Vectors for Targeted Alpha Particle Therapy

Efficient and specifically targeted carriers need to be developed in order to realize the potential and favorable properties of α emitters. A variety of conventional and novel drug-delivery systems have been investigated for these purposes: biological macromolecules (antibodies, antibody fragments), small molecule compounds (peptides, affibodies) and nanocarriers/nanoconstructs.

4.1. Small Molecules

4.1.1. MABG

[^{211}At]-*meta*-astatobenzylguanidine (^{211}At-MABG) was synthesized to improve the therapeutic effect for the treatment of malignant pheochromocytoma (PCC) and other diseases [76]. Compared with ^{131}I-MIBG, sufficient cellular uptake and suppression of tumor size after single administration of ^{211}At-MABG (555 kBq/head) have been reported [77]. A kit method for the high-level synthesis of ^{211}At-MABG was also developed [78].

4.1.2. Prostate-Specific Membrane Antigen (PSMA)

Clinical salvage therapy with ^{225}Ac-PSMA-617 was introduced for patients with advanced mCRPC in whom approved therapies had been ineffective. PSMA (prostate-specific membrane antigen) is a 750 amino acid type II transmembrane glycoprotein; after binding at the tumor cell surface the PSMA ligands are internalized allowing radioisotopes to be concentrated within the cell. A standard treatment activity of 100 kBq/kg administered every 8 weeks presents remarkable anti-tumor activity along with tolerable bystander effects and moderate hematological toxicity [4,5]. Figure 5 shows a patient case with impressive results of TAT in comparison with non-effective ^{177}Lu therapy. It was also shown, that [213]Bi labelled *PSMA* targeting agents induce DNA double-strand breaks in prostate cancer xenografts [79].

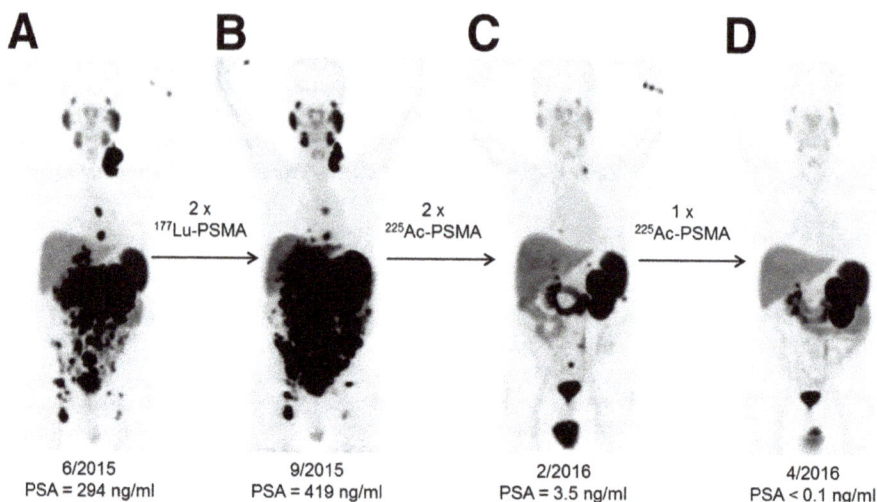

Figure 5. ^{68}Ga-PSMA-11 positron emission tomography (PET)/computed tomography (CT) scans of a patient comparing the initial tumor spread (**A**); restaging after 2 cycles of β⁻ emitting ^{177}Lu-PSMA-617 reveals progression (**B**). In contrast, restaging after second (**C**) and third (**D**) cycles of α emitting [225]Ac-PSMA-617 shows impressive response. This research was originally published in JNM. Kratochwil et al. [225]Ac-PSMA-617 for PSMA-Targeted α-Radiation Therapy of Metastatic Castration-Resistant Prostate Cancer. *J. Nucl. Med.* 2016, 57(12), 1941–1944. © by the Society of Nuclear Medicine and Molecular Imaging, Inc. [4].

4.1.3. Substance P

Clinical experience with the use of peptide carrier Substance P in TAT has recently been reported [80,81]. Patients with recurrent glioblastoma multiforme were treated with 1–7 doses of approx. 2 GBq [213]Bi-DOTA-Substance P or 1–4 doses of 10 MBq [225]Ac-DOTAGA-Substance P at two-month intervals. Favorable toxicity profile and prolonged median survival compared to standard therapy were observed.

4.2. Biomolecules—Antibodies

A detailed description of mAbs radiolabeling with α emitters has been recently given elsewhere [82,83]. Here we mention only some of the clinical and preclinical studies.

Actimab-A, which represents [225]Ac conjugated to lintuzumab (anti-CD33 mAb), demonstrated safety and efficacy against acute myeloid leukemia (AML) in two phase 1 trials. Total administered activities ranged from 37–148 kBq/kg and it was found that baseline peripheral blast count is a highly significant predictor of objective response [84]. The phase 2 trial is currently active at 16 clinical trial sites with patients with AML, age 60 and older, who are ineligible for standard induction chemotherapy [85].

[213]Bi-anti-EGFR-mAb radioimmunoconjugate was prepared by coupling [213]Bi and cetuximab via the chelating agent CHX-A"-DTPA. Intravesical instillation of 366–821 MBq of the [213]Bi-anti-EGFR-mAb in 40 mL of PBS was applied in recurrent bladder cancer patients revealing well-tolerated therapeutic efficacy [86].

The first-in-human clinical studies of ^{212}Pb-AR-RMX (AlphaMedix^{TM}, Houston, TX, USA) for therapy of neuroendocrine tumors were announced to have begun. The biodistribution and safety of this peptide derivative vehicle targeting SSTR2-(+) neuroendocrine cancer cells were clinically evaluated using ^{203}Pb-AR-RMX. No acute or delayed hematological or renal toxicity was observed [87,88].

Preclinical trials of ^{225}Ac-DOTA-anti-PD-L1-BC conjugate have demonstrated promising results in the radioimmunotherapeutic treatment of breast cancer. PD-L1, programmed cell Death Ligand 1, is part of an immune checkpoint system preventing autoimmunity. Anti-PD-L1 antibody (anti-PD-L1-BC) was coupled with p-SCN-Bn-DOTA, and the resulting DOTA-anti-PD-L1-BC conjugate was then labelled with ^{225}Ac in sodium acetate. According to the pilot therapeutic studies a single dose of 15 kBq of the ^{225}Ac-DOTA-anti-PD-L1-BC (3 mg/kg) increased median survival in a metastatic breast cancer mouse model [12].

8C3 mAb, a 2nd-generation murine antibody to melanin of the IgG isotype, was labelled with ^{188}Re or ^{213}Bi directly or via CHXA"-DTPA chelator, respectively to prepare a new agent for therapy of metastatic melanoma. There was statistically significant reduction of lesions in the lungs of mice treated with either 400 mCi ^{188}Re8C3 or 400 mCi ^{213}Bi-8C3 mAb without any undesirable side effects. The unlabeled mAb did not have any effect on the number of the lesions. A statistically significant difference between the ^{188}Re and the ^{213}Bi treatment was not observed [89].

The efficacy of IgC1k 35A7 mAb (anti-carcinoembryonic antigen, CEA) and trastuzumab (anti-HER2) labelled with ^{212}Pb was estimated *in vitro* and *in vivo* in the treatment of small-volume peritoneal carcinomatosis. A strong dose gradient was measured for ^{212}Pb-35A7 mAb; it was much more homogeneous for ^{212}Pb-trastuzumab. The heterogeneity in mAb distribution was found to be counterbalanced by the presence of bystander effects [90]. Trastuzumab was also labelled with ^{225}Ac and studied in a breast cancer spheroids model *in vitro* [91] and with ^{211}At in an athymic rat model with implanted MCF-7/HER2-18 breast carcinoma cells, in which the median survival almost doubled [92].

The small molecule of antibody fragment anti-HER2 2Rs15d Nb was studied as a vehicle of ^{225}Ac [93] and ^{211}At [94]. The labelling was performed via the bifunctional chelating agent p-SCN-Bn-DOTA for ^{225}Ac and three different prosthetic groups m-eATE, SGMAB, MSB for ^{211}At using random and site-specific labelling approaches. All prepared conjugates showed efficient degree of internalization in HER2 + SKOV-3 cells justifying their further *in vivo* evaluation.

Poly(ADP-ribose)polymerase-1 (PARP-1), the nuclear protein which exhibits the ability to target directly chromatin, was functionalized with ^{211}At for the therapy of high-risk neuroblastoma. The prepared ^{211}At-MM4 conjugate demonstrated cytotoxicity to several cell lines [95].

4.3. Macromolecules and Nanoconstructs

Conceptual differences in clinical translation of the above vehicles were pointed out. For instance, antibody conjugates target the cell surface and tend to have limited access to solid tumors [96], whereas radiolabeled peptides are more desirable due to straightforward chemical synthesis, versatility, easier radiolabeling, optimum clearance from the circulation, faster penetration and more uniform distribution into tissues, and also lower immunogenicity [97,98]. Nanoparticle-based systems have been designed to improve biodistribution, stability, specificity, pharmacological and targeting properties, daughter retention, as well as to exploit the theranostic approach [99–101].

Nanoparticles with two layers of cold LaPO$_4$ deposited on the core surface (LaPO$_4$ core and core +2 shells) were synthesized and labelled with either ^{223}Ra or ^{225}Ra/^{225}Ac. The NPs were additionally coated with GdPO$_4$ and gold shells demonstrating retention of both parents and daughters (over 27–35 days) without diminishing the tumoricidal properties of emitted α particles. Consequent conjugation of LaPO$_4$ NPs to 201b mAb, targeting trombomodulin in lung endothelium was carried out using a lipoamide polyethylene glycol (dPEG)-COOH linker. Efficacy of the NPs-antibody conjugate system was demonstrated on reduced EMT-6 lung colonies [102,103].

Novel nuclear-recoil-resistant carriers of ^{223}Ra based on hydroxyapatite were developed [17,104]. Two strategies were used to prepare the nanoconstructs: the surface and the intrinsic (volume) labelling.

High labelling yields as well acceptable *in vitro* and *in vivo* stabilities over the period of ^{223}Ra half-life make the developed nanoconstructs promising for targeted cancer therapy, e.g., bone matrix targeting [17]. Similarly, the ^{223}Ra labelled CaCO$_3$ microparticles were successfully tested in a mice model with ES-2 and SKOV3-luc intraperitoneal ovarian cancer xenografts resulting in considerably reduced tumor volume or a survival benefit [105].

The Au-S-PEG-Substance P (5-11) bioconjugates were proposed to utilize the formation of a strong bond between metallic gold and astatine for binding ^{211}At to the biomolecule. Gold NPs were conjugated with Substance P (5-11), neuropeptide fragment with high affinity to neurokinin type 1 receptors on the glioma cells, through HS-PEG-NHS linker. They were then labelled with ^{211}At by chemisorption on the gold surface. The radiobioconjugates were stable for 24 h in human serum and cerebrospinal fluid, exhibiting high toxicity to glioma cancer cells. However, only local drug application, not intravenous injection, was recommended because of their relatively large size and high hydrophilicity [57,106].

Substance P (5-11) (SP) was also used for functionalization of nanozeolite-A loaded with ^{223}Ra for targeting glioma cancer cells. The small (<5%) release of the daughter radionuclides from the prepared bioconjugate ^{223}Ra-A-silane-PEG-SP (5-11) and the ability of zeolite NPs to re-adsorption of recoiled ^{223}Ra decay products (as a molecular sieve and as a cation-exchanger) along with high receptor affinity toward NK-1 receptor expressing glioma cells *in vitro* make ^{223}Ra-A-silane-PEG-SP (5-11) promising tool for TAT [107]. Nevertheless, like the preceding vehicle it was not recommended for intravenous injection.

Nanocarriers composed of amphiphilic block copolymers, i.e., loaded polymersomes, make it possible to keep the recoiling ^{225}Ac daughters and causing complete destruction of spheroidal tumors. Nevertheless, more studies are necessary to evaluate the *in vivo* recoil-retention effectivity [108].

Nanocarriers in the form of lipid vesicles targeted to PSMA were labelled with ^{225}Ac and compared with to a PSMA-targeted radiolabeled antibody. It was found that targeted vesicles localize closer to the nucleus while antibodies localize near the plasma membrane. Targeted vesicles cause larger numbers of dsDNA breaks per nucleus of treated cells compared with radiolabeled mAb [109].

Interstitial vehicles in the form of pH-tunable liposomes encapsulating chelated ^{225}Ac were designed to enhance the penetration in solid tumors, which is usually limited for radionuclide carriers. The liposomes were composed of 21PC:DSPA:cholesterol(chol):DSPE-PEG:Rhd-lipid. In the slightly acidic tumor interstitium (7.4 > pH > 6.0) a pH-responsive mechanism on the liposome membrane results in the release of the encapsulated radioactivity [110]. This study together with refs. [4,5] actually supports the concept of DaRT therapy [75] in large solid tumors and metastases.

5. Summary

Targeted alpha-particle therapy is a very promising and effective therapeutical tool against cancer. This brief overview of recent developments shows great potential in solving partial pitfalls of this method mainly related to the nuclear-recoil effect. We speculate that two major strategies in TAT field are very likely to develop further—firstly, the use of single α particle emitters and/or carriers able to stop the spread of recoils labelled with chain α emitters; and, secondly, the use of carriers providing controlled release of chain α particle emitters (DaRT concept). While the former field would just apply already-known facts, the latter brings a relatively new concept in the TAT, with an overlap to immunologic signaling and cell death. Despite the many uncertainties and problems in TAT, e.g., concerning the proper dose targeting, it should be pointed out that successful treatment cases in animal models have already been reported for both strategies. Also, recent clinical trials showed that patient benefits prevailed over potential risks. Further research is, however, needed to clarify the dosimetry on all levels and to eliminate the unwanted spread of radioactive burden over the body and the induction of secondary malignancies. TAT should, therefore, become additional and equivalent tools in truly personalized medicine.

Acknowledgments: This work was funded by the Health Research Agency of the Czech Republic, grant No.: NV16-30544A, the Russian Foundation for Basic Research, and Moscow city government according to the research project No.: 15-33-70004 «mol_a_mos», the Technology Agency of the Czech Republic, grant No.: TJ01000334 and the EU & Ministry of Education Youth and Sports of the Czech Republic grant No.: CZ.02.1.01/0.0/0.0/15_003/0000464.

Author Contributions: Ján Kozempel, Martin Vlk and Olga Mokhodoeva wrote the paper and contributed equally.

Conflicts of Interest: The authors declare no conflict of interest. The funding sponsors had no role in the interpretation of data, in the writing of the manuscript, and in the decision to publish the results.

References

1. Song, H.; Senthamizhchelvan, S.; Hobbs, R.F.; Sgouros, G. Alpha Particle Emitter Radiolabeled Antibody for Metastatic Cancer: What Can We Learn from Heavy Ion Beam Radiobiology? *Antibodies* **2012**, *1*, 124–148. [CrossRef]

2. Borchardt, P.E.; Yuan, R.R.; Miederer, M.; McDevitt, M.R.; Scheinberg, D.A. Targeted Actinium-225 *In Vivo* Generators for Therapy of Ovarian Cancer. *Cancer Res.* **2003**, *63*, 5084–5090. [PubMed]

3. Jaggi, J.S.; Kappel, B.J.; McDevitt, M.R.; Sgouros, G.; Flombaum, C.D.; Cabassa, C.; Scheinberg, D.A. Efforts to control the errant products of a targeted *in vivo* generator. *Cancer Res.* **2005**, *65*, 4888–4895. [CrossRef] [PubMed]

4. Kratochwil, C.; Bruchertseifer, F.; Giesel, F.L.; Weis, M.; Verburg, F.A.; Mottaghy, F.; Kopka, K.; Apostolidis, C.; Habekorn, U.; Morgenstern, A. ^{225}Ac-PSMA-617 for PSMA-Targeted α-Radiation Therapy of Metastatic Castration-Resistant Prostate Cancer. *J. Nucl. Med.* **2016**, *57*, 1941–1944. [CrossRef] [PubMed]

5. Kratochwil, C.; Bruchertseifer, F.; Rathke, H.; Bronzel, M.; Apostolidis, C.; Weichert, W.; Haberkorn, U.; Giesel, F.L.; Morgenstern, A. Targeted α-Therapy of Metastatic Castration-Resistant Prostate Cancer with ^{225}Ac-PSMA-617: Dosimetry Estimate and Empiric Dose Finding. *J. Nucl. Med.* **2017**, *58*, 1624–1631. [CrossRef] [PubMed]

6. Woodward, J.; Kennel, S.J.; Stuckey, A.; Osborne, D.; Wall, J.; Rondinone, A.J.; Standaert, R.F.; Mirzadeh, S. LaPO$_4$ nanoparticles doped with actinium-225 that partially sequester daughter radionuclides. *Bioconj. Chem.* **2011**, *22*, 766–776. [CrossRef] [PubMed]

7. Kozempel, J.; Vlk, M. Nanoconstructs in Targeted Alpha-Therapy. *Rec. Pat. Nanomed.* **2014**, *4*, 71–76. [CrossRef]

8. Mokhodoeva, O.; Vlk, M.; Málková, E.; Kukleva, E.; Mičolová, P.; Štamberg, K.; Šlouf, M.; Dzhenloda, R.; Kozempel, J. Study of Ra-223 uptake mechanism by Fe$_3$O$_4$ nanoparticles: Towards new prospective theranostic SPIONs. *J. Nanopart. Res.* **2016**, *18*, 301. [CrossRef]

9. Piotrowska, A.; Leszczuk, E.; Bruchertseifer, F.; Morgenstern, A.; Bilewicz, A. Functionalized NaA nanozeolites labeled with Ra-224,Ra-225 for targeted alpha therapy. *J. Nanopart. Res.* **2013**, *15*, 2082. [CrossRef] [PubMed]

10. Máthé, D.; Szigeti, K.; Hegedűs, N.; Horváth, I.; Veres, D.S.; Kovács, B.; Szűcs, Z. Production and *in vivo* imaging of ^{203}Pb as a surrogate isotope for *in vivo* ^{212}Pb internal absorbed dose studies. *Appl. Radiat. Isot.* **2016**, *114*, 1–6. [CrossRef] [PubMed]

11. Wick, R.R. History and current uses of ^{224}Ra in ankylosing spondylitis and other diseases. *Environ. Int.* **1993**, *19*, 467–473. [CrossRef]

12. Nedrow, J.R.; Josefsson, A.; Park, S.; Back, T.; Hobbs, R.F.; Brayton, C.; Bruchertseifer, F.; Morgenstern, A.; Sgouros, G. Pharmacokinetics, microscale distribution, and dosimetry of alpha-emitter-labeled anti-PD-L1 antibodies in an immune competent transgenic breast cancer model. *EJNMI Res.* **2017**, *7*, 57. [CrossRef] [PubMed]

13. Al Darwish, R.; Staudacher, A.H.; Li, Y.; Brown, M.P.; Bezak, E. Development of a transmission alpha particle dosimetry technique using A549 cells and a Ra-223 source for targeted alpha therapy. *Med. Phys.* **2016**, *43*, 6145–6153. [CrossRef] [PubMed]

14. Ackerman, N.L.; Graves, E.E. The Potential for Cerenkov luminescence imaging of alpha emitting isotopes. *Phys. Med. Biol.* **2012**, *57*, 771–783. [CrossRef] [PubMed]

15. Jaggi, J.S.; Seshan, S.V.; McDevitt, M.R.; Sgouros, G.; Hyjek, E.; Scheinberg, D.A. Mitigation of radiation nephropathy after internal α-particle irradiation of kidneys. *Int. J. Radiat. Oncol. Biol. Phys.* **2006**, *64*, 1503–1512. [CrossRef] [PubMed]

16. Dekempeneer, Y.; Keyaerts, M.; Krasniqi, A.; Puttemans, J.; Muyldermans, S.; Lahoutte, T.; D'huyvetter, M.; Devoogdt, N. Targeted alpha therapy using short-lived alpha-particles and the promise of nanobodies as targeting vehicle. *Expert Opin. Biol. Ther.* **2016**, *16*, 1035–1047. [CrossRef] [PubMed]

17. Kozempel, J.; Vlk, M.; Malková, E.; Bajzíková, A.; Bárta, J.; Santos-Oliveira, R.; Malta Rossi, A. Prospective carriers of ^{223}Ra for targeted alpha particle therapy. *J. Radioanal. Nucl. Chem.* **2015**, *304*, 443–447. [CrossRef]

18. Kreyling, W.G.; Holzwarth, U.; Haberl, N.; Kozempel, J.; Hirn, S.; Wenk, A.; Schleh, C.; Schäffler, M.; Lipka, J.; Semmler-Behnke, M.; et al. Quantitative biokinetics of titanium dioxide nanoparticles after intravenous injection in rats: Part 1. *Nanotoxicology* **2017**, *11*, 434–442. [CrossRef] [PubMed]

19. Jekunen, A.; Kairemo, K.; Karnani, P. *In vivo* Modulators of Antibody Kinetics. *Acta Oncol.* **1996**, *35*, 267–271. [CrossRef] [PubMed]

20. McAlister, D.R.; Horwitz, E.P. Chromatographic generator systems for the actinides and natural decay series elements. *Radiochim. Acta* **2017**, *99*, 151–159. [CrossRef]

21. Sobolev, A.S.; Aliev, R.A.; Kalmykov, S.N. Radionuclides emitting short-range particles and modular nanotransporters for their delivery to target cancer cells. *Russ. Chem. Rev.* **2016**, *85*, 1011–1032. [CrossRef]

22. Steinber, E.P.; Stehney, A.F.; Stearns, C.; Spaletto, I. Production of ^{149}Tb in gold by high-energy protons and its use as an intensity monitor. *Nucl. Phys. A* **1968**, *113*, 265–271. [CrossRef]

23. Beyer, G.J.; Čomor, J.J.; Daković, M.; Soloviev, D.; Tamburella, C.; Hagebo, E.; Allan, B.; Dmitriev, S.N.; Zaitseva, N.G.; Starodub, G.Y.; et al. Production routes of the alpha emitting ^{149}Tb for medical application. *Radiochim. Acta* **2002**, *90*, 247–252. [CrossRef]

24. Lebeda, O.; Jiran, R.; Ráliš, J.; Štursa, J. A new internal target system for production of At-211 on the cyclotron U-120M. *Appl. Radiat. Isot.* **2005**, *63*, 49–53. [CrossRef] [PubMed]

25. Zalutsky, M.R.; Pruszynski, M. Astatine-211: Production and Availability. *Curr. Radiopharm.* **2011**, *4*, 177–185. [CrossRef] [PubMed]

26. Morgenstern, A.; Apostolidis, C.; Molinet, R.; Luetzenkirchen, K. Method for Producing Actinium-225. EP1610346 A1, 28 December 2005.

27. Apostolidis, C.; Molinet, R.; Rasmussen, G.; Morgenstern, A. Production of Ac-225 from Th-229 for targeted alpha therapy. *Anal. Chem.* **2005**, *77*, 6288–6291. [CrossRef] [PubMed]

28. Griswold, J.R.; Medvedev, D.G.; Engle, J.W.; Copping, R.; Fitzsimmons, J.M.; Radchenko, V.; Cooley, J.C.; Fassbender, M.E.; Denton, D.L.; Murphy, K.E.; et al. Large scale accelerator production of ^{225}Ac: Effective cross sections for 78-192 MeV protons incident on Th-232 targets. *Appl. Radiat. Isot.* **2016**, *118*, 366–374. [CrossRef] [PubMed]

29. Larsen, R.; Henriksen, G. The Preparation and Use of Radium-223 to Target Calcified Tissues for Pain Palliation, Bone Cancer Therapy, and Bone Surface Conditioning, WO 2000/40275. 13 July 2000.

30. Henriksen, G.; Hoff, P.; Alstad, J.; Larsen, R.H. ^{223}Ra for endoradiotherapeutic applications prepared from an immobilized ^{227}Ac/^{227}Th source. *Radiochim. Acta* **2001**, *89*, 661–666. [CrossRef]

31. Guseva, L.I.; Tikhomirova, G.S.; Dogadkin, N.N. Anion-exchange separation of radium from alkaline-earth metals and actinides in aqueous-methanol solutions of HNO_3. ^{227}Ac-^{223}Ra generator. *Radiochemistry* **2004**, *46*, 58–62. [CrossRef]

32. Shishkin, D.N.; Kupitskii, S.V.; Kuznetsov, S.A. Extraction generator of ^{223}Ra for nuclear medicine. *Radiochemistry* **2011**, *53*, 343–345. [CrossRef]

33. Schwarz, U.; Daniels, R. Novel Radiotherapeutic Formulations Containing 224Ra and a Method for Their Production. WO 2002/015943, 28 February 2002.

34. Šebesta, F.; Starý, J. A generator for preparation of carrier-free ^{224}Ra. *J. Radioanal. Chem.* **1974**, *21*, 151–155. [CrossRef]

35. Larsen, R.H. Radiopharmaceutical Solutions with Advantageous Properties. WO 2016/135200, 1 September 2016.

36. Ziegler, J.F. SRIM-2013 Code. Available online: http://www.srim.org/ (accessed on 11 November 2017).

37. Frenvik, J.O.; Dyrstad, K.; Kristensen, S.; Ryan, O.B. Development of separation technology for the removal of radium-223 from targeted thorium conjugate formulations. Part I: Purification of decayed thorium-227 on cation exchange columns. *Drug Dev. Ind. Pharm.* **2017**, *43*, 225–233. [CrossRef] [PubMed]

38. Frenvik, J.O.; Dyrstad, K.; Kristensen, S.; Ryan, O.B. Development of separation technology for the removal of radium-223 from targeted thorium conjugate formulations. Part II: Purification of targeted thorium conjugates on cation exchange columns. *Drug Dev. Ind. Pharm.* **2017**, *43*, 1440–1449. [CrossRef] [PubMed]

39. Ivanov, K.P.; Kalinina, M.K.; Levkovich, Y.I. Blood flow velocity in capillaries of brain and muscles and its physiological significance. *Microvasc. Res.* **1981**, *22*, 143–155. [CrossRef]

40. Maheshwari, V.; Dearling, J.L.J.; Treves, S.T.; Packard, A.B. Measurement of the rate of copper(II) exchange for ^{64}Cu complexes of bifunctional chelators. *Inorg. Chim. Acta* **2012**, *393*, 318–323. [CrossRef]

41. Chakravarty, R.; Chakraborty, S.; Ram, R.; Vatsa, R.; Bhusari, P.; Shukla, J.; Mittal, B.R.; Dash, A. Detailed evaluation of different ^{68}Ge/^{68}Ga generators: An attempt toward achieving efficient ^{68}Ga radiopharmacy. *J. Label. Compd. Radiopharm.* **2016**, *59*, 87. [CrossRef] [PubMed]

42. Notni, J.; Plutnar, J.; Wester, H.J. Bone-seeking TRAP conjugates: Surprising observations and their implications on the development of gallium-68-labeled bisphosphonates. *EJNMMI Res.* **2012**, *2*, 13. [CrossRef] [PubMed]

43. Holub, J.; Meckel, M.; Kubíček, V.; Rösch, F.; Hermann, P. Gallium(III) complexes of NOTA-bis (phosphonate) conjugates as PET radiotracers for bone imaging. *Contrast Media Mol. Imaging* **2015**, *10*, 122–134. [CrossRef] [PubMed]

44. Chang, C.A.; Liu, Y.L.; Chen, C.Y.; Chou, X.M. Ligand Preorganization in Metal Ion Complexation: Molecular Mechanics/Dynamics, Kinetics, and Laser-Excited Luminescence Studies of Trivalent Lanthanide Complex Formation with Macrocyclic Ligands TETA and DOTA. *Inorg. Chem.* **2001**, *40*, 3448–3455. [CrossRef] [PubMed]

45. Chan, H.S.; de Blois, E.; Konijnenberg, M.; Morgenstern, A.; Bruchertseifer, F.; Breeman, W.; de Jong, M. Optimizing labeling conditions of ^{213}Bi-somatostatin analogs for receptor-mediated processes in preclinical models. *J. Nucl. Med.* **2014**, *55* (Suppl. 1), 1179.

46. Ryan, O.B.; Cuthbertson, A.; Herstad, G.; Grant, D.; Bjerke, R.M. Development of effective chelators for Th-227 to be used in targeted thorium conjugates. In Proceedings of the 10th International Symposium on Targeted Alpha Therapy, Kanazawa, Japan, 30 May–1 June 2017; p. 57.

47. Notni, J.; Pohle, K.; Wester, H.J. Comparative gallium-68 labeling of TRAP-, NOTA-, and DOTA-peptides: Practical consequences for the future of gallium68-PET. *EJNMMI Res.* **2012**, *2*, 28. [CrossRef] [PubMed]

48. Simeček, J.; Hermann, P.; Wester, H.J.; Notni, J. How is ^{68}Ga-labeling of macrocyclic chelators influenced by metal ion contaminants in ^{68}Ge/^{68}Ga generator eluates? *ChemMedChem* **2013**, *8*, 95–103. [CrossRef] [PubMed]

49. Kratochwil, C.; Giesel, F.L.; Bruchertseifer, F.; Mier, W.; Apostolidis, C.; Boll, R.; Murphy, K.; Haberkom, U.; Morgenstern, A. ^{213}Bi-DOTATOC receptor-targeted alpha-radionuclide therapy induces remission in neuroendocrine tumours refractory to beta radiation: A first-in-human experience. *Eur. J. Nucl. Med. Mol. Imaging* **2014**, *41*, 2106–2119. [CrossRef] [PubMed]

50. Sathekge, M.; Knoesen, O.; Meckel, M.; Modiselle, M.; Vorster, M.; Marx, S. ^{213}Bi-PSMA-617 targeted alpha-radionuclide therapy in metastatic castration-resistant prostate cancer. *Eur. J. Nucl. Med. Mol. Imaging* **2017**, *44*, 1099–1100. [CrossRef] [PubMed]

51. Müller, C.; Reber, J.; Haller, S.; Dorrer, H.; Köster, U.; Johnston, K.; Zhernosekov, K.; Türler, A.; Schibli, R. Folate Receptor Targeted Alpha-Therapy Using Terbium-149. *Pharmaceuticals* **2014**, *7*, 353–365. [CrossRef] [PubMed]

52. Müller, C.; Vermeulen, C.; Köster, U.; Johnston, K.; Türler, A.; Schibli, R.; Van der Meulen, N.P. Alpha-PET with terbium-149: Evidence and perspectives for radiotheragnostics. *EJNMMI Radiopharm. Chem.* **2016**, *1*, 5. [CrossRef]

53. Beyer, G.J.; Miederer, M.; Vranješ-Durić, S.; Čomor, J.J.; Künzi, G.; Hartley, O.; Senekowitsch-Schmidtke, R.; Soloviev, D.; Buchegger, F. The ISOLDE Collaboration. Targeted alpha therapy *in vivo*: Direct evidence for single cancer cell kill using ^{149}Tb-rituximab. *Eur. J. Nucl. Med. Mol. Imaging* **2004**, *31*, 547–554. [CrossRef] [PubMed]

54. Hartman, K.B.; Hamlin, D.K.; Wilbur, D.S.; Wilson, L.J. ^{211}AtCl@US-Tube Nanocapsules: A New Concept in Radiotherapeutic-Agent Design. *Small* **2007**, *3*, 1496–1499. [CrossRef] [PubMed]

55. Kučka, J.; Hrubý, M.; Koňák, Č.; Kozempel, J.; Lebeda, O. Astatination of nanoparticles containing silver as possible carriers of ^{211}At. *Appl. Radiat. Isot.* **2006**, *64*, 201–206. [CrossRef] [PubMed]

56. Leszczuk, E.; Piotrowska, A.; Bilewicz, A. Modified TiO$_2$ nanoparticles as carries for At-211. *J. Label. Compd. Radiopharm.* **2013**, *56*, S242.

57. Dziawer, L.; Koźmiński, P.; Męczyńska-Wielgosz, S.; Pruszyński, M.; Łyczko, M.; Wąs, B.; Celichowski, G.; Grobeny, J.; Jastrzębsky, J.; Bilewicz, A. Gold nanoparticle bioconjugates labelled with 211At for targeted alpha therapy. *RSC Adv.* **2017**, *7*, 41024–41032. [CrossRef]

58. Chang, M.-Y.; Seideman, J.; Sofou, S. Enhanced Loading Efficiency and Retention of ^{225}Ac in Rigid Liposomes for Potential Targeted Therapy of Micrometastases. *Bioconj. Chem.* **2008**, *19*, 1274–1282. [CrossRef] [PubMed]

59. Maeda, H.; Bharate, G.Y.; Daruwalla, J. Polymeric drugs for efficient tumor-targeted drug delivery based on EPR-effect. *Eur. J. Pharm. Biopharm.* **2009**, *71*, 409–419. [CrossRef] [PubMed]

60. Bae, Y.H.; Park, K. Targeted drug delivery to tumors: Myths, reality and possibility. *J. Control. Release* **2011**, *153*, 198–205. [CrossRef] [PubMed]

61. Baidoo, K.E.; Yong, K.; Brechbiel, M.W. Molecular Pathways: Targeted α-Particle Radiation Therapy. *Clin. Cancer Res.* **2013**, *19*, 530–537. [CrossRef] [PubMed]

62. De Kruijff, R.M.; Wolterbeek, H.T.; Denkova, A.G. A Critical Review of Alpha Radionuclide Therapy—How to Deal with Recoiling Daughters? *Pharmaceuticals* **2015**, *8*, 321–336. [CrossRef] [PubMed]

63. Chan, H.S.; Konijnenberg, M.W.; de Blois, E.; Koelewijn, S.; Baum, R.P.; Morgenstern, A.; Bruchertseifer, F.; Breeman, W.A.; de Jong, M. Influence of tumour size on the efficacy of targeted alpha therapy with ^{213}Bi-[DOTA^{0},Tyr^{3}]-octreotate. *EJNMMI Res.* **2016**, *6*, 6–15. [CrossRef] [PubMed]

64. Song, E.Y.; Abbas Rizvi, S.M.; Qu, C.F.; Raja, C.; Brechbiel, M.W.; Morgenstern, A.; Apostolidis, C.; Allen, B.J. Pharmacokinetics and toxicity of ^{213}Bi-labeled PAI2 in preclinical targeted alpha therapy for cancer. *Cancer Biol. Ther.* **2007**, *6*, 898–904. [CrossRef] [PubMed]

65. Nedrow, J.R.; Josefsson, A.; Park, S.; Hobbs, R.F.; Bruchertseifer, F.; Morgenstern, A.; Sgouros, G. Reducing renal uptake of free ^{213}Bi associated with the decay of ^{225}Ac-labeled radiopharmaceuticals. In Proceedings of the 10th International Symposium on Targeted Alpha Therapy, Kanazawa, Japan, 30 May–1 June 2017; p. 67.

66. Hanadate, S.; Washiyama, K.; Yoshimoto, M.; Matsumoto, H.; Tsuji, A.; Higashi, T.; Yoshii, Y. Oral administration of barium sulfate reduces radiation exposure to the large intestine during alpha therapy with radium-223 dichloride. *J. Nucl. Med.* **2017**, *58* (Suppl. l), 1030.

67. Edem, P.E.; Fonslet, J.; Kjaer, A.; Herth, M.; Severin, G. *In vivo* Radionuclide Generators for Diagnostics and Therapy. *Bioinorg. Chem. Appl.* **2016**, 6148357. [CrossRef] [PubMed]

68. Hindorf, C.; Chittenden, S.; Aksnes, A.K.; Parker, C.; Flux, G.D. Quantitative imaging of ^{223}Ra-chloride (Alpharadin) for targeted alpha-emitting radionuclide therapy of bone metastases. *Nucl. Med. Commun.* **2012**, *33*, 726–732. [CrossRef] [PubMed]

69. Robertson, A.K.H.; Ramogida, C.F.; Rodriguez-Rodriguez, C.; Blinder, S.; Kunz, P.; Sossi, V.; Schaffer, P. Multi-isotope SPECT imaging of the Ac-225 decay chain: Feasibility studies. *Phys. Med. Biol.* **2017**, *62*, 4406–4420. [CrossRef] [PubMed]

70. Bäck, T.; Jacobsson, L. The α-Camera: A Quantitative Digital Autoradiography Technique Using a Charge-Coupled Device for *Ex Vivo* High-Resolution Bioimaging of α-Particles. *J. Nucl. Med.* **2010**, *51*, 1616–1623. [CrossRef] [PubMed]

71. Altman, M.B.; Wang, S.J.; Whitlock, J.L.; Roeske, J.C. Cell detection in phase-contrast images used for alpha-particle track-etch dosimetry: A semi-automated approach. *Phys. Med. Biol.* **2005**, *50*, 305–318. [CrossRef] [PubMed]

72. Muggiolu, G.; Pomorski, M.; Claverie, G.; Berthet, G.; Mer-Calfati, C.; Saada, S.; Devès, G.; Simon, M.; Seznec, H.; Barberet, P. Single α-particle irradiation permits real-time visualization of RNF8 accumulation at DNA damaged sites. *Sci. Rep.* **2017**, *7*, 41764. [CrossRef] [PubMed]

73. Gholami, Y.; Zhu, X.; Fulton, R.; Meikle, S.; El-Fakhri, G.; Kuncic, Z. Stochastic simulation of radium-223 dichloride therapy at the sub-cellular level. *Phys. Med. Biol.* **2015**, *60*, 6087–6096. [CrossRef] [PubMed]

74. Roeske, J.C.; Stinchcomb, T.G. The average number of alpha-particle hits to the cell nucleus required to eradicate a tumour cell population. *Phys. Med. Biol.* **2006**, *51*, N179–N186. [CrossRef] [PubMed]

75. Lazarov, E.; Arazi, L.; Efrati, M.; Cooks, T.; Schmidt, M.; Keisari, Y.; Kelson, I. Comparative *in vitro* microdosimetric study of murine- and human-derived cancer cells exposed to alpha particles. *Radiat. Res.* **2012**, *177*, 280–287. [CrossRef] [PubMed]

76. Batra, V.; Ranieri, P.; Makvandi, M.; Tsang, M.; Hou, C.; Li, Y.; Vaidyanathan, G.; Pryma, D.A.; Maris, J.M. Development of meta-[^{211}At]astatobenzylguanidine ([^{211}At]MABG) as an alpha particle emitting systemic targeted radiotherapeutic for neuroblastoma. *Cancer Res.* **2015**, *75* (Suppl. 15), 1610. [CrossRef]

77. Ohshima, Y.; Watanabe, S.; Tsuji, A.; Nagatsu, K.; Sakashima, T.; Sugiyama, A.; Harada, Y.; Waki, A.; Yoshinaga, K.; Ishioka, N. Therapeutic efficacy of α-emitter meta-[211]At-astato-benzylguanidine (MABG) in a pheochromocytoma model. *J. Nucl. Med.* **2016**, *57* (Suppl. 2), 468.

78. Vaidyanathan, G.; Affleck, D.J.; Alston, K.L.; Zhao, X.-G.; Hens, M.; Hunter, D.H.; Babich, J.; Zalutsky, M.R. A Kit Method for the High Level Synthesis of [[211]At]MABG. *Bioorg. Med. Chem.* **2007**, *15*, 3430–3436. [CrossRef] [PubMed]

79. Nonnekens, J.; Chatalic, K.L.S.; Molkenboer-Kuenen, J.D.M.; Beerens, C.E.M.T.; Bruchertseifer, F.; Morgenstern, A.; Veldhoven-Zweistra, J.; Schottelius, M.; Wester, H.-J.; van Gent, D.C.; et al. [213]Bi-Labeled Prostate-Specific Membrane Antigen-Targeting Agents Induce DNA Double-Strand Breaks in Prostate Cancer Xenografts. *Cancer Biother. Radiopharm.* **2017**, *32*, 67–73. [CrossRef] [PubMed]

80. Krolicki, L.; Bruchertseifer, F.; Kunikowska, J.; Koziara, H.; Królicki, B.; Jakuciński, M.; Pawlak, D.; Apostolidis, C.; Rola, R.; Merlo, A.; et al. Targeted alpha therapy of glioblastoma multiforme: Clinical experience with [213]Bi- and [225]Ac-Substance P. In Proceedings of the 10th International Symposium on Targeted Alpha Therapy, Kanazawa, Japan, 30 May–1 June 2017; p. 24.

81. Cordier, D.; Krolicki, L.; Morgenstern, A.; Merlo, A. Targeted Radiolabeled Compounds in Glioma Therapy. *Semin. Nucl. Med.* **2016**, *46*, 243–249. [CrossRef] [PubMed]

82. Marcu, L.; Bezak, E.; Allen, B.J. Global comparison of targeted alpha vs targeted beta therapy for cancer: *In vitro, in vivo* and clinical trials. *Crit. Rev. Oncol. Hematol.* **2018**, *123*, 7–20. [CrossRef] [PubMed]

83. Aghevlian, S.; Boyle, A.J.; Reilly, R.M. Radioimmunotherapy of cancer with high linear energy transfer (LET) radiation delivered by radionuclides emitting α-particles or Auger electrons. *Adv. Drug Deliv. Rev.* **2017**, *109*, 102–118. [CrossRef] [PubMed]

84. Berger, M.; Jurcic, J.; Scheinberg, D. Efficacy of Ac-225-labeled anti-CD33 antibody in acute myeloid leukemia (AML) correlates with peripheral blast count. In Proceedings of the 10th International Symposium on Targeted Alpha Therapy, Kanazawa, Japan, 30 May–1 June 2017; p. 22.

85. Actinium Pharmaceuticals Provides Update on Actimab-A Phase 2 Clinical Trial for Patients with Acute Myeloid Leukemia. Available online: https://ir.actiniumpharma.com/press-releases/detail/247 (accessed on 31 December 2017).

86. Autenrieth, M.E.; Horn, T.; Kurtz, F.; Nguyen, K.; Morgenstern, A.; Bruchertseifer, F.; Schwaiger, M.; Blechert, M.; Seidl, C.; Senekowitsch-Schmidtke, R.; et al. Intravesical radioimmunotherapy of carcinoma in situ of the urinary bladder after BCG failure. *Urol. A* **2017**, *56*, 40–43. [CrossRef] [PubMed]

87. Tworowska, I.; Stallons, T.; Saidi, A.; Wagh, N.; Rojas-Quijano, F.; Jurek, P.; Kiefer, G.; Torgue, J.; Delpassand, E. Pb[203]-AR-RMX conjugates for image-guided TAT of neuroendocrine tumors (NETs). In Proceedings of the American Association for Cancer Research Annual Meeting 2017, Washington, DC, USA, 1–5 April 2017. [CrossRef]

88. RadioMedix and AREVA Med Announce Initiation of Phase 1 Clinical Trial of AlphaMedixTM, a Targeted Alpha Therapy for Patients with Neuroendocrine Tumors. Available online: http://radiomedix.com/news/radiomedix-and-areva-med-announce-initiation-of-phase-1-clinical-trial-of-alphamedixtm-a-targeted-alpha-therapy-for-patients-with-neuroendocrine-tumors (accessed on 31 December 2017).

89. Dadachova, E.; Morgenstern, A.; Bruchertseifer, F.; Rickles., D.J. Radioimmunotherapy with novel IgG to melanin and its comparison with immunotherapy. *J. Nucl. Med.* **2017**, *58* (Suppl. 1), 1036.

90. Boudousq, V.; Bobyk, L.; Busson, M.; Garambois, V.; Jarlier, M.; Charalambatou, P.; Pèlegrin, A.; Paillas, S.; Chouin, N.; Quenet, F.; et al. Comparison between Internalizing Anti-HER2 mAbs and Non-Internalizing Anti-CEA mAbs in Alpha-Radioimmunotherapy of Small Volume Peritoneal Carcinomatosis Using [212]Pb. *PLoS ONE* **2013**, *8*, e69613. [CrossRef] [PubMed]

91. Ballangrud, Å.M.; Yang, W.-H.; Palm, S.; Enmon, R.; Borchardt, P.E.; Pellegrini, V.A.; McDevitt, M.R.; Scheinberg, D.A.; Sgouros, G. Alpha-Particle Emitting Atomic Generator (Actinium-225)-Labeled Trastuzumab (Herceptin) Targeting of Breast Cancer Spheroids. *Clin. Cancer Res.* **2004**, *10*, 4489–4497. [CrossRef] [PubMed]

92. Boskovitz, A.; McLendon, R.E.; Okamura, T.; Sampson, J.H.; Bigner, D.D.; Zalutsky, M.R. Treatment of HER2-positive breast carcinomatous meningitis with intrathecal administration of α-particle-emitting [211]At-labeled trastuzumab. *Nucl. Med. Biol.* **2009**, *36*, 659–669. [CrossRef] [PubMed]

93. Pruszyński, M.; D'Huyvetter, M.; Cędrowska, E.; Lahoutte, T.; Bruchertseifer, F.; Morgenstern, A. Preclinical evaluation of anti-HER2 2Rs15d nanobody labeled with ^{225}Ac. In Proceedings of the 10th International Symposium on Targeted Alpha Therapy, Kanazawa, Japan, 30 May–1 June 2017; p. 34.

94. Dekempeneer, Y.; D'Huyvetter, M.; Aneheim, E.; Xavier, C.; Lahoutte, T.; Bäck, T.; Jensen, H.; Caveliers, V.; Lindegren, S. Preclinical evaluation of astatinated nanobodies for targeted alpha therapy. In Proceedings of the 10th International Symposium on Targeted Alpha Therapy, Kanazawa, Japan, 30 May–1 June 2017; p. 35.

95. Puentes, L.; Xu, K.; Hou, C.; Mach, R.H.; Maris, J.M.; Pryma, D.A.; Makvandi, M. Targeting PARP-1 to deliver alpha-particles to cancer chromatin. In Proceedings of the American Association for Cancer Research Annual Meeting 2017, Washington, DC, USA, 1–5 April 2017. [CrossRef]

96. Carrasquillo, J.A. Alpha Radionuclide Therapy: Principles and Applications to NETs. In *Diagnostic and Therapeutic Nuclear Medicine for Neuroendocrine Tumors*; Pacak, K., Taïeb, D., Eds.; Humana Press: Cham, Switzerland, 2017; pp. 429–445.

97. Norain, A.; Dadachova, E. Targeted Radionuclide Therapy of Melanoma. *Semin. Nucl. Med.* **2016**, *46*, 250–259. [CrossRef] [PubMed]

98. Basu, S.; Banerjee, S. Envisaging an alpha therapy programme in the atomic energy establishments: The priorities and the nuances. *Eur. J. Nucl. Med. Mol. Imaging* **2017**, *44*, 1244–1246. [CrossRef] [PubMed]

99. Koziorowski, J.; Stanciu, A.E.; Gomez-Vallejo, V.; Llop, J. Radiolabeled nanoparticles for cancer diagnosis and therapy. *Anticancer Agents Med. Chem.* **2017**, *17*, 333–354. [CrossRef] [PubMed]

100. Beeler, E.; Gabani, P.; Singh, O.M. Implementation of nanoparticles in therapeutic radiation oncology. *J. Nanopart. Res.* **2017**, *19*, 179. [CrossRef]

101. Drude, N.; Tienken, L.; Mottaghy, F.M. Theranostic and nanotheranostic probes in nuclear medicine. *Methods* **2017**, *130*, 14–22. [CrossRef] [PubMed]

102. McLaughlin, M.F.; Robertson, D.; Pevsner, P.H.; Wall, J.S.; Mirzadeh, S.; Kennel, S.J. LnPO$_4$ Nanoparticles Doped with Ac-225 and Sequestered Daughters for Targeted Alpha Therapy. *Cancer Biother. Radiopharm.* **2014**, *29*, 34–41. [CrossRef] [PubMed]

103. Rojas, J.V.; Woodward, J.D.; Chen, N.; Rondinone, A.J.; Castano, C.H.; Mirzadeh, S. Synthesis and characterization of lanthanum phosphate nanoparticles as carriers for Ra-223 and Ra-225 for targeted alpha therapy. *Nucl. Med. Biol.* **2015**, *42*, 614–620. [CrossRef] [PubMed]

104. Salberg, G.; Larsen, R. Alpha-Emitting Hydroxyapatite Particles. WO 2005/079867, 1 September 2015.

105. Westrøm, S.; Bønsdorff, T.B.; Bruland, Ø.; Larsen, R. Therapeutic Effect of α-Emitting Ra-Labeled Calcium Carbonate Microparticles in Mice with Intraperitoneal Ovarian Cancer. *Transl. Oncol.* **2018**, *11*, 259–267. [CrossRef] [PubMed]

106. Ostrowski, S.; Majkowska-Pilip, A.; Bilewicz, A.; Dobrowolski, J.C. On Au$_n$At clusters as potential astatine carriers. *RSC Adv.* **2017**, *7*, 35854–35857. [CrossRef]

107. Piotrowska, A.; Męczyńska-Wielgosz, S.; Majkowska-Pilip, A.; Koźmiński, P.; Wójciuk, G.; Cędrowska, E.; Bruchertseifer, F.; Morgenstern, A.; Kruszewski, M.; Bilewicz, A. Nanozeolite bioconjugates labeled with ^{223}Ra for targeted alpha therapy. *Nucl. Med. Biol.* **2017**, *47*, 10–18. [CrossRef] [PubMed]

108. De Kruijff, R.M.; Drost, K.; Thijssen, L.; Morgenstern, A.; Bruchertseifer, F.; Lathouwers, D.; Wolterbeek, H.T.; Denkova, A.G. Improved ^{225}Ac daughter retention in LnPO$_4$ containing polymersomes. *Appl. Radiat. Isot.* **2017**, *128*, 183–189. [CrossRef] [PubMed]

109. Zhu, C.; Bandekar, A.; Sempkowski, M.; Banerjee, S.R.; Pomper, M.G.; Bruchertseifer, F.; Morgenstern, A.; Sofou, S. Nanoconjugation of PSMA-targeting ligands enhances perinuclear localization and improves efficacy of delivered alpha-particle emitters against tumor endothelial analogues. *Mol. Cancer Ther.* **2016**, *15*, 106–113. [CrossRef] [PubMed]

110. Zhu, C.; Sempkowski, M.; Holleran, T.; Linz, T.; Bertalan, T.; Josefsson, A.; Bruchertseifer, F.; Morgenstern, A.; Sofou, S. Alpha-particle radiotherapy: For large solid tumors diffusion trumps targeting. *Biomaterials* **2017**, *130*, 67–75. [CrossRef] [PubMed]

molecules

MDPI

Article

Influence of Storage Temperature on Radiochemical Purity of ^{99m}Tc-Radiopharmaceuticals

Licia Uccelli [1,2,*], Alessandra Boschi [1], Petra Martini [3,4], Corrado Cittanti [1,2], Stefania Bertelli [2], Doretta Bortolotti [2], Elena Govoni [2], Luca Lodi [2], Simona Romani [2], Samanta Zaccaria [2], Elisa Zappaterra [2], Donatella Farina [2], Carlotta Rizzo [1], Melchiore Giganti [1] and Mirco Bartolomei [2]

[1] Morphology, Surgery and Experimental Medicine Department, University of Ferrara, Via L. Borsari, 46, 44121 Ferrara (FE), Italy; alessandra.boschi@unife.it (A.B.); corrado.cittanti@unife.it (C.C.); carlotta.rizzo@student.unife.it (C.R.); ggm@unife.it (M.G.)
[2] Nuclear Medicine Unit, University Hospital, Via Aldo Moro, 8, 44124 Ferrara (FE), Italy; s.bertelli@ospfe.it (S.B.); d.bortolotti@ospfe.it (D.B.); e.govoni@ospfe.it (E.G.); l.lodi@ospfe.it (L.L.); s.romani@ospfe.it (S.R.); s.zaccaria@ospfe.it (S.Z.); e.zappaterra@ospfe.it (E.Z.); d.farina@ospfe.it (D.F.); m.bartolomei@ospfe.it (M.B.)
[3] Physics and Heart Science Department, University of Ferrara, Via Giuseppe Saragat, 1, 44122 Ferrara (FE), Italy; petra.martini@unife.it
[4] Legnaro National Laboratories, Italian National Institute for Nuclear Physics (LNL-INFN), Viale dell'Università, 2, 35020 Legnaro (PD), Italy
* Correspondence: ccl@unife.it; Tel.: +39-0532-237462

Received: 26 January 2018; Accepted: 14 March 2018; Published: 15 March 2018

Abstract: The influence of effective room temperature on the radiochemical purity of ^{99m}Tc-radiopharmaceuticals was reported. This study was born from the observation that in the isolators used for the preparation of the ^{99m}Tc-radiopharmaceuticals the temperatures can be higher than those reported in the commercial illustrative leaflets of the kits. This is due, in particular, to the small size of the work area, the presence of instruments for heating, the continuous activation of air filtration, in addition to the fact that the environment of the isolator used for the ^{99m}Tc-radiopharmaceuticals preparation and storage is completely isolated and not conditioned. A total of 244 ^{99m}Tc-radiopharmaceutical preparations (seven different types) have been tested and the radiochemical purity was checked at the end of preparation and until the expiry time. Moreover, we found that the mean temperature into the isolator was significantly higher than 25 °C, the temperature, in general, required for the preparation and storage of ^{99m}Tc-radiopharmaceuticals. Results confirmed the radiochemical stability of radiopharmaceutical products. However, as required in the field of quality assurance, the impact that different conditions than those required by the manufacturer on the radiopharmaceuticals quality have to be verified before human administration.

Keywords: ^{99m}Tc-radiopharmaceuticals; radiochemical purity; radiopharmaceuticals quality control

1. Introduction

Radiopharmaceuticals labelled with technetium-99m (^{99m}Tc) have been, and still are, by far the most used conventional agents for single photon emission computed tomography (SPECT). These radiopharmaceuticals are prepared directly in hospital sites, using commercial lyophilized kit formulations, by introducing the ^{99}Mo/^{99m}Tc generator-eluted ^{99m}Tc-pertechnetate into the kit and following the manufacturer's instruction contained in the package leaflet. The instructions concern not only the reconstitution procedure of the kit formulation, but also the storage conditions and the associated quality control procedures. Environmental factors, such as temperature, humidity, and/or

Molecules **2018**, *23*, 661

light, intrinsic factors, such as chemical-physical properties of the reagents and excipients present in the kit formulations, pharmaceutical composition, quality of the generator eluted [^{99m}Tc]NaTcO$_4$, etc., could compromise the radiochemical purity of technetium-99m radiopharmaceuticals [1–4]. In particular, these factors could cause the formation of chemical impurities inducing phenomena such as hydrolysis and oxidation-reduction reactions. Among the impurities, residual ^{99m}Tc-pertechnetate, as consequence of non-reduction, and the reduced hydrolyzed technetium-99m [^{99m}Tc]TcO$_2$, due to the pertechnetate reduction, but not the complexation of the technetium-99m [5], lead to consequent radiochemical purity (RCP) loss of the final radiopharmaceutical. The effect of many listed factors on RCP of some ^{99m}Tc-radiopharmaceuticals have been investigated in the past, e.g., the technetium-99 presence excess in solution [2]; the humidity influence on radiochemical purity of [^{99m}Tc]Tc-ECD and [^{99m}Tc]Tc-MIBI [1]; the saline storage condition, used for the preparation of [^{99m}Tc]Tc-HMPAO [6]; and the effect of increased temperature on the labelling efficiency of some ^{99m}Tc-radiopharmaceutilcals [7]. In particular, Maksin et al. exposed some compounds, such as pyrophosphate (PyP), dicarboxypropane diphosphonate (DPD), dimercaptosuccinate (DMS), and Sn-colloid, at 37 °C for different time intervals before labelling with ^{99m}Tc-pertechnetate. They found that, only for ^{99m}Tc-Sn-colloids, the increased temperature exposure had no effect on the expected labelling yield in the routine quality control.   Taking into account these results and in particular, considering that in our nuclear medicine hospital unit the preparation of ^{99m}Tc-radiopharmaceuticals takes place inside a not conditioned isolator and, therefore, the temperature could be significantly higher than required by manufacturer, we decided to investigate the effect of the temperature on the radiochemical purity of radiopharmaceuticals at the end of the preparation and during the storage before injection. In fact, while the temperature required by manufacturers for the production of ^{99m}Tc-radiopharmaceuticals is variable, the radiolabelled kits storage temperature value has to be <25 °C. Furthermore, the cited works [6,7] analyzed different aspects from this that we have investigated.

The aim of this work was to investigate the effect of storage temperature on the radiochemical purity of Ceretec (GE Healthcare, Milan, Italy), MAASOL (GE Healthcare, Milan, Italy), Nanocoll (GE Healthcare, Milan, Italy), Renocis (IBA molecular Italy S.r.l., Milan, Italy), Technemibi (Mallinckrodt, Petten, The Netherlands), Technescan HDP (Mallinckrodt, Petten, The Netherlands) and Technescan MAG3 (Mallinckrodt, Petten, The Netherlands).

2. Materials and Methods

All materials used and the methods followed are described in the manufacturer's instructions. These instructions represent the reference for operators in the production and quality control of radiopharmaceuticals. All tools and equipment described in the following sections were initially qualified (installation qualification, operational qualification, and performance qualification) and periodically re-qualified.

2.1. ^{99m}Tc-Radiopharmaceuticals Preparation

^{99}Mo/^{99m}Tc Drytec generators (25 GBq of ^{99}Mo) were purchased by GE (GE Healthcare S.r.l, Milan, Italy). Generator elution and ^{99m}Tc-radiopharmaceuticals synthesis from kit formulation were conducted into a class A laminar flow isolator (Murphil-Tc, MecMurphil S.r.l, Ferrara, Italy, Figure 1), placed in a class D room.

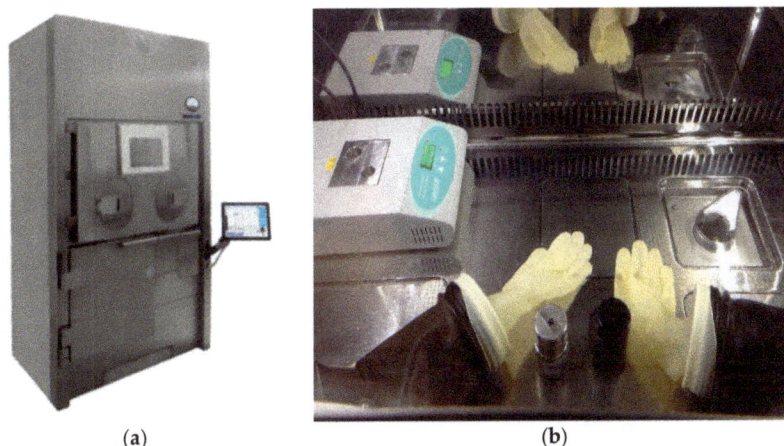

(a)   (b)

Figure 1. Murphil-Tc (MecMurphil): the isolator (**a**) and the work area inside the isolator (**b**).

Freeze-dried kits (Table 1) have been stored until their use in a temperature-controlled refrigerator (Medical Proget, KW control, marketed by MecMurphil S.r.l, Ferrara, Italy).

Table 1. Freeze-dried kits analyzed and the temperature required by the manufacturer for the storage of lyophilized kits and radiopharmaceutical preparations.

Name	Radiopharmaceuticals	Kit Storage Conditions [1]	Radiopharmaceuticals Storage Conditions [1]
Ceretec (GE Healthcare)	[^{99m}Tc]Tc-HMPAO	<25 °C (1 year)	<25 °C (30 min)
MAASOL (GE Healthcare)	[^{99m}Tc]Tc-albumin colloid	2–8 °C (2 year)	<25 °C (6 h)
Nanocoll (GE Healthcare)	[^{99m}Tc]Tc-nanocolloid	2–8 °C (1 year)	<25 °C (6 h)
Renocis (IBA molecular)	[^{99m}Tc]Tc-DMSA	2–8 °C (1 year)	15–25 °C (8h)
Technemibi (Mallinckrodt)	[^{99m}Tc]Tc-MIBI	<25 °C (2 year)	<25 °C (10 h)
Technescan HDP (Mallinckrodt)	[^{99m}Tc]Tc-HDP	<25 °C (2 year)	<25 °C (8 h)
Technescan MAG3 (Mallinckrodt)	[^{99m}Tc]Tc-MAG3	2–8 °C (1 year)	<25 °C (8 h)

[1] The temperature and time values of storage have been obtained from the most recently approved package inserts.

Kit reconstitution was performed according to the methods described in the package insert included within the commercial kits. In order to evaluate radiochemical purity in the worst case, the volumes and the activities used during the reconstitution of the kit were the maximum possible. When required by the synthesis process, e.g., in the [^{99m}Tc]Tc-MAG3 and [^{99m}Tc]Tc-MIBI production (Table 2), a digital dry heater (AccuBlockTM, LabNet, purchased by MecMurphil S.r.l, Ferrara, Italy) was used. After labelling with ^{99m}Tc-pertechnetate, the radiopharmaceuticals solutions were stored inside the isolator until the complete fractionation. Continuous temperature monitoring inside the isolator was performed using calibrated graphic recorder KT621 (Dickson, provided by the University Hospital of Ferrara). The temperature was recorded on a 101 mm diameter paper disc using a red pen (Figure 2).

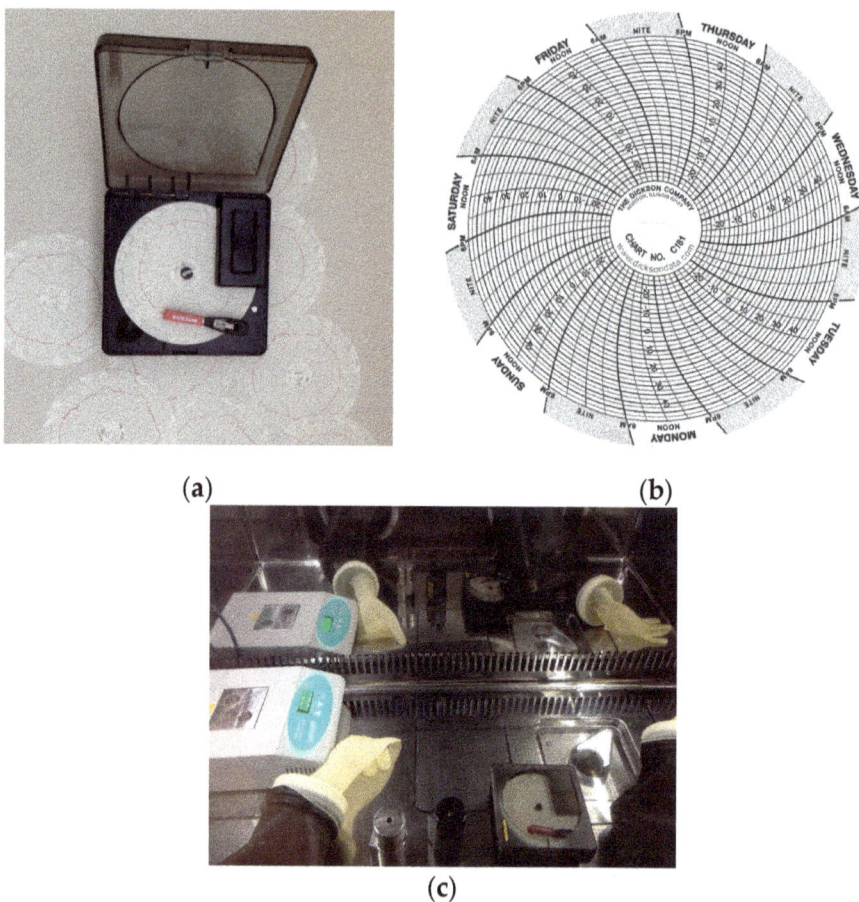

Figure 2. Graphic recorder (**a**,**b**) diagram disks (Dickson); and (**c**) recorder position in the work area.

A preliminary temperature registration, in order to ascertain the temperature fluctuation, was made for 24 h/7 days for three weeks inside the isolator (hot cell closed): on the outer edge of the paper disc are indicated the different days of the week divided into three hours-groups; the temperatures, printed on the disk in the range from −30 °C to +50 °C, are then detected by a graduated scale with a precision of ±2.5 °C. Every Friday, the paper disk was replaced and the data were collected and registered into datasheets. The same temperature control has been continuously performed during all the radiochemical purity three months monitoring.

2.2. ^{99m}Tc-Radiopharmaceuticals Radiochemical Purity Assessment

The radiochemical purity (RCP) of radiopharmaceuticals was evaluated immediately after the preparation (time = 0), in the middle and at the end of the expiry time indicated by the manufacturer (Table 1). The radiochemical purity was measured using methods specified by manufacturer, with the exception of Technescan (Mallinckrodt) for which the following chromatographic system was used [8]: mobile phase, 54/45/1 (physiological/methanol/glacial acetic acid, *v/v/v*) and stationary phase, RP-18 plates (Merck, Serono, Roma, Italy). The radioactivity distribution's determination was performed by a scanning radio-chromatography detection system for thin layer chromatography (Cyclone instrument equipped with a phosphor imaging screen and an OptiQuant image analysis software; PerkinElmer, Waltham, MA, USA). The solvents were obtained from Sigma Aldrich (Sigma-Aldrich, Milan, Italy); only methanol was obtained from VWR (VWR International, Fontanay sous Bois, France); MilliQ (18.2 MΩ) water was obtained from a Direct Q® system (Millipore, Darmstadt, Germany); physiological solution Fresenius KABI was obtained from Medexitalia (Medexitalia, Roma, Italy). All mobile phases used for quality control of radiopharmaceuticals were stored in a controlled and ventilated refrigerator (Antiscintilla 400 ECT-F-TOUCH, Fiocchetti, marketed by MecMurphil S.r.l, Ferrara, Italy); the stationary phases were stored in a special dehumidified container. ITLC-SG and ITLC-SA plates were obtained from Agilent Technology (Folsom, CA, USA); Whatman n.1 (Merck, Serono, Roma, Italy). Table 2 summarizes all of the experimental variables.

Table 2. Summary of the experimental conditions. We chose to use, for each preparation, the maximum activity and the maximum volume allowed by the manufacturer.

Compound	Synthesis	Quality Control System	Quality Control Time
Ceretec (GE Healthcare) [^{99m}Tc]Tc-HMPAO $N = 24$	Volume: 5 mL Activity: 1.10 GBq [1] Incubation Time: 30 s Package leaflet: 03/2016	TLC-SA Methylethylketone (MEK) —————————— TLC-SA physiological solution	0 min, 15 min, 30 min
MAASOL (GE Healthcare) [^{99m}Tc] Tc-albumin colloid $N = 24$	Volume: 8 mL Activity: 2.96 GBq Incubation Time: 5 min Package leaflet: 03/2015	ITLC-SG Methanol:water (85:15, v:v)	0 h, 3 h, 6 h
Nanocoll (GE Healthcare) [^{99m}Tc]Tc-nanocolloid $N = 80$	Volume: 5 mL Activity: 5.55 GBq Incubation Time: 30 min Package leaflet: 11/2015	TLC-SA Methanol:water (85:15, v:v)	0 h, 3 h, 6 h
Renocis (IBA molecular) [^{99m}Tc]Tc-DMSA $N = 8$	Volume: 6 mL Activity: 3.70 GBq Incubation Time: 10 min Package leaflet: 05/2011	Whatman n.1 MEK	0 h, 4 h, 8 h
Technemibi (Mallinckrodt) [^{99m}Tc]Tc-MIBI $N = 24$	Volume: 3 mL Activity: 11.10 GBq Incubation Time: 10 min, 100 °C Package leaflet: 10/2014	Baker flex aluminum oxide Ethanol >95%	0 h, 5 h, 10 h
Technescan HDP (Mallinckrodt) [^{99m}Tc]Tc-HDP $N = 60$	Volume: 10 mL Activity: 11.10 GBq Incubation Time: 30 s Package leaflet: 08/2017	TLC-SG 13.6% sodium acetate —————————— TLC-SG MEK	0 h, 4 h, 8 h
Technescan MAG3 (Mallinckrodt) [^{99m}Tc]Tc-MAG3 $N = 24$	Volume: 10 mL Activity: 2.96 GBq [2] Time: 10 min, 120 °C Package leaflet: 02/2017	RP-18 54/45/1 (saline/methanol/glacial acetic acid)	0 h, 4 h, 8 h

[1] Fresh sodium pertechnetate (not older than 2 h) obtained by generator eluted not more than 24 h before. [2] Fresh sodium pertechnetate obtained by generator (generator not older than 1 week from manufacturing) eluted not more than 24 h before. Where not specified, the incubation temperature during the synthesis, inside the isolator, was detected by our measurement. Finally, the production process of ^{99m}Tc-radiopharmaceuticals is constantly subjected to the Fill average test.

3. Results and Discussion

The illustrative leaflets of the commercial kits used for the preparation of ^{99m}Tc-radiopharmaceuticals, set out in detail all the preparations conditions, the quality controls to be executed, the values of the radiochemical purity and storage conditions, such as the temperature to which the radiopharmaceutical must be preserved before injection.

The small size of the work area, the presence of instruments for heating, the continuous activation of air filtration, the internal lighting, in addition to the fact that the environment of the isolator, used for the preparation and holding of ^{99m}Tc-radiopharmaceuticals, is completely isolated and not conditioned (Figure 1), could affect the temperature in the work area. At the moment, to our knowledge, there are no conditioning-equipped isolators, for nuclear medicine, available on the market even though the nuclear medicine community has chorally underlined this problem to the hot cells producers.

These potential discrepancies of temperatures and its influence on the radiochemical purity of radiopharmaceuticals have been deeply investigated in this work. The first objective was to verify the real temperature inside the isolator, with a calibrated instrument (Figure 1). The temperature required by manufacturers for the production of radiopharmaceuticals is variable, but the storage value must be <25 °C. Data were collected for three months (7 days/24 h) and the results are summarized in Table 3. The data collected through the graphical temperature recorder were firstly transcribed and subdivided by time slots in an Excel sheet. From these data, it was possible to determine the temperature average values and error, on different days of the week, divided by time slots.

Table 3. Average and standard error of temperature detected on the work surface inside the isolator.

Day [1]	Temperature °C (8:00–12:00)	Temperature °C (12:00–18:00)
Monday	29.0 ± 1.4	30.0 ± 1.4
Tuesday	30.0 ± 1.4	30.0 ± 1.4
Wednesday	30.0 ± 1.4	29.0 ± 1.4
Thursday	30.0 ± 1.4	31.0 ± 1.4
Friday	29.0 ± 1.4	30.0 ± 1.4
Saturday/Sunday	23.0 ± 1.4	23.0 ± 1.4

[1] The peak of activity (production and quality control of radiopharmaceuticals) is from Monday to Friday from 8:00 to 12:00 a.m.; Saturday and Sunday the Nuclear Medicine Unit is closed, therefore we consider 23 °C as the standard temperature during the non-activity period (isolator-off). The weekly radiopharmaceuticals production program is standardized and is repeated with minimal variations.

The results show a significant difference in temperatures compared to the reference standard (<25 °C). In the afternoon the instrument used to heat is off, but, despite this, the temperature, on average, results higher than 25 °C; this fact can be explained, most likely, by the operation of the isolator ventilation systems, which causes air heating. Establishing that the internal temperature of the isolator differed from that required for storage, we evaluated how much this temperature difference could affect the RCP of radiopharmaceuticals during the storage period, which also includes the fractionation of the preparation to be administered to the patients. A total of 244 ^{99m}Tc-radiopharmaceutical preparations (seven different types), performed in our nuclear medicine unit, have been tested in this work. Table 4 shows the RCP values of the preparations, reported as the mean and standard deviation of the individual values found during the three months of the test run. Along with these values, the temperature mean values detected during the RCP determination were added.

The results show that the temperatures are well above the reference standards but the RCP values are always higher than those established by manufacturers. The overall analysis of the results showed no anomalies with respect to the radiopharmaceutical quality of the products intended for administration to the patients at the time of preparation or at the limit of their stability. Moreover, all "media fill" tests confirmed the sterility of the radiopharmaceutical products.

Table 4. RCP values expressed as % media ± SD and temperature values expressed as average ± standard error.

Radiopharmaceuticals	RCP-1 (%) [1]	RCP-2 (%) [2]	RCP-3 (%) [3]	RCP (%) Expected Value
Ceretec (GE Healthcare) [99mTc]Tc-HMPAO N = 24	98.1 (±1.0) 31.0 °C (±0.5)	94.8 (±1.1) 31.0 °C (±0.5)	94.5 (±2.1) 31.0 °C (±0.5)	≥80%
MAASOL (GE Healthcare) [99mTc]Tc-albumin colloid N = 24	100.0 (±0.0) 32.0 °C (±0.5)	100.0 (±0.0) 32.0 °C (±0.5)	100.0 (±0.0) 30.0 °C (±0.5)	≥95%
Nanocoll (GE Healthcare) [99mTc]Tc-nanocolloid N = 80	99.10 (±0.09) 29.0 °C (±0.3)	100.0 (±0.0) 29.0 °C (±0.3)	99.3 (±1.2) 29.0 °C (±0.3)	≥95%
Renocis (IBA molecular) [99mTc]Tc-DMSA N = 8	100.0 (±0.9) 33.0 °C (±0.9)	100.0 (±0.9) 29.0 °C (±0.9)	100.0 (±0.9) 29.0 °C (±0.9)	≥95%
Technemibi (Mallinckrodt) [99mTc]Tc-MIBI N = 24	98.4 (±0.2) 29.0 °C (±0.5)	99.1 (±0.7) 29.0 °C (±0.5)	98.9 (±0.8) 29.0 °C (±0.5)	≥94%
Technescan HDP (Mallinckrodt) [99mTc]Tc-HDP N = 60	100.0 (±1.2) 33.0 °C (±0.3)	98.8 (±1.6) 33.0 °C (±0.3)	98.9 (±1.8) 31.0 °C (±0.3)	≥95%
Technescan MAG3 (Mallinckrodt) [99mTc]Tc-MAG3 N = 24	100.0 (±0.0) 30.0 °C (±0.5)	99.5 (±0.5) 30.0 °C (±0.5)	99.4 (±0.5) 29.0 °C (±0.5)	≥94%

[1] RCP-1 = value at the end of the labelling and temperature at the time of test; [2] RCP-2 = value determined in the middle between the labelling and the expiry time and temperature at the time of test; [3] RCP-3 = value determined at the expiry time and temperature at the time of test; expected value is the standard value required by the manufacturer for administration to patients.

4. Conclusions

The radiochemical purity of radiopharmaceuticals plays a key role in the protection of the patient to which the minimum undesirable exposure must be ensured. The evaluation of the impact that different conditions may have on the radiopharmaceuticals quality, it is crucial. The results of this work confirmed that the temperature variability, measured inside the isolator, does not have any effect on the radiochemical purity of radiopharmaceutical products until the expiry time. However, as required in the field of quality assurance, an assessment of the impact that conditions, other than those required by the manufacturer have on the radiopharmaceuticals for human use, is needed.

Acknowledgments: Products are furnished by University Hospital of Ferrara.

Author Contributions: Licia Uccelli and Alessandra Boschi conceived and designed the experiments and wrote the paper; Petra Martini and Corrado Cittanti analyzed the data; Stefania Bertelli, Doretta Bortolotti, Elena Govoni, Luca Lodi, Simona Romani, Samanta Zaccaria, Elisa Zappaterra, Donatella Farina, and Carlotta Rizzo performed the experiments; and Mirco Bartolomei and Melchiore Giganti contributed reagents/materials/analysis tools and revised the paper.

Conflicts of Interest: The authors declare no conflict of interest and declare that have not received funding, contracts or other forms of personal or institutional funding, with companies whose products are mentioned in the text.

References

1. Vasconcelos dos Santos, E.; Liane de Oliveira, M.; Eudes do Nascimento, J. Influence of humidity on radiochemical purity of ^{99m}Tc-ECD and ^{99m}Tc-Sestamibi. *Radioanal. Nucl. Chem.* **2015**, *306*, 751–755. [CrossRef]
2. Uccelli, L.; Boschi, A.; Pasquali, M.; Duatti, A.; Di Domenico, G.; Pupillo, G.; Esposito, J.; Giganti, M.; Taibi, A.; Gambaccini, M. Influence of the Generator in-Growth Time on the Final Radiochemical Purity and Stability of ^{99m}Tc Radiopharmaceuticals. *Sci. Technol. Nucl. Install.* **2013**, *2013*, 7. [CrossRef]
3. Norenberg, J.P.; Vaidya, M.P.; Hladik, W.B., III; Pathak, D.R.; Born, J.L.; Anderson, T.L.; Carroll, T.R. The effect of selected preparation variables on the radiochemical purity of ^{99m}Tc-Sestamibi. *J. Nucl. Med. Technol.* **2005**, *33*, 34–41. [PubMed]
4. Sanchez, C.; Zimmer, M.; Cutrera, P.; McDonald, N.; Spies, S. Radiochemical purity and stability of generic Tc-99m Sestamibi Kits. *J. Nucl. Med.* **2009**, *50*, 22–35.
5. Vallabhajosula, S.; Killeen, R.P.; Osborne, J.R. Altered Biodistribution of Radiopharmaceuticals: Role of Radiochemical/Pharmaceutical Purity, Physiological, and Pharmacologic Factors. *Semin. Nucl. Med.* **2010**, *40*, 220–241. [CrossRef] [PubMed]
6. Uccelli, L.; Martini, P.; Pasquali, M.; Boschi, A. Radiochemical purity and stability of ^{99m}Tc-HMPAO in routine Preparations. *J. Radioanal. Nucl. Chem.* **2017**, *314*, 1177–1181. [CrossRef]
7. Maksin, T.; Djokib, D.; Jankovik, D. Effect of Increased Temperature on Labelling Efficiency of Some ^{99m}Tc-Radiopharmaceuticals. *Label. Compd. Radiopharm.* **2001**, *44* (Suppl. 1), S636–S638. [CrossRef]
8. Zolle, I. *Technetium-99m Pharmaceuticals: Preparation and Quality Control in Nuclear Medicine*; Springer: Berlin, Germany, 2017; ISBN 978-3-540-33990-8.

Sample Availability: Samples of the compounds are not available from the authors.

molecules

MDPI

Article

Synthesis of ^{11}C-Labelled Ureas by Palladium(II)-Mediated Oxidative Carbonylation

Sara Roslin [1,*], Peter Brandt [1], Patrik Nordeman [1], Mats Larhed [2], Luke R. Odell [1] and Jonas Eriksson [1]

[1] Organic Pharmaceutical Chemistry, Department of Medicinal Chemistry, BMC, Uppsala University, Box 574, SE-751 23 Uppsala, Sweden; peter.brandt@orgfarm.uu.se (P.B.); patrik.nordeman@akademiska.se (P.N.); luke.odell@orgfarm.uu.se (L.R.O.); jonas.p.eriksson@akademiska.se (J.E.)

[2] Science for Life Laboratory, Department of Medicinal Chemistry, BMC, Uppsala University, Box 574, SE-751 23 Uppsala, Sweden; mats.larhed@orgfarm.uu.se

* Correspondence: sara.roslin@orgfarm.uu.se; Tel.: +46-018-4714282

Received: 25 September 2017; Accepted: 4 October 2017; Published: 10 October 2017

Abstract: Positron emission tomography is an imaging technique with applications in clinical settings as well as in basic research for the study of biological processes. A PET tracer, a biologically active molecule where a positron-emitting radioisotope such as carbon-11 has been incorporated, is used for the studies. Development of robust methods for incorporation of the radioisotope is therefore of the utmost importance. The urea functional group is present in many biologically active compounds and is thus an attractive target for incorporation of carbon-11 in the form of [11]Ccarbon monoxide. Starting with amines and [11]Ccarbon monoxide, both symmetrical and unsymmetrical ^{11}C-labelled ureas were synthesised via a palladium(II)-mediated oxidative carbonylation and obtained in decay-corrected radiochemical yields up to 65%. The added advantage of using [11]Ccarbon monoxide was shown by the molar activity obtained for an inhibitor of soluble epoxide hydrolase (247 GBq/μmol–319 GBq/μmol). DFT calculations were found to support a reaction mechanism proceeding through an ^{11}C-labelled isocyanate intermediate.

Keywords: carbon-11; ^{11}C-labelling; urea; carbonylation; positron emission tomography; carbon monoxide

1. Introduction

Positron emission tomography (PET) is a non-invasive imaging technique used to visualise and study biological processes in vivo by use of a molecular probe, a PET tracer, where a positron-emitting radioisotope has been incorporated. PET has found extensive use in clinical applications such as oncology, neurology and cardiology [1–3]. PET also serves as a useful technique in drug development, where PET offers the possibility to study the distribution, kinetics and target occupancy of potential drugs in vivo [4–6].

A radioisotope commonly used in PET is carbon-11 (^{11}C), with a half-life of 20.4 min. The natural abundance of carbon in biologically active molecules makes carbon-11 an appealing isotope to incorporate in PET tracers. The short half-life offers the possibility to perform several scans in one patient during a day but it also puts time-restraints on the production of the ^{11}C-labelled PET tracer. Efficient incorporation of carbon-11 is therefore a key step for a successful production. The labelling reaction should preferably be performed as the last step in the synthetic route and be a fast, high-yielding and robust method.

Carbon-11 can be incorporated in the PET tracer via a carbonylation reaction using [11]Ccarbon monoxide ([11]CCO). There are numerous carbonyl-containing biologically active molecules, thus the ^{11}C-carbonylative labelling reaction has great potential in PET tracer development but the minute amounts (typically 10–100 nmol), the short physical half-life and the low solubility of [11]CCO in organic solvents pose particular challenges [7–9]. To achieve sufficiently fast reaction kinetics, high reagent concentrations are needed, something that can be accomplished by confining the [^{11}C]CO in low volume reaction vessels [8]. However, conventional bubbling of [^{11}C]CO into the reaction mixture in this setting typically results in low recovery due to the low solubility of [^{11}C]CO and efficient purging of the gas phase as the carrier gas is vented. This problem has been approached in different ways, for example using high-pressure autoclave reactors [10–12], [^{11}C]CO-trapping agents [13–18], microfluidic reactors [19–21] and an ambient pressure system [22]. The ambient pressure system, developed by Eriksson et al., uses xenon as a carrier gas instead of helium or nitrogen. Most of the carrier gas is absorbed by the reaction solvent, thus precluding the need for high-pressure autoclave reactors or [^{11}C]CO-trapping reagents.

The urea moiety is a structural motif with a long history in medicinal chemistry and is found in various drugs and biologically active compounds [23]. By offering possibilities for hydrogen bonding, modulation of physicochemical properties and unique binding modes, therapeutic agents acting as inhibitors at targets such as protein kinases, Hepatitis C NS3 protease and methionyl–tRNA synthetase all contain the urea functional group (Figure 1) [23–26]. With the incorporation of the urea motif in aspiring drugs, there is a rationale for developing a simple and reliable method for incorporation of carbon-11 into the urea functional group.

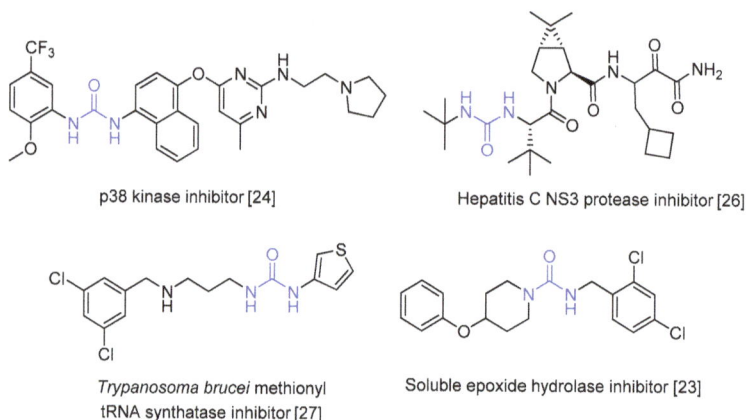

p38 kinase inhibitor [24]

Hepatitis C NS3 protease inhibitor [26]

Trypanosoma brucei methionyl
tRNA synthatase inhibitor [27]

Soluble epoxide hydrolase inhibitor [23]

Figure 1. Biologically active ureas.

[^{11}C]Urea and ^{11}C-labelled urea-derivatives have traditionally been synthesised from [^{11}C]phosgene ([^{11}C]COCl$_2$) [27–31] but also from [^{11}C]cyanide [32–34], [^{11}C]carbon dioxide ([^{11}C]CO$_2$) [35–41] and [^{11}C]CO [14,42–49]. The different methods come with their own unique limitations. Production of [^{11}C]COCl$_2$ is rather complicated, and the method is burdened by low product molar activities (A_m) and is less suited for the labelling of unsymmetrical ureas. [^{11}C]HCN also suffers from low A_m and the production requires a series of chemical transformations, hence the long reaction times. Methods utilising [^{11}C]CO$_2$ and [^{11}C]CO offer improved A_m and fewer or no subsequent chemical transformations. Cyclotron-produced [^{11}C]CO$_2$ can be utilised directly but fixation agents and drying agents are needed in addition to the amine/amines to be incorporated in the ^{11}C-labelled urea. Great care must be taken to make sure that all agents used are freed of atmospheric CO$_2$ to avoid isotopic dilution and reduced A_m. The A_m reached with [^{11}C]CO$_2$-fixation methods have been in the range of 25–148 GBq/μmol [36–38,41,50]. High A_m is of particular importance when imaging a less abundant target, especially in the central nervous system [51,52].

Different approaches have been employed for the synthesis of ^{11}C-labelled ureas from [^{11}C]CO and amines. The first method published was a selenium-mediated carbonylation for the synthesis of a number of symmetrical and unsymmetrical ^{11}C-labelled ureas, where secondary amines were difficult to employ [42]. Rhodium(I) has mainly been used in the synthesis of unsymmetrical ^{11}C-labelled ureas and necessitates the use of an azide as precursor, which, according to Doi et al., converts to a nitrene intermediate and subsequently to the ^{11}C-labelled isocyanate when reacting with [^{11}C]CO. [43–48]. In contrast to the case with Rh(I), palladium(II)-mediated ^{11}C-urea syntheses can use amines as sole precursors. Kealey et al. reported the use of a Cu(I)scorpionate complex for the trapping of [^{11}C]CO for a Pd(II)-mediated formation of symmetrical and unsymmetrical ^{11}C-labelled ureas [49]. Primary, aliphatic amines performed well as substrates, whereas anilines were found to be more challenging. Since no ^{11}C-labelled ureas were isolated nor any A_m determined, the practical utility of the method was difficult to assess.

To address some of the issues with the related methods and to improve the access to ^{11}C-labelled ureas, we here report on Pd(II)-mediated oxidative ^{11}C-carbonylation of amines for the synthesis of symmetrical and unsymmetrical ^{11}C-labelled ureas. The developed protocol utilised [^{11}C]CO, with xenon as a carrier gas, a palladium source and amines for the isolation of 14 ^{11}C-labelled ureas. Additionally, to demonstrate the advantage of using [^{11}C]CO to reduce isotopic dilution, the A_m was determined for an inhibitor of soluble epoxide hydrolase (sEH, Figure 1).

2. Results and Discussion

The minute amounts of [^{11}C]CO available for reaction and the requisite for a finished synthesis within 2–3 half-lives of carbon-11 set the framework for a transition-metal-mediated ^{11}C-carbonylation. We have previously used the xenon system for ambient pressure carbonylations and demonstrated its feasibility in synthesising amides [53,54] and sulfonyl carbamates [55]. The report by Kealey et al. as well as our own observations that ^{11}C-labelled ureas can form as byproducts in the synthesis of ^{11}C-labelled amides, especially when a Pd(II) source is used as a pre-catalyst for the aminocarbonylation, sparked our interest in exploring the Pd(II)-mediated formation of ^{11}C-labelled ureas [14].

The investigation into the synthesis of symmetrical ^{11}C-labelled ureas began with using benzylamine (1) as a model amine and Pd(Xantphos)Cl$_2$ as a Pd source [56,57]. Initially, 56% of the [^{11}C]CO was converted to non-volatile ^{11}C-labelled compounds and the selectivity for [^{11}C]-*N,N'*-dibenzylurea **2** was >99%, giving a radiochemical yield (RCY) of 55% calculated from the conversion and product selectivity (Table 1, entry 1) [58].

Table 1. Optimisation of reaction conditions for synthesis of symmetrical ^{11}C-labelled urea.

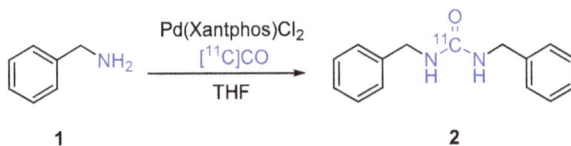

Entry	T (°C)	Time (min)	Conversion [a] (%)	Product Selectivity [b] (%)	RCY [c] (%)
1	120	5	56 ± 2.2	>99	55 ± 2.1 (3)
2	120	10	83 ± 3.9	>99	82 ± 3.9 (3)
3	150	5	66 ± 4.5	96 ± 2.6	63 ± 4.3 (3)
4	150	10	90 ± 2.5	97 ± 2.1	87 ± 3.4 (3)
5 [d]	120	10	66 ± 1.0	>99	65 ± 1.0 (2)

Conditions: **1** (30 µmol), Pd(Xantphos)Cl$_2$ (4 µmol), THF (400 µL). [a] Percentage of [^{11}C]CO converted to non-volatile products. Decay-corrected. [b] Percentage of product formed, assessed by analytical HPLC of crude reaction mixture, after volatiles were purged. [c] Radiochemical yield, calculated from the conversion and product selectivity. Number of experiments in brackets. [d] 10 µmol of **1**.

With these encouraging results, the reaction time and temperature were altered in order to improve the conversion. Extending the reaction time to 10 min (entry 2) increased the conversion to 83% and returned a RCY of 82%. Raising the reaction temperature to 150 °C (entry 3) whilst keeping the reaction time at 5 min did not improve the reaction to the same extent as prolonging the reaction time (66% conversion and 63% RCY). When heating the reaction at 150 °C for 10 min (entry 4), the conversion and RCY were further improved to 90% and 87%, respectively. The gain in RCY was minor for entry 4 as compared to entry 2 and we therefore decided to continue with the conditions as in entry 2 to avoid unwanted side reactions caused by the high temperature. In a final experiment, the amount of **1** was lowered to 10 μmol to test whether the conversion of [^{11}C]CO and the selectivity could be retained (entry 5). The conversion dropped somewhat but **2** was still obtained in 65% RCY. Further experiments were conducted using 30 μmol of amine.

Next, the scope for symmetrical ^{11}C-urea formation was investigated (Table 2). Symmetrical ureas are not as abundant in medicinal chemistry as unsymmetrical ureas. There are, however, examples of bioactive, symmetrical ureas such as [(7-amino(2-naphthyl)sulfonyl]phenylamines derivatives that have been shown to activate insulin receptor tyrosine kinases or sulfonylated naphthyl urea derivatives inhibiting protein arginine methyl transferases [59,60]. The products were isolated by semi-preparative HPLC purification and the radiochemical yields are based on the amount of [^{11}C]CO transferred to the reaction vial [58]. Primary, aliphatic amines (**2**, **3** and **4**) were found to be very good substrates and the products were isolated in good radiochemical yields and in >99% radiochemical purity (RCP). The gain in conversion that was seen when increasing the reaction time from 5 min to 10 min manifested itself as a gain in RCY for **2** (41% compared with 65%).

Table 2. Scope for symmetrical ^{11}C-labelled ureas.

$$RNH_2 \xrightarrow[\text{THF, 120 °C, 10 min}]{Pd(Xantphos)Cl_2,\ [^{11}C]CO} R{\underset{H}{\overset{O}{\underset{N}{\parallel}}}}^{11}C{\underset{H}{\overset{}{\underset{N}{}}}}R$$

2-5

Compound	^{11}C-Labelled Urea	Conversion [a] (%)	RCY [b] (%)	RCP [c] (%)
2		81 ± 5	65 ± 1 41 [d]	>99
3		67 ± 4	40 ± 6	>99
4		71 ± 2	48 ± 4	>99
5		15 ± 1	4 ± 1	>99

Conditions: Amine (30 μmol), Pd(Xantphos)Cl$_2$ (4 μmol), THF (400 μL). All experiments were performed in duplicate. [a] Percentage of [^{11}C]CO converted to non-volatile products. Decay-corrected. [b] Radiochemical yield. Based on the ^{11}C-labelled product obtained after semi-preparative HPLC and amount of [^{11}C]CO collected in the reaction vial. Decay-corrected. [c] Radiochemical purity. Determined by analytical HPLC of the isolated ^{11}C-labelled product. [d] 5 min reaction time, single experiment.

Urea **2** was isolated with 0.44 GBq 34 min after end of nuclide production (EOB) following a 5-min reaction, whereas a 10 min reaction time returned **2** with 0.83–0.88 GBq 33–36 min after EOB. Another example of a symmetrical urea in medicinal chemistry is *N*,*N*′-dicyclohexylurea, which has been identified as a potent inhibitor of sEH (K_i 30 nM), a discovery that sparked interest in the urea as a scaffold for designing sEH inhibitors [61]. Notably, [^{11}C]*N*,*N*′-dicyclohexylurea (**4**) was isolated in a high RCY (48%). Aniline was found to be a sluggish substrate and **5** was only isolated in 4% RCY.

This is in line with previous studies using either a Pd(II)-source and high concentrations of aniline (90 µmol) or [^{11}C]CO$_2$-fixation methods using varying amounts of aromatic amines [35,36,49,50]. Rh(I)-mediated synthesis of aromatic ^{11}C-labelled ureas has returned higher RCYs, albeit starting from an aromatic azide rather than an aromatic amine [44,45]. When piperidine **6** was used as substrate, the tetra-substituted urea was not detected.

For optimisation of the synthesis of unsymmetrical ^{11}C-labelled ureas, **7** was chosen as the model compound (Table 3). Starting from the same conditions as in the synthesis of symmetrical ^{11}C-labelled ureas, **7** was obtained in 42% RCY based on a conversion of 53% and a product selectivity of 79% (entry 1) [58]. The ratio of symmetrical (**2**) versus unsymmetrical (**7**) formation was 12:88. Initial optimisation was performed by changing the Pd-source (entries 2–5) [14,49]. Although the Pd-species in entries 2–5 gave higher conversions than Pd(Xantphos)Cl$_2$, the product selectivity was lower and consequently the RCYs as well (7–32%). Using DMF as solvent was unfavorable for the product selectivity (entry 6). An investigation of the temperature influence (entries 7 and 8), showed that a lowering of the temperature to 80 °C, which has been used in Rh-based ^{11}C-labelled urea syntheses, retained the conversion whilst losing product selectivity [43]. On the other hand, heating the reaction to 150 °C improved the product selectivity but, because of the lower conversion, the yield was in the same range as in entry 1 (41% in entry 8 and 42% in entry 1). The reaction temperature was therefore kept at 120 °C.

Table 3. Optimisation of reaction conditions for synthesis of unsymmetrical ^{11}C-labelled urea.

Entry	Catalyst	T (°C)	6 (Equiv.)	Conversion [a] (%)	Product Selectivity [b] (%)	2:7 [c]	RCY [d] (%)
1	Pd(Xantphos)Cl$_2$	120	1	53 ± 5.6	79 ± 2.9	12:88	42 ± 5.9 (3)
2	Pd(PPh$_3$)$_2$Cl$_2$	120	1	69 ± 4.1	46 ± 3.6	16:84	32 ± 3.2 (3)
3	Pd(OAc)$_2$ + dppf	120	1	95 ± 3.5	13 ± 3.5	11:89	12 ± 4.0 (2)
4	Pd(OAc)$_2$ + dppp	120	1	75 ± 2.5	21 ± 0.5	23:77	16 ± 0.5 (2)
5	Pd(OAc)$_2$ + Xantphos	120	1	67 ± 9	10 ± 1	23:77	7 ± 1.5 (2)
6 [e]	Pd(Xantphos)Cl$_2$	120	1	43 ± 1.7	49 ± 3.6	16:84	21 ± 1.6 (3)
7	Pd(Xantphos)Cl$_2$	80	1	57 ± 9.2	44 ± 6.8	9:91	26 ± 8.5 (3)
8	Pd(Xantphos)Cl$_2$	150	1	44 ± 11	87 ± 4.3	9:91	41 ± 6.2 (4)
9	Pd(Xantphos)Cl$_2$	120	2	46 ± 4.3	63 ± 2.2	9:91	29 ± 3.3 (3)
10	Pd(Xantphos)Cl$_2$	120	5	58 ± 1.7	71 ± 3.6	2:98	42 ± 3.1 (3)
11 [f]	Pd(Xantphos)Cl$_2$	120	1	67 ± 1.7	89 ± 3.3	7:93	60 ± 3.4 (3)
12 [g]	Pd(PPh$_3$)$_4$	120	1	93	-	-	-

Conditions: **1** (30 µmol), **2** (30 µmol), catalyst ([Pd] 4 µmol + ligand 4 µmol), THF (400 µL). 5 min reaction time unless otherwise stated. [a] Percentage of [^{11}C]CO converted to non-volatile products, after purge. Decay-corrected. [b] Percentage of product formed, assessed by analytical HPLC of crude reaction mixture, after purge. [c] Product ratio of **2** to **7**, assessed by analytical HPLC of crude reaction mixture. [d] Radiochemical yield, calculated from the conversion and product selectivity. Number of experiments in brackets. [e] DMF as solvent. [f] 10 min reaction time. [g] Single experiment.

Next, increasing the amount of piperidine (**6**) was investigated (entries 9 and 10). Not surprisingly, using five equivalents of **6** gave almost sole formation of unsymmetrical ^{11}C-labelled urea **7** (entry 10) whereas having two equivalents of **6** did not improve the product ratio to the same extent (entry 9). The RCY was markedly lower in entry 9 (29%) compared to entry 1 and entry 10 (both 42%), because of both lower conversion and inferior product selectivity. As no improvement in yield was gained by using five equivalents of **6**, the reaction time was altered next. Heating the reaction for 10 min enhanced the conversion, the product selectivity and the ratio of **2** to **7** formed, with a 60% RCY (entry 11). A final experiment, with Pd(PPh$_3$)$_4$, supported the reaction to be Pd(II)-mediated as neither **2** nor **7** formed with the Pd(0)-source. The conditions in entry 11 were continued with for investigation of the scope for synthesis of unsymmetrical ^{11}C-labelled ureas, including sEH inhibitor **19** (Table 4).

Compounds **7–19** were isolated by semi-preparative HPLC and the radiochemical yields are based on the amount of [^{11}C]CO transferred to the reaction vial [58]. Aliphatic amines were isolated in good RCYs ranging from 12% to 41% (**7–10**). Another demonstration of the gain in measured radioactivity of the product with a prolonged reaction time is seen for **7**, where a 5 min reaction time led to a RCY of 17% and 0.15 GBq isolated 37 min from EOB, whilst the 10 min reaction time gave 41% in RCY and 0.47–0.57 GBq isolated after 36–45 min from EOB. 2-(2-Aminoethyl)pyridine proved to be a challenging substrate and **9** was isolated in 12% RCY. Here, an unknown ^{11}C-labelled byproduct formed along with product **9**. The byproduct was not observed when amine **6** was removed. In the synthesis of **10**, steric hindrance was introduced in the primary amine, which resulted in a lower RCY compared to **7** and **8**.

Table 4. Scope for unsymmetrical ^{11}C-labelled ureas.

Compound	^{11}C-Labelled Urea	Conversion [a] (%)	RCY [b] (%)	RCP [c] (%)
7		65 ± 0 39 [d] 59 [e]	41 ± 6 17 [d] 31 [e]	98 ± 1
8		55 ± 4 60 [e]	23 ± 1 14 [e]	>99
9		20 ± 0	12 ± 1	>99
10		35 ± 6	14 ± 4	>99
11		60 ± 3 [f] 59 ± 1 70 [e]	7 ± 2 [f] 12 ± 0 [g] 6 [e]	80 ± 7 [f] 97 ± 2 [g] 88 [e]
12		66 ± 3 74 [e]	9 ± 1 21 [e]	99 ± 1
13		67 ± 9 [f] 88 [e]	8 ± 1 [f] 28 [e]	>99
14		63 ± 0 83 [e]	5 ± 1 6 [e]	>99

Table 4. *Cont.*

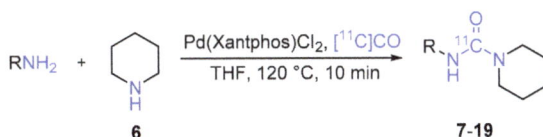

Compound	¹¹C-Labelled Urea	Conversion [a] (%)	RCY [b] (%)	RCP [c] (%)
15		27 ± 1 90 [e]	1 ± 0 Trace [e]	90 ± 10 -
16		12 ± 2	Trace	-
17		51 ± 0	Trace	-
18		66 ± 8	-	-
19		74	41 ± 7 [f]	99 ± 0 [f]

Conditions as in entry 11, Table 2. All experiments were performed in duplicate unless otherwise stated. [a] Percentage of [¹¹C]CO converted to non-volatile products. Decay-corrected. [b] Radiochemical yield. Based on the ¹¹C-labelled product obtained after semi-preparative HPLC and amount of [¹¹C]CO collected in the reaction vial. Decay-corrected. [c] Radiochemical purity. Determined by analytical HPLC of the isolated ¹¹C-labelled product. [d] 5 min reaction time, one experiment. [e] 10 equiv. of 1-butanol added, single experiment. [f] Average of three experiments. [g] 3 equiv. of aniline used.

A set of aniline derivatives were also synthesised (**11–14**) and, similar to **5**, were found to be less reactive than their aliphatic counterparts and were isolated in RCYs varying from 5% to 12%. In the reaction with unfunctionalised aniline, an impurity was present after the purification and **11** could only be isolated with 80% RCP. When three equivalents of aniline were used, the RCY increased from 7% to 12% and the RCP reached 97%. Substitution in the 4-position of the aromatic ring was explored and methoxy- and fluoro-substituents were equally tolerated (**12** and **13**, 8–9%), whereas the nitro group gave a slightly lower RCY of 5% (**14**). It should be noted that the conversion for the reactions with aniline derivatives were in the range of 60–67% and the product selectivities varied from 33% to 71%. The true outcome of the reaction only became fully apparent after isolation of the respective product. Thus, the radiochemical yields estimated for a non-isolated ¹¹C-labelled product should be interpreted with care as it may not fully correspond to the radiochemical yield of the isolated product. Next, cyclic ¹¹C-labelled urea **15** was synthesised but isolated in a very low RCY (1%). An unknown ¹¹C-labelled byproduct formed during the reaction. This byproduct was not formed when the reference was synthesised (even when using Pd(Xantphos)Cl₂). ¹¹C-Labelled sulfonylureas have been synthesised from the corresponding sulfonyl azide in a Rh(I)-mediated reaction but here, synthesis of ¹¹C-labelled sulfonyl ureas **16** and **17** did not result any in isolable product [43]. To probe whether the trace formation of **16** was related to the poor nucleophilicity of aniline, the amine was changed to **6** to aid in the plausible ^{11}C-sulfonyl isocyanate formation and subsequent product formation to sulphonylurea **17**. However, as seen in Table 4, this change did not improve the reaction outcome.

Ethanol has been found to increase the conversion of ^{11}C]phosgene-derived ammonium ^{11}C]isocyanate to ^{11}C]urea [31]. Similarly, different alcoholic additives were found to accelerate the conversion of ammonium cyanate to urea [62]. Therefore, 10 equivalents of 1-butanol were added to the reaction mixture of **7**, **8** and **11–15**. Compounds **12** and **13**, were isolated with improved RCY (21% and 28%) whereas the RCY of **7** and **8**, synthesised from an aliphatic primary amine and **6**, were not improved (31% and 14%). Hence, the addition of 1-butanol could be most beneficial when a poorly nucleophilic primary amine is used. However, the RCY of **11**, **14** and **15** were not improved thus making the hypothesis of nucleophilicity less certain. Of note is that the ^{11}C-labelled byproduct formed during the synthesis of **15**, was isolated in a RCY of 72% when 1-butanol was added.

Lactams has previously been synthesised under Pd(II)-catalysed conditions, however, with the conditions employed here, in particular the short reaction time and small amount of ^{11}C]CO, ^{11}C-labelled lactam **18** was not formed [63]. A ^{11}C-labelled byproduct was formed, which in part accounts for the relatively high ^{11}C]CO-conversion (66%). The byproduct was hypothesised from LC/MS to be ^{11}C]N-benzyl-N-ethylbenzamide, formed by scrambling of a phenyl group from the Pd-ligand. A subsequent synthesis of N-benzyl-N-ethylbenzamide from benzoic acid and N-ethyl benzylamine and matching of LC retention times confirmed the identity. When the reaction mixture was preheated, to aid in dissolution of reactants, ^{11}C]N-benzyl-N-ethylbenzamide was formed in 25% RCY, whereas the RCY dropped to 4% when the reaction mixture was not preheated. Lastly, sEH inhibitor **19** was synthesised in a good 41% RCY [23]. To further demonstrate the utility of the method, with respect to molar activity, was the A_m determined for **19** in two experiments. When starting from 17.8 GBq of ^{11}C]CO, product **19** was isolated with 1.9 GBq and in a A_m of 247 GBq/µmol, 43 min after EOB. In the second experiment, when starting with 12.8 GBq of ^{11}C]CO, **20** could be isolated with 2.1 GBq and a A_m of 319 GBq/µmol, 41 min from EOB.

In Scheme 1, hypothetical paths **A–E** through which the reaction can proceed are presented. A carbamoylpalladium species has been proposed by Hiwatari et al. to form as an intermediate in the reaction [64]. Thus, hypothesising the same species will form in the present reaction with ^{11}C]CO, ^{11}C-labelled carbamoylpalladium species **20** can generate ^{11}C-labelled isocyanate **21** through the deprotonation of **20** by an external base such as in path **A**. Alternatively, **21** can be formed by a β-elimination to give a hydridopalladium(II) complex **23**, which subsequently can reductively eliminate HCl to yield Pd(0)-complex **22** (path **B**). The elimination depicted in path **A** has found experimental support in kinetic studies where the disappearance of a carbamoylpalladium complex was found to have a first-order dependence on the concentration of a tertiary amine [64]. Attack of a second amine on **21** will furnish product **24**. Carbamoyl chloride **25** is proposed to form in path **C** after a reductive elimination from **20**. As in **A** and **B**, attack of another amine gives **24**. In paths **D** and **E**, the second amine reacts directly with **20**, either through initial coordination to Pd (complex **26**, path **D**) or through a direct attack on the carbonyl (path **E**).

Path **A**, and path **B** to some extent, could explain the low RCY of **5** if the weakly basic aniline cannot deprotonate **20** and/or is slow to react with **21**. However, the low RCYs of **11–14** where **6** with ease should be able to both deprotonate **20** and subsequently trap **21** to form product cannot be rationalised in the same way. Although paths **C–E** are feasible on paper, previous experimental studies support an isocyanate intermediate, i.e., path **A** or **B**, on the basis that secondary amines does not furnish the corresponding tetra-substituted urea under the conditions explored [63–65]. Further support of path **A** or **B** is that the ratio of symmetric ^{11}C-labelled urea **2** to unsymmetrical ^{11}C-labelled urea **7** was here found to be approximately 1:9 and the same ratio was found when **1** and **6** were allowed to react with benzyl isocyanate. Furthermore, when performing the reaction with only 4-fluoroaniline and Pd(Xantphos)Cl$_2$, was the presence of ^{11}C]4-fluorophenyl isocyanate confirmed by analytical HPLC and co-injection with 4-fluorophenyl isocyanate. Amine **6** was thereafter added and the reaction was heated for another 10 min. A second analysis revealed that **13** had formed. However, the mere presence of the ^{11}C-labelled isocyanate does not preclude the reaction from proceeding through an alternative mechanism.

To lend further support to path **A** in Scheme 1, DFT-calculations were performed for the urea formation from CO and *N*-methyl amine. Although a solvent model for THF was used, the accuracy of the energies will suffer from the variation in charge for the palladium complexes on the reaction path. However, it was reasoned that the calculations would be accurate enough to give a rough estimate of the stability of intermediates along the reaction path (Scheme 2).

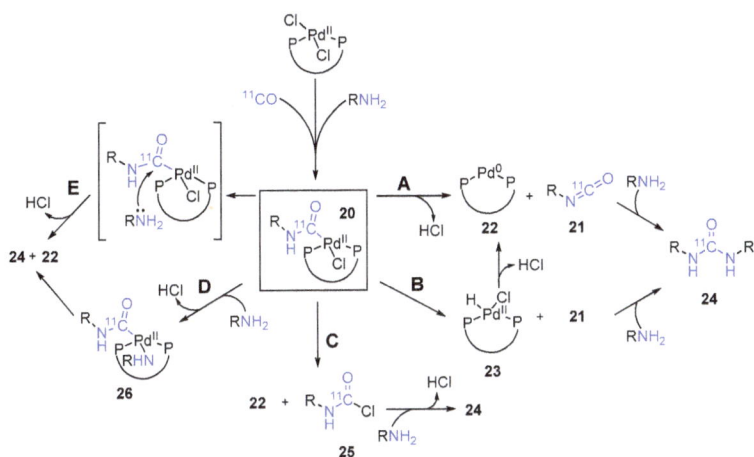

Scheme 1. Hypothetical reaction paths for ^{11}C-labelled urea formation.

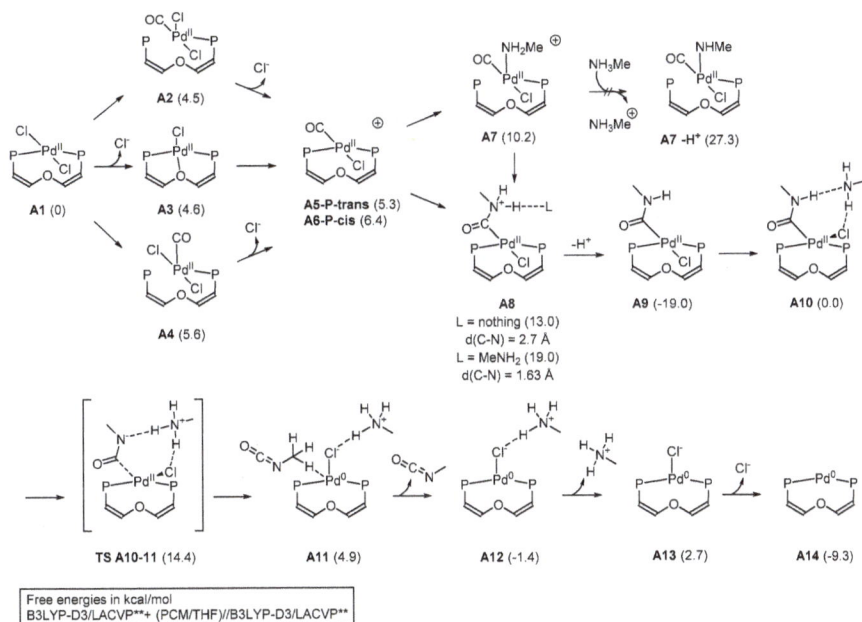

Free energies in kcal/mol
B3LYP-D3/LACVP**+ (PCM/THF)//B3LYP-D3/LACVP**

Scheme 2. Calculated free energies (kcal/mol) of intermediates along path **A** in Scheme 1 showing that the path is energetically feasible.

As shown in Scheme 2, addition of CO to the initial palladium dichloro complex **A1** does not involve any dramatic energy changes and addition of a primary amine is also feasible. Deprotonation of the palladium-bound amine is not likely, however. Instead, the amine can attack the carbonyl directly forming a labile *N*-protonated carbamoyl species. In the gas-phase, the C–N bond varies between 2.07 Å and 2.70 Å depending on whether the cation can be stabilised by the phenylphosphine or not. Upon addition of a stabilising amine to **A8**, the C–N bond shortens to 1.63 Å. The carbamoyl species **A9**, formed by deprotonation of **A8**, is the most stable intermediate on the reaction path. Base-assisted deprotonation of this species leads to the formation of an isocyanate and the palladium(0)-complex **A14** in accordance with the first-order dependence on the concentration of a tertiary amine reported by Hiwatari et al. [64].

3. Materials and Methods

3.1. General Chemistry Information

Reagents and solvents were obtained from Sigma-Aldrich (St. Louis, MO, USA) and Fischer (Pittsburgh, PA, USA) and used without further purification. Thin layer chromatography (TLC) was performed on aluminium sheets pre-coated with silica gel 60 F_{254} (0.2 mm, Merck KGaA, Darmstadt, Germany). Column chromatography was performed using silica gel 60 (40–63 µm, Merck KGaA, Darmstadt, Germany). Carbon-11 was prepared by the $^{14}N(p,\alpha)^{11}C$ nuclear reaction using 17 MeV protons produced by a Scanditronix MC-17 Cyclotron at PET Centre, Uppsala University Hospital, and obtained as [^{11}C]carbon dioxide. The target gas used was nitrogen (AGA Nitrogen 6.0) containing 0.05% oxygen (AGA Oxygen 4.8). The [^{11}C]CO$_2$ was transferred to the low-pressure xenon system in a stream of helium gas and concentrated before reduction to [^{11}C]CO, over zinc heated to 400 °C [22,54]. A column with Ascarite was used to trap residual [^{11}C]CO$_2$. Before transferring the concentrated [^{11}C]CO into the reaction vial, through a transfer needle placed in the capped reaction vial, the carrier gas was changed from helium to xenon (>99.9%, 1.5 mL/min). No venting needle was required during the transfer and, when completed, the transfer needle was removed and the reaction vial was lowered into a heating block. Analytical reversed phase HPLC-MS was performed on a Dionex Ultimate 3000 system using 0.05% HCOOH in water and 0.05% HCOOH in acetonitrile as mobile phase with MS detection, equipped with a C18 (Phenomenex Kinetex SB-C18 (4.8 × 50 mm)) column using a UV diode array detector. Purity determinations were performed using C18 (Phenomenex Kinetex SB-C18 (4.8 × 50 mm)) column and Biphenyl (Phenomenex Kinetex Biphenyl (2.6 µm, 4.6 × 50 mm) column, with UV detection at 214 nm or 254 nm. Semi-preparative reversed phase HPLC was performed on a Gilson–Finnigan ThermoQuest AQA system equipped with Gilson GX-271 system equipped with a C18 (Macherey-Nagel Nucleodur HTec (5 µm, 125 × 21 mm)) using 0.1% TFA in water and 0.1% TFA in acetonitrile as eluent. Purifications of ^{11}C-labelled ureas were performed on a VWR La Prep Sigma system with a LP1200 pump, 40D UV detector, a Bioscan flowcount radiodetector. The identities, concentration and radiochemical purities of the purified ^{11}C-labeled ureas were determined with either a VWR Hitachi Elite LaChrom system (L-2130 pump, L-2200 autosampler, L-2300 column oven, L-2450 diode array detector in series with a Bioscan β$^+$-flowcount radiodetector) or an Elite LaChrom VWR International (LaPrep P206 pump, an Elite LaChrom L-2400 UV detector in series with a Bioscan β$^+$-flowcount detector), 8.1 mM ammonium carbonate (aq.) and acetonitrile mobile phase and Merck Chromolith Performance RP-18e column (4.6 × 100 mm) using isotopically unmodified compounds as references. NMR spectra were recorded on a Varian Mercury plus spectrometer (^{1}H at 399.8 MHz, ^{13}C at 100.5 MHz and ^{19}F at 376.5 MHz) at ambient temperature. Chemical shifts (δ) are reported in ppm, indirectly referenced to tetrametylsilane (TMS) via the residual solvent signal (^{1}H: CHCl$_3$ δ 7.26, CD$_2$HOD δ 3.31, (CHD$_2$)(CD$_3$)SO δ 2.50, (CHD$_2$)(CD$_3$)CO δ 2.05 ^{13}C: CDCl$_3$ δ 77.2, CD$_3$OD δ 49.0, (CHD$_2$)(CD$_3$)SO δ 39.5, (CD$_3$)$_2$CO δ 29.8, 206.3.

3.2. Synthesis of Starting Materials

3.2.1. *tert*-Butyl 4-hydroxypiperidine-1-carboxylate [66] CAS: 109384-19-2

4-Hydroxypiperdine (3.1 mmol) was dissolved in THF (10 mL), to which 10% NaOH (aq., 1.5 mL) was added. Di-*tert*-butyl dicarbonate (1.2 equiv.) was slowly added to the reaction mixture. After completion of the reaction, the reaction mixture was extracted with 3 × 15 mL chloroform. The combined organic phases were washed with 15 mL Brine, dried over Na_2SO_4, filtered and concentrated in vacuo. Flash column chromatography followed, using 15% ethanol in ethyl acetate as eluent. Isolated as a white solid in 95% purity and used without further purification. ^{1}H-NMR (400 MHz, $CDCl_3$) δ 3.91–3.76 (m, 3H), 3.07–2.96 (m, 2H), 1.89–1.80 (m, 2H), 1.68 (d, *J* = 4.2 Hz, 1H), 1.50–1.46 (m, 1H), 1.45 (s, 9H), 1.43–1.39 (m, 1H).

3.2.2. *tert*-Butyl 4-phenoxypiperidine-1-carboxylate [66] CAS: 155989-69-8

tert-Butyl 4-hydroxypiperidine-1-carboxylate (3.1 mmol), phenol (1.0 equiv.) and PPh_3 (1.0 equiv.) were dissolved in anhydrous THF (16 mL) before addition of diisopropyl azodicarboxylate (1.0 equiv.). The reaction mixture was stirred at room temperature for 18 h. The reaction mixture was concentrated before purification with flash column chromatography, using 7:1 *i*-hexane:ethyl acetate as eluent. Isolated as a white solid (498 mg, 59%). R_f = 0.47 (7:1 *i*-hexane:ethyl acetate). ^{1}H-NMR (400 MHz, $CDCl_3$) δ 7.31–7.27 (m, 2H), 6.99–6.88 (m, 3H), 4.46 (hept, *J* = 3.5 Hz, 1H), 3.75–3.64 (m, 2H), 3.39–3.26 (m, 2H), 1.98–1.86 (m, 2H), 1.83–1.69 (m, 2H), 1.47 (s, 9H). ^{13}C-NMR (101 MHz, $CDCl_3$) δ 157.3, 155.0, 129.7, 121.2, 116.3, 79.7, 72.3, 40.9, 30.7, 28.5. MS-EI *m/z* 277.1.

3.2.3. 4-Phenoxypiperidine [66] CAS: 3202-33-3

tert-Butyl 4-phenoxypiperidine-1-carboxylate (453 mg) was dissolved in dichloromethane (10 mL). TFA (1.4 mL) was added and the mixture was stirred at room temperature for 2.5 h. The reaction mixture was concentrated and re-dissolved in dichloromethane (20 mL), followed by washing with 1.5 M NaOH (20 mL). The water phase was extracted with 2 × 20 mL DCM. The combined organic phases were dried over Na_2SO_4, filtered and concentrated. Purification with flash column chromatography using 5% MeOH and 1% triethylamine (TEA) in DCM Product was isolated as a white solid (290 mg, 96%). R_f = 0.18 (5% MeOH, 1% TEA in DCM). ^{1}H-NMR (400 MHz, $CDCl_3$) δ 7.30–7.25 (m, 2H), 6.96–6.89 (m, 3H), 4.39 (hept, *J* = 3.9 Hz, 1H), 3.20–3.12 (m, 2H), 2.80–2.72 (m, 2H), 2.24 (br. s, 1H), 2.08–1.97 (m, 2H), 1.75–1.65 (m, 2H). ^{13}C-NMR (101 MHz, $CDCl_3$) δ 157.4, 129.6, 121.0, 116.3 73.1, 43.8, 32.2. HPLC purity > 99%.

3.3. Synthesis of Reference Compounds

3.3.1. General Procedure for Synthesis of Reference Compounds via an Oxidative Carbonylation

The reaction was performed in a double-chamber system [67,68]. Amine/amines, $Pd(OAc)_2$ (0.05 equiv.) and $Cu(OAc)_2$ (0.5 equiv.) were added to the reaction chamber and dissolved in 1,4-dioxane (2 mL). In the CO-chamber was $Mo(CO)_6$ (200 mg) dissolved in 1,4-dioxane (2 mL). After capping of the system, 1,8-diazabicyclo[5.4.0]undec-7-ene was added to the CO-chamber. The double-chamber system was positioned in a Dry–Syn heating block and heated to 95 °C. After completion of the reaction, the reaction mixture was filtered through a short silica plug before flash column chromatography.

3.3.2. *N,N'*-Dibenzylurea [69] CAS: 1466-67-7

Synthesised according to the general procedure from benzylamine (1 mmol). Purified with flash column chromatography, using 1% MeOH and 1% AcOH in DCM as eluent. Isolated as a white powder (74 mg, 31%). R_f = 0.23 (1% MeOH and 1% acetic acid in DCM). ^{1}H-NMR (400 MHz, DMSO-d_6) δ

7.34–7.18 (m, 10H), 6.46 (t, J = 6.1 Hz, 2H), 4.23–4.20 (m, 4H). ^{13}C-NMR (101 MHz, CDCl$_3$ and MeOD) δ 159.1, 139.3, 128.4, 127.2, 127.0, 44.0. MS-ESI [M + H$^+$] = m/z 241.3. HPLC purity > 99%.

3.3.3. *N*-(2-(Pyridin-2-yl)ethyl)piperidine-1-carboxamide CAS: 1710806-84-0

Synthesised according to the general procedure from 2-(2-aminoethyl)pyridine (1.0 mmol) and piperidine (3 equiv.). Purified with flash column chromatography, using 1% MeOH and 1% AcOH in DCM as eluent. Isolated as a solid (49 mg, 21%). R_f = 0.16 (ethyl acetate + 1% TEA). ^{1}H-NMR (400 MHz, (CD$_3$)$_2$CO) δ 8.51–8.47 (m, 1H), 7.67 (td, J = 7.7, 1.9 Hz, 1H), 7.24 (d, J = 7.8 Hz, 1H), 7.20–7.15 (m, 1H), 6.04 (br. s, 1H), 3.54–3.47 (m, 2H), 3.33–3.27 (m, 4H), 2.94 (t, J = 7.0 Hz, 2H), 1.60–1.53 (m, 2H), 1.49–1.41 (m, 4H). ^{13}C-NMR (101 MHz, (CD$_3$)$_2$CO) δ 161.2, 158.3, 150.0, 137.1, 124.0, 122.1, 45.4, 41.3, 41.2, 39.1, 39.0, 26.5, 25.4. HRMS calc: 234.1606 found: 234.1614. HPLC purity > 99%.

3.3.4. 3,4-Dihydroquinazolin-2(1H)-one [70] CAS: 66655-67-2

Synthesised according to the general procedure from 2-aminobenzylamine (1.5 mmol). Purified twice with flash column chromatography, using 5% MeOH, 1% TEA in DCM and 1% TEA in ethyl acetate, respectively, followed by semi-preparative purification. Isolated as a white powder (26 mg, 12%). R_f = 0.74 (5% MeOH, 1% TEA in DCM). ^{1}H-NMR (400 MHz, (CD$_3$)$_2$SO) δ 8.98 (br. s, 1H), 7.12–7.04 (m, 2H), 6.84 (td, J = 7.5, 1.2 Hz, 1H), 6.79–6.73 (m, 2H), 4.29 (s, 2H). ^{13}C-NMR (101 MHz, (CD$_3$)$_2$SO) δ 154.6, 138.1, 127.6, 125.6, 120.9, 118.1, 113.5, 42.5. MS-ESI [M + H$^+$] = m/z 149.1. HPLC purity > 99%.

3.3.5. 2-Ethylisoindolin-1-one [71] CAS: 23967-95-5

Synthesised according to the general procedure from *N*-ethylbenzylamine (1 mmol). Purified with flash column chromatography, using 3:1 i-hexane:ethyl acetate + 1% TEA. Isolated as a yellow liquid (49 mg, 31%). R_f = 0.25 (3:1 i-hexane:ethyl acetate + 1% TEA). ^{1}H-NMR (400 MHz, CDCl$_3$/CD$_3$OD) δ 7.77 (d, J = 7.5 Hz, 1H), 7.54–7.49 (m, 1H), 7.47–7.41 (m, 2H), 4.40 (s, 2H), 3.64 (q, J = 7.3 Hz, 2H), 1.25 (t, J = 7.3 Hz, 3H). ^{13}C-NMR (101 MHz, CDCl$_3$/CD$_3$OD) δ 141.4, 132.7, 131.7, 128.4, 123.6, 123.0, 49.8, 37.4, 13.6 (Carbonyl carbon missing). MS-ESI [M + H$^+$] = m/z 162.1. HPLC purity > 99%.

3.3.6. *N*-(2,4-Dichlorobenzyl)-4-phenoxypiperidine-1-carboxamide CAS: 950645-62-2

Synthesised according to the general procedure from 2,4-dichlorobenzylamine (0.43 mmol) and 4-phenoxypiperidine (2 equiv.). Purified with flash column chromatography, using 1% MeOH and 1% AcOH in DCM as eluent, followed by semi-preparative chromatography. Isolated as a white powder (47 mg, 29%). R_f = 0.53 (5% MeOH, 1% TEA in DCM). ^{1}H-NMR (400 MHz, (CD$_3$)$_2$CO) δ 7.46–7.42 (m, 2H), 7.33 (dd, J = 8.2, 2.2 Hz, 1H), 7.31–7.24 (m, 2H), 7.00–6.94 (m, 2H), 6.92 (tt, J = 7.3, 1.1 Hz, 1H), 4.62 (tt, J = 7.6, 3.7 Hz, 1H), 4.45 (d, J = 5.0 Hz, 2H), 3.86–3.75 (m, 2H), 3.41–3.30 (m, 2H), 2.03–1.95 (m, 2H), 1.72–1.61 (m, 2H).^{13}C-NMR (101 MHz, (CD$_3$)$_2$CO) δ 157.6, 157.5, 137.2, 133.3, 132.5, 130.3, 129.6, 128.6, 127.1, 120.8, 116.1, 72.2, 41.6, 41.2, 30.7. HRMS calc: 379.0980 found: 379.0999. HPLC purity > 99%.

3.3.7. General Procedure for the Synthesis of Reference Compounds from an Isocyanate

The isocyanate derivative (1 mmol) was dissolved in DCM (2 mL) and cooled to $-10\ ^\circ$C before piperidine (1 equiv.) in DCM (2 mL) was added under N$_2$. The reaction was stirred under N$_2$ for 30 min and then let to warm room temperature and stir until completion. The reaction mixture was concentrated in vacuo and recrystallised.

3.3.8. *N*-Benzylpiperidine-1-carboxamide [72] CAS: 39531-35-6

Synthesised according to the general procedure from benzyl isocyanate. Recrystallised in petroleum ether and isolated as colourless crystals (87 mg, 40%). R_f = 0.33 (3:1 i-hexane:ethyl acetate + 1% TEA). Melting point: 99–100 $^\circ$C [73]. ^{1}H-NMR (400 MHz, (CD$_3$)$_2$CO) δ 7.32–7.24 (m, 4H), 7.21–7.16

(m, 1H), 6.27 (s, 1H), 4.35 (d, J = 5.8 Hz, 2H), 3.40–3.35 (m, 4H), 1.63–1.55 (m, 2H), 1.53–1.44 (m, 4H). ^{13}C-NMR (101 MHz, (CD$_3$)$_2$CO) δ 158.3, 142.3, 128.9, 128.2, 127.3, 45.6, 44.9, 26.6, 25.4. MS-ESI [M + H$^+$] = m/z 219.4. HPLC purity > 99%.

3.3.9. *N*-Butylpiperidine-1-carboxamide CAS: 1461-79-6

Synthesised according to the general procedure from butyl isocyanate. Recrystallised in petroleum ether and isolated as colourless crystals (150 mg, 81%). R_f = 0.15 (3:1 *i*-hexane:ethyl acetate + 1% TEA). Melting point: 60–62 °C. ^{1}H-NMR (400 MHz, (CD$_3$)$_2$CO) δ 5.73 (s, 1H), 3.35–3.26 (m, 4H), 3.16–3.09 (m, 2H), 1.60–1.52 (m, 2H), 1.51–1.39 (m, 6H), 1.37–1.24 (m, 2H), 0.88 (t, J = 7.3 Hz, 3H). ^{13}C-NMR (101 MHz, (CD$_3$)$_2$CO) δ 158.4, 45.5, 41.0, 40.9, 33.4, 33.4, 26.5, 25.4, 20.7, 14.2. HRMS calc: 185.1654 found: 185.1653. HPLC purity > 99%.

3.3.10. *N*-Isopropylpiperidine-1-carboxamide CAS: 10581-04-1

Synthesised according to the general procedure from isopropyl isocyanate. Recrystallised in petroleum ether and isolated as white crystals (100 mg, 58%). Melting point: 114–121 °C. ^{1}H-NMR (400 MHz, (CD$_3$)$_2$SO) δ 6.02 (d, J = 7.6 Hz, 1H), 3.80–3.66 (m, 1H), 3.25–3.20 (m, 4H), 1.56–1.46 (m, 2H), 1.44–1.34 (m, 4H), 1.03 (d, J = 6.6 Hz, 6H). ^{13}C-NMR (101 MHz, (CD$_3$)$_2$CO) δ 156.7, 44.2, 41.6, 25.3, 24.2, 23.0. HRMS calc: 171.1497 found: 171.1504. HPLC purity > 99%.

3.3.11. *N*-Phenylpiperidine-1-carboxamide [72] CAS: 2645-36-5

Synthesised according to the general procedure from phenyl isocyanate. Recrystallised in ethyl acetate and isolated as colorless crystals (173 mg, 84%). R_f = 0.37 (3:1 *i*-hexane:ethyl acetate + 1% TEA). Melting point: 168–171 °C. ^{1}H-NMR (400 MHz, (CD$_3$)$_2$CO) δ 7.84 (br. s, 1H), 7.54–7.49 (m, 2H), 7.23–7.17 (m, 2H), 6.91 (tt, J = 7.4, 1.2 Hz, 1H), 3.50–3.44 (m, 4H), 1.67–1.59 (m, 2H), 1.58–1.49 (m, 4H). ^{13}C-NMR (101 MHz, (CD$_3$)$_2$CO) δ 155.7, 141.9, 129.1, 122.4, 120.2, 45.8, 26.6, 25.3. MS-ESI [M + H$^+$] = m/z 205.4. HPLC purity > 99%.

3.3.12. *N*-(4-Methoxyphenyl)piperidine-1-carboxamide CAS: 2645-37-6

Synthesised according to the general procedure from 4-methoxyphenyl isocyanate. Recrystallised in petroleum ether and isolated as colourless crystals (183 mg, 78%). R_f = 0.49 (3:1 *i*-hexane:ethyl acetate + 1% TEA). Melting point: 124–125 °C. ^{1}H-NMR (400 MHz, (CD$_3$)$_2$CO) δ 7.68 (s, 1H), 7.43–7.37 (m, 2H), 6.82–6.76 (m, 2H), 3.73 (s, 3H), 3.50–3.39 (m, 4H), 1.65–1.58 (m, 2H), 1.57–1.49 (m, 4H). ^{13}C-NMR (101 MHz, (CD$_3$)$_2$CO) δ 155.9, 135.0, 122.1, 122.0, 114.3, 55.6, 45.8, 26.6, 25.3. HRMS calc: 235.1447 found: 235.1450. HPLC purity > 99%.

3.3.13. *N*-(4-Fluorophenyl)piperidine-1-carboxamide CAS: 60465-12-5

Synthesised according to the general procedure from 4-fluorophenyl isocyanate. Recrystallised in ethyl acetate and isolated as colourless crystals (188 mg, 84%). R_f = 0.39 (3:1 *i*-hexane:ethyl acetate + 1% TEA). Melting point: 179–184 °C. ^{1}H-NMR (400 MHz, (CD$_3$)$_2$CO) δ 7.90 (br. s, 1H), 7.55–7.48 (m, 2H), 7.02–6.94 (m, 2H), 3.50–3.42 (m, 4H), 1.67–1.58 (m, 2H), 1.57–1.48 (m, 4H). ^{13}C-NMR (101 MHz, (CD$_3$)$_2$CO) δ 158.8 (d, J = 238.2 Hz), 155.8, 138.2 (d, J = 2.6 Hz), 121.9 (d, J = 7.5 Hz), 115.4 (d, J = 22.3 Hz), 45.7, 26.6, 25.3. ^{19}F-NMR (376 MHz, (CD$_3$)$_2$CO) δ -119.3 – -130.5. HRMS calc: 223.1247 found: 223.1245. HPLC purity > 99%.

3.3.14. *N*-(4-Nitrophenyl)piperidine-1-carboxamide [63] CAS: 2589-20-0

Synthesised according to the general procedure from 4-nitrophenyl isocyanate. Recrystallised in petroleum ether and ethyl acetate and isolated as yellow crystals (223 mg, 89%). R_f = 0.39 (3:1 *i*-hexane:ethyl acetate + 1% TEA). Melting point: 159–162 °C. ^{1}H-NMR (400 MHz, (CD$_3$)$_2$CO) δ 8.51–8.47 (m, 1H), 7.67 (td, J = 7.7, 1.9 Hz, 1H), 7.24 (d, J = 7.8 Hz, 1H), 7.20–7.15 (m, 1H), 6.04 (br. s,

1H), 3.54–3.47 (m, 2H), 3.33–3.27 (m, 4H), 2.94 (t, *J* = 7.0 Hz, 2H), 1.60–1.53 (m, 2H), 1.49–1.41 (m, 4H). ^{13}C-NMR (101 MHz, (CD$_3$)$_2$CO) δ 161.2, 158.3, 150.0, 137.1, 124.0, 122.1, 45.4, 41.3, 41.2, 39.1, 39.0, 26.5, 25.4. MS-ESI [M + H$^+$] = *m/z* 250.2. HPLC purity > 99%.

3.3.15. *N*-Tosylpiperidine-1-carboxamide CAS: 23730-08-7

Synthesised according to the general procedure from *p*-toluenesulfonyl isocyanate. Recrystallised in petroleum ether and isolated as white crystals (144 mg, 51%). *R$_f$* = 0.85 (5% MeOH, 1% TEA in DCM). Melting point: 133–137 °C. ^{1}H-NMR (400 MHz, (CD$_3$)$_2$SO) δ 10.75 (br. s, 1H), 7.76 (d, *J* = 8.2 Hz, 2H), 7.36 (d, *J* = 8.1 Hz, 2H), 3.29–3.23 (m, 4H), 2.38 (s, 3H), 1.54–1.46 (m, 2H), 1.43–1.35 (m, 4H). ^{13}C-NMR (101 MHz, (CD$_3$)$_2$CO) δ 151.4, 143.4, 138.7, 129.0, 128.2, 44.9, 25.6, 24.1, 20.6. HRMS calc: 283.1116 found: 283.1103. HPLC purity > 99%.

3.4. *General Procedure for the Synthesis and Analysis of ^{11}C-Labelled Ureas*

Pd(Xantphos)Cl$_2$ (4 μmol) and amine/amines (30 μmol) were added to an oven-dried, conical glass vial followed by freshly distilled tetrahydrofuran (400 μL). [^{11}C]CO was transferred to the capped reaction vial and the radioactivity was measured to determine the starting amount of [^{11}C]CO. The reaction was heated at 120 °C for 10 min. When finished, the radioactivity was measured to confirm that no radioactive material had escaped during heating. The vial was purged with N$_2$ to remove unreacted [^{11}C]CO and, possibly, volatile labelled compounds formed during the reaction. The radioactivity was measured after the purge followed by either analytical HPLC for product selectivity determination or semi-preparative HPLC purification for ^{11}C-labelled product isolation. Purification was performed using either column C1 = Phenomenex Kinetex C18 (5 μm, 150 × 10.0 mm) column, C2 = Reprosil–Pur Basic C18 (5 μm, 150 × 10.0 mm) column or C3 = Gemini NX C18 (5 μm, 250 × 10.0 mm) column with ammonium formate buffer 50 mM (pH 3.5) (A) and acetonitrile (B) as eluents. Run time was 20 min with flow 5 mL/min followed by flushing the column with 100% B. After isolation and a final radioactivity measurement of the ^{11}C-labelled product, an aliquot was analysed to determine radiochemical purity and the identity of the ^{11}C-labelled product was confirmed using the isotopically unmodified product as reference. The analytical method was the same for all compounds, 10–90% acetonitrile in 10 min (flow 2 mL/min). Molar activity determinations were based on a calibration curve, constructed with isotopically unmodified **20**. For full definitions and calculations, see the supporting materials.

3.4.1. [*carbonyl*-^{11}C]*N,N'*-Dibenzylurea **2**

Synthesised according to the general procedure (three experiments). (1) Purification method: 40% B. Column: C1. Starting from 3.4 GBq, 0.88 GBq was isolated at 33 min from EOB; (2) Purification method: 35% B Column: C1. Starting from 3.4 GBq, 0.83 GBq was isolated at 36 min from EOB; (3) Purification method: 45% B Column: C3. Starting from 2.7 GBq, 0.44 GBq was isolated at 34 min from EOB. Analytical HPLC R$_t$ = 5.1 min.

3.4.2. [*carbonyl*-^{11}C]*N,N'*-Dipropylurea **3**

Synthesised according to the general procedure (two experiments). (1) Purification method: 0–70% B Column: C1. Starting from 2.2 GBq, 0.24 GBq was isolated at 42 min from EOB; (2) Purification method: 0–70% B Column: C1. Starting from 6.7 GBq, 0.62 GBq was isolated at 53 min from EOB. Analytical HPLC R$_t$ = 2.6 min.

3.4.3. [*carbonyl*-^{11}C]*N,N'*-Dicyclohexylurea **4**

Synthesised according to the general procedure (two experiments). (1) Purification method: 40% B Column: C1. Starting from 3.4 GBq, 0.53 GBq was isolated at 37 min from EOB; (2) Purification method:

35% B Column: C1. Starting from 3.8 GBq, 0.60 GBq was isolated at 41 min from EOB. Analytical HPLC R_t = 5.6 min.

3.4.4. [*carbonyl*-^{11}C]*N,N'*-Diphenylurea 5

Synthesised according to the general procedure (two experiments). (1) Purification method: 35% B Column: C1. Starting from 3.0 GBq, 0.028 GBq was isolated at 50 min from EOB; (2) Purification method: 35% B Column: C1. Starting from 6.7 GBq, 0.059 GBq was isolated at 43 min from EOB. Analytical HPLC R_t = 5.5 min.

3.4.5. [*carbonyl*-^{11}C]*N*-Benzylpiperidine-1-carboxamide 7

Synthesised according to the general procedure (four experiments). (1) Purification method: 35% B Column: C1. Starting from 4.9 GBq, 0.47 GBq was isolated at 45 min from EOB; (2) Purification method: 35% B Column: C1. Starting from 3.0 GBq, 0.57 GBq was isolated at 36 min from EOB; (3) 5 min reaction time. Purification method: 35% B Column: C3. Starting from 2.6 GBq, 0.15 GBq was isolated at 37 min from EOB; (4) 10 equiv. of 1-butanol added. Purification method: 35% B Column: C3. Starting from 3.2 GBq, 0.29 GBq was isolated at 42 min from EOB. Analytical HPLC R_t = 4.9 min.

3.4.6. [*carbonyl*-^{11}C]*N*-Butylpiperidine-1-carboxamide 8

Synthesised according to the general procedure (three experiments). (1) Purification method: 20–50% Column: C1. Starting from 3.7 GBq, 0.25 GBq was isolated at 44 min from EOB; (2) Purification method: 20–50% Column: C1. Starting from 2.9 GBq, 0.17 GBq was isolated at 46 min from EOB; (3) 10 equiv. of 1-butanol added. Purification method: 20–50% Column: C3. Starting from 2.8 GBq, 0.093 GBq was isolated at 49 min from EOB. Analytical HPLC R_t = 4.5 min.

3.4.7. [*carbonyl*-^{11}C]*N*-(2-(Pyridin-2-yl)ethyl)piperidine-1-carboxamide 9

Synthesised according to the general procedure (two experiments). (1) Purification method: 20–50% B Column: C1. Starting from 3.3 GBq, 0.14 GBq was isolated at 36 min from EOB; (2) Purification method: 20% B Column: C1. Starting from 3.7 GBq, 0.13 GBq was isolated at 39 min from EOB. Analytical HPLC R_t = 3.8 min.

3.4.8. [*carbonyl*-^{11}C]*N*-Isopropylpiperidine-1-carboxamide 10

Synthesised according to the general procedure (two experiments). (1) Purification method: 15–50% B Column: C1. Starting from 3.2 GBq, 0.11 GBq was isolated at 39 min from EOB; (2) Purification method: 15–50% B Column: C1. Starting from 3.2 GBq, 0.19 GBq was isolated at 38 min from EOB. Analytical HPLC R_t = 3.3 min.

3.4.9. [*carbonyl*-^{11}C]*N*-Phenylpiperidine-1-carboxamide 11

Synthesised according to the general procedure (six experiments). (1) Purification method: 35% B Column: C1. Starting from 3.6 GBq, 0.19 GBq was isolated at 40 min from EOB; (2) Purification method: 30% B Column: C1. Starting from 3.1 GBq, 0.094 GBq was isolated at 42 min from EOB; (3) 3 equiv. aniline. Purification method: 20–50% B Column: C1. Starting from 3.6 GBq, 0.051 GBq was isolated at 40 min from EOB; (4) 3 equiv. aniline. Purification method: 20–50%% B Column: C1. Starting from 3.2 GBq, 0.0.058 GBq was isolated at 40 min from EOB; (5) 3 equiv. aniline. Purification method: 20–50% B Column: C1. Starting from 4.3 GBq, 0.16 GBq was isolated at 42 min from EOB; (6) 10 equiv. of 1-butanol. Purification method: 20–50% B Column: C1. Starting from 2.7 GBq, 0.026 GBq was isolated at 62 min from EOB. Analytical HPLC R_t = 4.7 min.

3.4.10. [*carbonyl*-¹¹C]*N*-(4-Methoxyphenyl)piperidine-1-carboxamide **12**

Synthesised according to the general procedure (three experiments). (1) Purification method: 20–50% B Column: C1. Starting from 3.2 GBq, 0.084 GBq was isolated at 41 min from EOB; (2) Purification method: 20–50% B Column: C1. Starting from 1.8 GBq, 0.052 GBq was isolated at 43 min from EOB; (3) 10 equiv. 1-butanol. Purification method: 20–50% B Column: C3. Starting from 2.9 GBq, 0.16 GBq was isolated at 43 min from EOB. Analytical HPLC R_t = 4.5 min.

3.4.11. [*carbonyl*-¹¹C]*N*-(4-Fluorophenyl)piperidine-1-carboxamide **13**

Synthesised according to the general procedure (four experiments). (1) Purification method: 35% B Column: C1. Starting from 3.7 GBq, 0.072 GBq was isolated at 46 min from EOB; (2) Purification method: 35% B Column: C1. Starting from 3.6 GBq, 0.086 GBq was isolated at 37 min from EOB; (3) Purification method: 35% B Column: C1. Starting from 21.7 GBq, 0.61 GBq was isolated at 37 min from EOB; (4) 10 equiv. 1-butanol. Purification method: 35% B Column: C3. Starting from 4.3 GBq, 0.38 GBq was isolated at 40 min from EOB. Analytical HPLC R_t = 4.8 min.

3.4.12. [*carbonyl*-¹¹C]*N*-(4-Nitrophenyl)piperidine-1-carboxamide **14**

Synthesised according to the general procedure (three experiments). (1) Purification method: 40% B Column: C1. Starting from 3.3 GBq, 0.028 GBq was isolated at 49 min from EOB; (2) Purification method: 40% B Column: C1. Starting from 3.6 GBq, 0.059 GBq was isolated at 39 min from EOB; (3) 10 equiv. 1-butanol. Purification method: 40% B Column: C3. Starting from 2.7 GBq, 0.007 GBq was isolated at 96 min from EOB. Analytical HPLC R_t = 5.4 min.

3.4.13. [*carbonyl*-¹¹C]3,4-Dihydroquinazolin-2(1H)-one **15**

Synthesised according to the general procedure (three experiments). (1) Purification method: 20% B Column: C1. Starting from 3.4 GBq, 0.007 GBq was isolated at 43 min from EOB; (2) Purification method: 20% B Column: C1. Starting from 3.1 GBq, 0.018 GBq was isolated at 34 min from EOB. Analytical HPLC R_t = 2.7 min.

3.4.14. [*carbonyl*-¹¹C]*N*-(2,4-Dichlorobenzyl)-4-phenoxypiperidine-1-carboxamide **19**

Synthesised according to the general procedure (three experiments). (1) Purification method: 60% B Column: C1. Starting from 3.2 GBq, 0.46 GBq was isolated at 33 min from EOB; (2) Purification method: 60% B Column: C2. Starting from 17.8 GBq, 1.9 GBq was isolated at 43 min from EOB; (3) Purification method: 60% B Column: C3. Starting from 12.8 GBq, 2.1 GBq was isolated at 41 min from EOB. Analytical HPLC R_t = 7.4 min.

3.5. Computational Details

The density functional theory calculations were performed in Jaguar version 9.5, release 11, Schrodinger, Inc., New York, NY, USA, 2016 [74]. To facilitate the calculations, *N*-methyl amine was used as a minimal primary amine. The geometries were optimised using the LACVP** basis set [75] and the B3LYP-D3 a posteriori-corrected functional [76]. Vibrational analyses were performed to characterise the stationary points identified and to calculate zero point energies. The contributions to the free energies were calculated at a temperature of 393.15 K (120 °C) using the B3LYP-D3/LACVP** geometries. For the chloride anion, an entropy of 40.97 cal K^{-1}·mol^{-1} was used [77]. Solvation energies were calculated using B3LYP-D3/LACVP**+ and the standard Poisson-Boltzmann continuum solvation model [78] with parameters suitable for THF using the B3LYP-D3/LACVP** geometries.

4. Conclusions

An important aspect of PET tracer development is the possibility to label aspiring tracers, to enable their preclinical and subsequent clinical evaluation. The labelling method should give the labelled compound with high enough radioactivity and purity to enable the study. The palladium(II)-mediated oxidative carbonylation of aliphatic and aromatic amines presented herein is a facile and simple method for the synthesis of ^{11}C-labelled ureas. In total, 14 symmetrical and unsymmetrical ^{11}C-labelled ureas were synthesised and isolated using only [^{11}C]CO, a Pd-source and amines as reaction components. DFT-calculations and selectivity experiments supported a reaction proceeding via a ^{11}C-labelled isocyanate. The reaction outcome was largely dependent on the amine nucleophilicity, with aliphatic amines performing better than aromatic amines as substrates. However, with the addition of 1-butanol to the reaction mixture, the radiochemical yields of two aromatic ^{11}C-labelled ureas were improved. Not all aniline derivatives benefitted from the alcoholic additive, but the addition presents a viable approach when using challenging aromatic amines in ^{11}C-urea synthesis. Finally, the advantage of using [^{11}C]CO to achieve high molar activity, as opposed to other ^{11}C-labelling synthons, was apparent as the isolation of sEH inhibitor **19** in a good 41% radiochemical yield and with high molar activity (247 GBq/μmol–319 Gbq/μmol).

Supplementary Materials: The following is available online: Definitions and molar activity calculations, ^{1}H and ^{13}C-NMR chromatograms of all synthesised reference compounds, HPLC chromatograms of isolated ^{11}C-compounds, DFT calculated structures of intermediates on path **A**.

Acknowledgments: We gratefully acknowledge the financial support from the Disciplinary domain of Medicine and Pharmacy, Uppsala University and the King Gustaf V and Queen Victoria Freemason Foundation.

Author Contributions: S.R. synthesised the reference compounds and starting material, designed the experiments, performed labelling experiments and drafted the manuscript. P.B. performed, analysed and summarised the DFT-calculations. P.N. designed the experiments and performed labelling experiments. L.O. designed the experiments and contributed with reference compounds. J.E. designed the experiments and set up the purification method. M.L. designed the experiments. All authors contributed to the critical discussion and presentation of the results. All authors read, commented on and approved the final manuscript.

Conflicts of Interest: The authors declare no conflict of interest.

References and Note

1. Wood, K.A.; Hoskin, P.J.; Saunders, M.I. Positron emission tomography in oncology: A review. *Clin. Oncol.* **2007**, *19*, 237–255. [CrossRef] [PubMed]
2. Rocchi, L.; Niccolini, F.; Politis, M. Recent imaging advances in neurology. *J. Neurol.* **2015**, *262*, 2182–2194. [CrossRef] [PubMed]
3. Boutagy, N.E.; Sinusas, A.J. Recent advances and clinical applications of PET cardiac autonomic nervous system imaging. *Curr. Cardiol. Rep.* **2017**, *19*. [CrossRef] [PubMed]
4. Langer, O. Use of PET Imaging to evaluate transporter-mediated drug-drug interactions. *J. Clin. Pharmacol.* **2016**, *56*, S143–S156. [CrossRef] [PubMed]
5. Papadimitriou, L.; Smith-Jones, P.M.; Sarwar, C.M.S.; Marti, C.N.; Yaddanapudi, K.; Skopicki, H.A.; Gheorghiade, M.; Parsey, R.; Butler, J. Utility of positron emission tomography for drug development for heart failure. *Am. Heart J.* **2016**, *175*, 142–152. [CrossRef] [PubMed]
6. Declercq, L.D.; Vandenberghe, R.; Van Laere, K.; Verbruggen, A.; Bormans, G. Drug development in Alzheimer's disease: The contribution of PET and SPECT. *Front. Pharmacol.* **2016**, *7*. [CrossRef] [PubMed]
7. Colquhoun, H.M.; Thompson, D.J.; Twigg, M.W. *Carbonylation—Direct Synthesis of Carbonyl Compounds*; Springer Science + Business Media: New York, NY, USA, 1991; pp. 1–281.
8. Rahman, O. Carbon monoxide in labeling chemistry and positron emission tomography tracer development: Scope and limitations. *J. Label. Compd. Radiopharm.* **2015**, *58*, 86–98. [CrossRef] [PubMed]
9. Purwanto; Deshpande, R.M.; Chaudhari, R.V.; Delmas, H. Solubility of hydrogen, carbon monoxide, and 1-octene in various solvents and solvent mixtures. *J. Chem. Eng. Data* **1996**, *41*, 1414–1417.
10. Kihlberg, T.; Långström, B. Method and Apparatus for Production and Use of [^{11}C]Carbon Monoxide in Labeling Synthesis. Patent WO 02/102711 A1, 2002.

11. Hostetler, E.D.; Burns, H.D. A remote-controlled high pressure reactor for radiotracer synthesis with [11C]carbon monoxide. *Nucl. Med. Biol.* **2002**, *29*, 845–848. [CrossRef]
12. Kihlberg, T.; Långström, B. Biologically active 11C-labeled amides using palladium-mediated reactions with aryl halides and [11C]carbon monoxide. *J. Org. Chem.* **1999**, *64*, 9201–9205. [CrossRef]
13. Dahl, K.; Schou, M.; Amini, N.; Halldin, C. Palladium-mediated [11C]carbonylation at atmospheric pressure: A general method using xantphos as supporting ligand. *Eur. J. Org. Chem.* **2013**, 1228–1231. [CrossRef]
14. Kealey, S.; Miller, P.W.; Long, N.J.; Plisson, C.; Martarello, L.; Gee, A.D. Copper(i)scorpionate complexes and their application in palladium-mediated [11C]carbonylation reactions. *Chem. Commun.* **2009**, *25*, 3696–3698. [CrossRef] [PubMed]
15. Dahl, K.; Schou, M.; Rahman, O.; Halldin, C. Improved yields for the palladium-mediated C-11-carbonylation reaction using microwave technology. *Eur. J. Org. Chem.* **2014**, *2*, 307–310. [CrossRef]
16. Audrain, H.; Martarello, L.; Gee, A.; Bender, D. Utilisation of [11C]-labelled boron carbonyl complexes in palladium carbonylation reaction. *Chem. Commun.* **2004**, *5*, 558–559. [CrossRef] [PubMed]
17. Nordeman, P.; Friis, S.D.; Andersen, T.L.; Audrain, H.; Larhed, M.; Skrydstrup, T.; Antoni, G. Rapid and Efficient Conversion of 11CO2 to 11CO through silacarboxylic acids: Applications in Pd-mediated carbonylations. *Chem. Eur. J.* **2015**, *21*, 17601–17604. [CrossRef] [PubMed]
18. Taddei, C.; Bongarzone, S.; Dheere, A.K.H.; Gee, A.D. [11C]CO2 to [11C]CO conversion mediated by [11C]silanes: A novel route for [11C]carbonylation reactions. *Chem. Commun.* **2015**, *51*, 11795–11797. [CrossRef] [PubMed]
19. Miller, P.W.; Long, N.J.; De Mello, A.J.; Vilar, R.; Audrain, H.; Bender, D.; Passchier, J.; Gee, A. Rapid multiphase carbonylation reactions by using a microtube reactor: Applications in positron emission tomography 11C-radiolabeling. *Angew. Chem. Int. Ed.* **2007**, *46*, 2875–2878. [CrossRef] [PubMed]
20. Kealey, S.; Plisson, C.; Collier, T.L.; Long, N.J.; Husbands, S.M.; Martarello, L.; Gee, A.D. Microfluidic reactions using [11C]carbon monoxide solutions for the synthesis of a positron emission tomography radiotracer. *Org. Biomol. Chem.* **2011**, *9*, 3313–3319. [CrossRef] [PubMed]
21. Dahl, K.; Schou, M.; Ulin, J.; Sjöberg, C.-O.; Farde, L.; Halldin, C. 11C-carbonylation reactions using gas–liquid segmented microfluidics. *RSC Adv.* **2015**, *5*, 88886–88889. [CrossRef]
22. Eriksson, J.; Van Den Hoek, J.; Windhorst, A.D. Transition metal mediated synthesis using [11C]CO at low pressure—A simplified method for 11C-carbonylation. *J. Label. Compd. Radiopharm.* **2012**, *55*, 223–228. [CrossRef]
23. Jagtap, A.D.; Kondekar, N.B.; Sadani, A.A.; Chern, J.-W. Ureas: Applications in drug design and development. *Curr. Med. Chem.* **2017**, *24*, 622–651. [CrossRef] [PubMed]
24. Dumas, J.; Smith, R.A.; Lowinger, T.B. Recent developments in the discovery of protein kinase inhibitors from the urea class. *Curr. Opin. Drug Discov. Dev.* **2004**, *7*, 600–616.
25. Venkatraman, S.; Bogen, L.; Arasappan, A.; Bennett, F.; Chen, K.; Jao, E.; Liu, Y.; Lovey, R.; Hendrata, S.; Huang, Y.; et al. Orally bioavailable hepatitis C virus NS3 protease inhibitor: A potential therapeutic agent for the treatment of hepatitis C infection. *J. Med. Chem.* **2015**, *49*, 6074–6086. [CrossRef] [PubMed]
26. Shibata, S.; Gillespie, J.R.; Ranade, R.M.; Koh, C.Y.; Kim, J.E.; Laydbak, J.U.; Zucker, F.H.; Hol, W.G.J.; Verlinde, C.L.M.J.; Buckner, F.S.; et al. Urea-based inhibitors of trypanosoma brucei methionyl-trna synthetase: Selectivity and in vivo characterization. *J. Med. Chem.* **2012**, *55*, 6342–6351. [CrossRef] [PubMed]
27. Asakawa, C.; Ogawa, M.; Fujinaga, M.; Kumata, K.; Xie, L.; Yamasaki, T.; Yui, J.; Fukumura, T.; Zhang, M.R. Utilization of [11C]phosgene for radiosynthesis of N-(2-{3-[3,5-bis(trifluoromethyl)]phenyl[11C]ureido}ethyl) glycyrrhetinamide, an inhibitory agent for proteasome and kinase in tumors. *Bioorg. Med. Chem. Lett.* **2012**, *22*, 3594–3597. [CrossRef] [PubMed]
28. Dollé, F.; Martarello, L.; Bramoullé, Y.; Bottlaender, M.; Gee, A.D. Radiosynthesis of carbon-11-labelled GI181771, a new selective CCK-A agonist. *J. Label. Compd. Radiopharm.* **2005**, *48*, 501–513. [CrossRef]
29. Roeda, D.; Westera, G. The Synthesis of Some 11C-labelled antiepileptic drugs with potential utility as radiopharmaceuticals: Hydantoins and Barbiturates. *Int. J. Appl. Radiat. Isot.* **1981**, *32*, 843–845. [CrossRef]
30. Lemoucheux, L.; Rouden, J.; Sobrio, F.; Lasne, M. Debenzylation of tertiary amines using phosgene or triphosgene: An efficient and rapid procedure for the preparation of carbamoyl chlorides and unsymmetrical ureas. Application of carbon-11 chemistry. *J. Org. Chem.* **2003**, *68*, 7289–7297. [CrossRef] [PubMed]

31. Roeda, D.; Dollé, F. [^{11}C]Phosgene: A versatile reagent for radioactive carbonyl insertion into medicinal radiotracers for positron emission tomography. *Curr. Top. Med. Chem.* **2010**, *10*, 1680–1700. [CrossRef] [PubMed]

32. Emran, A.M.; Boothe, T.E.; Finn, R.D.; Vora, M.M.; Kothari, P.J. Preparation of ^{11}C-urea from no-carrier-added ^{11}C-cyanide. *Int. J. Appl. Radiat. Isot.* **1983**, *34*, 1013–1014. [CrossRef]

33. Boothe, T.E.; Emran, A.L.M.; Kothari, J. Use of ^{11}C as a tracer for studying the synthesis of [^{11}C]urea from [^{11}C]cyanide. *Int. J. Appl. Radiat. Isot.* **1985**, *36*, 141–144. [CrossRef]

34. Bera, R.K.; Hartman, N.G.; Jay, M. Continuous production of [C-11] urea for medical application. *Appl. Radiat. Isot.* **1991**, *42*, 407–409. [CrossRef]

35. Van Tilburg, E.W.; Windhorst, A.D.; Van Der Mey, M.; Herscheid, J.D.M. One-pot synthesis of [^{11}C]ureas via triphenylphosphinimines. *J. Label. Compd. Radiopharm.* **2006**, *49*, 321–330. [CrossRef]

36. Schirbel, A.; Holschbach, M.H.; Coenen, H.H. N.C.A.[^{11}C]CO$_2$ as a safe substitute for phosgene in the carbonylation of primary amines. *J. Label. Compd. Radiopharm.* **1999**, *42*, 537–551. [CrossRef]

37. Hicks, J.W.; Wilson, A.A.; Rubie, E.A.; Woodgett, J.R.; Houle, S.; Vasdev, N. Towards the preparation of radiolabeled 1-aryl-3-benzyl ureas: Radiosynthesis of [^{11}C-*carbonyl*] AR-A014418 by [^{11}C]CO$_2$ fixation. *Bioorg. Med. Chem. Lett.* **2012**, *22*, 2099–2101. [CrossRef] [PubMed]

38. Hicks, J.W.; Parkes, J.; Tong, J.; Houle, S.; Vasdev, N.; Wilson, A.A. Radiosynthesis and ex vivo evaluation of [^{11}C-*carbonyl*]carbamate- and urea-based monoacylglycerol lipase inhibitors. *Nucl. Med. Biol.* **2014**, *41*, 688–694. [CrossRef] [PubMed]

39. Dheere, A.; Bongarzone, S.; Taddei, C.; Yan, R.; Gee, A. Synthesis of ^{11}C-labelled symmetrical ureas via the rapid incorporation of [^{11}C]CO$_2$ into aliphatic and aromatic amines. *Synlett* **2015**, *26*, 2257–2260.

40. Chakraborty, P.K.; Mangner, T.J.; Chugani, H.T. The synthesis of no-carrier-added [^{11}C]urea from [^{11}C]carbon dioxide and application to [^{11}C]uracil synthesis. *Appl. Radiat. Isot.* **1997**, *48*, 619–621. [CrossRef]

41. Dahl, K.; Collier, T.L.; Chang, R.; Zhang, X.; Sadovski, O.; Liang, S.H.; Vasdev, N. "In-Loop" [^{11}C]CO$_2$-fixation: Prototype and proof-of-concept. *J. Label. Compd. Radiopharm.* **2017**. [CrossRef] [PubMed]

42. Kihlberg, T.; Karimi, F.; Långström, B. [^{11}C]carbon monoxide in selenium-mediated synthesis of ^{11}C-carbamoyl compounds. *J. Org. Chem.* **2002**, *67*, 3687–3692. [CrossRef] [PubMed]

43. Åberg, O.; Långström, B. Synthesis of substituted [^{11}C]ureas and [^{11}C]sulphonylureas by Rh(I)-mediated carbonylation. *J. Label. Compd. Radiopharm.* **2011**, *54*, 38–42. [CrossRef]

44. Ilovich, O.; Åberg, O.; Långström, B.; Mishania, E. Rhodium-mediated [^{11}C]carbonylation: A library of N-phenyl-N'-{4-(4-quinolyloxy)-phenyl]-[^{11}C]urea derivatives as potential PET angiogenic probes. *J. Label. Compd. Radiopharm.* **2009**, *52*, 151–157. [CrossRef]

45. Poot, A.J.; van der Wildt, B.; Stigter-van Walsum, M.; Rongen, M.; Schuit, R.C.; Hendrikse, N.H.; Eriksson, J.; van Dongen, G.A.M.S.; Windhorst, A.D. [^{11}C]Sorafenib: Radiosynthesis and preclinical evaluation in tumor-bearing mice of a new TKI-PET tracer. *Nucl. Med. Biol.* **2013**, *40*, 488–497. [CrossRef] [PubMed]

46. Doi, H.; Barletta, J.; Suzuki, M.; Noyori, R.; Watanabe, Y. Synthesis of ^{11}C-labelled N,N'-diphenylurea and ethyl phenylcarbamate by a rhodium-promoted carbonylation via [^{11}C]isocyanatebenzene using phenyl azide and [^{11}C]carbon monoxide. *Org. Biomol. Chem.* **2004**, *2*, 3063–3066. [CrossRef] [PubMed]

47. Barletta, J.; Karimi, F.; Långström, B. Synthesis of [^{11}C-*carbonyl*]hydroxyureas by a rhodium-mediated carbonylation reaction using [^{11}C]carbon monoxide. *J. Label. Compd. Radiopharm.* **2006**, *49*, 429–436. [CrossRef]

48. Dahl, K.; Itsenko, O.; Rahman, O.; Ulin, J.; Sjöberg, C.O.; Sandblom, P.; Larsson, L.A.; Schou, M.; Halldin, C. An evaluation of a high-pressure ^{11}CO carbonylation apparatus. *J. Label. Compd. Radiopharm.* **2015**, *58*, 220–225. [CrossRef] [PubMed]

49. Kealey, S.; Husbands, S.M.; Bennacef, I.; Gee, A.D.; Passchier, J. Palladium-mediated oxidative carbonylation reactions for the synthesis of ^{11}C-radiolabelled ureas. *J. Label. Compd. Radiopharm.* **2014**, *57*, 202–208. [CrossRef] [PubMed]

50. Wilson, A.A.; Garcia, A.; Houle, S.; Sadovski, O.; Vasdev, N. Synthesis and application of isocyanates radiolabeled with carbon-11. *Chem. Eur. J.* **2011**, *17*, 259–264. [CrossRef] [PubMed]

51. Gómez-vallejo, V.; Gaja, V.; Koziorowski, J.; Llop, J. *Positron Emission Tomography—Current Clinical and Research Aspects*; Hsieh, C.-H., Ed.; InTech: Rijeka, Croatia, 2012; p. 183.

52. Henriksen, G.; Drzezga, A. *Small Animal Imaging*; Kiessling, F., Pichler, B.J., Eds.; Springer: Berlin, Germany, 2011; pp. 499–513.

53. Roslin, S.; Rosa, M.D.; Deuther-Conrad, W.; Eriksson, J.; Odell, L.R.; Antoni, G.; Brust, P.; Larhed, M. Synthesis and in vitro evaluation of 5-substituted benzovesamicol analogs containing *N*-substituted amides as potential positron emission tomography tracers for the vesicular acetylcholine transporter. *Bioorg. Med. Chem.* **2017**, *25*, 5095–5106. [CrossRef] [PubMed]

54. Chow, S.Y.; Odell, L.R.; Eriksson, J. Low-pressure radical ^{11}C-aminocarbonylations of alkyl iodides via thermal initiation. *Eur. J. Org. Chem.* **2016**, *2016*, 5980–5989. [CrossRef]

55. Stevens, M.Y.; Chow, S.Y.; Estrada, S.; Eriksson, J.; Asplund, V.; Orlova, A.; Mitran, B.; Antoni, G.; Larhed, M.; Åberg, O.; et al. Synthesis of ^{11}C-labeled sulfonyl carbamates through a multicomponent reaction employing sulfonyl azides, alcohols, and [^{11}C]CO. *ChemistryOpen* **2016**, *5*, 566–573. [CrossRef] [PubMed]

56. Martinelli, J.R.; Watson, D.A.; Freckmann, D.M.M.; Barder, T.E.; Buchwald, S.L. Palladium-catalyzed carbonylation reactions of aryl bromides at atmospheric pressure: A general system based on xantphos. *J. Org. Chem.* **2008**, *73*, 7102–7107. [CrossRef] [PubMed]

57. Andersen, T.L.; Friis, S.D.; Audrain, H.; Nordeman, P.; Antoni, G.; Skrydstrup, T. Efficient ^{11}C-carbonylation of isolated aryl palladium complexes for PET: Application to challenging radiopharmaceutical synthesis. *J. Am. Chem. Soc.* **2015**, *137*, 1548–1555. [CrossRef] [PubMed]

58. For more information, see Supplementary Materials.

59. Lum, R.T.; Cheng, M.; Cristobal, C.P.; Goldfine, I.D.; Evans, J.L.; Keck, J.G.; Macsata, R.W.; Manchem, V.P.; Matsumoto, Y.; Park, S.J.; et al. Design, synthesis, and structure-activity relationships of novel insulin receptor tyrosine kinase activators. *J. Med. Chem.* **2008**, *51*, 6173–6187. [CrossRef] [PubMed]

60. Cheng, D.; Yadav, N.; King, R.W.; Swanson, M.S.; Weinstein, E.J.; Bedford, M.T. Small molecule regulators of protein arginine methyltransferases. *J. Biol. Chem.* **2004**, *279*, 23892–23899. [CrossRef] [PubMed]

61. Shen, H.C.; Hammock, B.D. Discovery of inhibitors of soluble epoxide hydrolase: A target with multiple potential therapeutic indications discovery of inhibitors of soluble epoxide hydrolase: A target with multiple potential therapeutic indications RY800-C114 department of medici. *J. Med. Chem.* **2012**, *55*, 1789–1808. [CrossRef] [PubMed]

62. Walker, J.; Kay, S.A. XLVIII—Velocity of urea formation in aqueous alcogol. *J. Chem. Soc. Trans.* **1897**, *71*, 489–508. [CrossRef]

63. Orito, K.; Miyazawa, M.; Nakamura, T.; Horibata, A.; Ushito, H.; Nagasaki, H.; Yuguchi, M.; Yamashita, S.; Yamazaki, T.; Tokuda, M. Pd(OAc)$_2$-Catalyzed carbonylation of amines. *J. Org. Chem.* **2006**, *71*, 5951–5958. [CrossRef] [PubMed]

64. Hiwatari, K.; Kayaki, Y.; Okita, K.; Ukai, T.; Shimizu, I.; Yamamoto, A. Selective oxidative carbonylation of amines to oxamides and ureas catalyzed by palladium complexes. *Bull. Chem. Soc. Jpn.* **2004**, *77*, 2237–2250. [CrossRef]

65. Aresta, M.; Giannoccaro, P.; Tommasi, I.; Dibenedetto, A.; Maria, A.; Lanfredi, M.; Ugozzoli, F.; Lanfredi, A.M.M.; Ugozzoli, F. Synthesis and solid state and solution characterization of mono- and di-(η [1]-C) carbamoyl-palladium complexes. New efficient palladium-catalyzed routes to carbamoyl chlorides: Key intermediates to isocyanates, carbamic esters, and ureas. *Organometallics* **2000**, *19*, 3879–3889. [CrossRef]

66. Kiesewetter, D.O.; Eckelman, W.C. Utility of azetidinium methanesulfonates for radiosynthesis of 3-[^{18}F]fluoropropyl amines. *J. Label. Compd. Radiopharm.* **2004**, *47*, 953–969. [CrossRef]

67. Hermange, P.; Lindhardt, A.T.; Taaning, R.H.; Bjerglund, K.; Lupp, D.; Skrydstrup, T. Ex situ generation of stoichiometric and substoichiometric [12]CO and [13]CO and its efficient incorporation in palladium catalyzed aminocarbonylations. *J. Am. Chem. Soc.* **2011**, *133*, 6061–6071. [CrossRef] [PubMed]

68. Nordeman, P.; Odell, L.R.; Larhed, M. Aminocarbonylations employing Mo(CO)$_6$ and a bridged two-vial system: Allowing the use of nitro group substituted aryl iodides and aryl bromides. *J. Org. Chem.* **2012**, *77*, 11393–11398. [CrossRef] [PubMed]

69. Guan, Z.H.; Lei, H.; Chen, M.; Ren, Z.H.; Bai, Y.; Wang, Y.Y. Palladium-catalyzed carbonylation of amines: Switchable approaches to carbamates and *N,N'*-disubstituted ureas. *Adv. Synth. Catal.* **2012**, *354*, 489–496. [CrossRef]

70. Paz, J.; Pérez-Balado, C.; Iglesias, B.; Muñoz, L. Carbon dioxide as a carbonylating agent in the synthesis of 2-oxazolidinones, 2-oxazinones, and cyclic ureas: Scope and limitations. *J. Org. Chem.* **2010**, *75*, 3037–3046. [CrossRef] [PubMed]

71. Das, S.; Addis, D.; Knöpke, L.R.; Bentrup, U.; Junge, K.; Brückner, A.; Beller, M. Selective catalytic monoreduction of phthalimides and imidazolidine-2,4-diones. *Angew. Chem. Int. Ed.* **2011**, *50*, 9180–9184. [CrossRef] [PubMed]

72. Lee, S.H.; Matsushita, H.; Clapham, B.; Janda, K.D. The direct conversion of carbamates to ureas using aluminum amides. *Tetrahedron* **2004**, *60*, 3439–3443. [CrossRef]

73. Matsumura, Y.; Satoh, Y.; Onomura, O.; Maki, T. A New method for synthesis of unsymmetrical ureas using electrochemically prepared trifluoroethyl carbamates. *J. Org. Chem.* **2000**, *65*, 1549–1551. [CrossRef] [PubMed]

74. Bochevarov, A.D.; Harder, E.; Hughes, T.F.; Greenwood, J.R.; Braden, D.A.; Philipp, D.M.; Rinaldo, D.; Halls, M.D.; Zhang, J.; Friesner, R.A. Jaguar: A high-performance quantum chemistry software program with strengths in life and materials sciences. *Int. J. Quantum Chem.* **2013**, *113*, 2110–2142. [CrossRef]

75. Hay, P.J.; Wadt, W.R. Ab initio effective core potentials for molecular calculations. Potentials for K to Au including the outermost core orbitals. *J. Chem. Phys.* **1985**, *82*, 299. [CrossRef]

76. Grimme, S.; Antony, J.; Ehrlich, S.; Krieg, H. A consistent and accurate ab initio parametrization of density functional dispersion correction (DFT-D) for the 94 elements H-Pu. *J. Chem. Phys.* **2010**, *132*, 154104. [CrossRef] [PubMed]

77. Chase, M.W., Jr.; Davies, C.A.; Downey, J.R., Jr.; Frurip, D.J.; McDonald, R.A.; Syverud, A.N. Janaf Thermochemical Tables 3rd ed. *J. Phys. Chem. Ref. Data* **1985**, *14*, 718–1856.

78. Marten, B.; Kim, K.; Cortis, C.; Friesner, R.A.; Murphy, R.B.; Ringnalda, M.N.; Sitkoff, D.; Honig, B. New model for calculation of solvation free energies: Correction of self-consistent reaction field continuum dielectric theory for short-range hydrogen-bonding effects. *J. Phys. Chem.* **1996**, *100*, 11775–11788. [CrossRef]

Sample Availability: Samples of the reference compounds **2**, **7–15** and **17–19** are available from the authors.

MDPI

Review

Advances in the Development of PET Ligands Targeting Histone Deacetylases for the Assessment of Neurodegenerative Diseases

Tetsuro Tago and Jun Toyohara *

Research Team for Neuroimaging, Tokyo Metropolitan Institute of Gerontology, 35-2 Sakae-cho, Itabashi-ku, Tokyo 173-0015, Japan; tago@pet.tmig.or.jp
* Correspondence: toyohara@pet.tmig.or.jp; Tel.: +81-3-3964-3241; Fax: +81-3-3964-1148

Received: 1 January 2018; Accepted: 29 January 2018; Published: 31 January 2018

Abstract: Epigenetic alterations of gene expression have emerged as a key factor in several neurodegenerative diseases. In particular, inhibitors targeting histone deacetylases (HDACs), which are enzymes responsible for deacetylation of histones and other proteins, show therapeutic effects in animal neurodegenerative disease models. However, the details of the interaction between changes in HDAC levels in the brain and disease progression remain unknown. In this review, we focus on recent advances in development of radioligands for HDAC imaging in the brain with positron emission tomography (PET). We summarize the results of radiosynthesis and biological evaluation of the HDAC ligands to identify their successful results and challenges. Since 2006, several small molecules that are radiolabeled with a radioisotope such as carbon-11 or fluorine-18 have been developed and evaluated using various assays including in vitro HDAC binding assays and PET imaging in rodents and non-human primates. Although most compounds do not readily cross the blood-brain barrier, adamantane-conjugated radioligands tend to show good brain uptake. Until now, only one HDAC radioligand has been tested clinically in a brain PET study. Further PET imaging studies to clarify age-related and disease-related changes in HDACs in disease models and humans will increase our understanding of the roles of HDACs in neurodegenerative diseases.

Keywords: positron emission tomography; histone deacetylase; radioligand; imaging; neurodegenerative disease

---

1. Introduction

Epigenetics is defined as mitotically and/or meiotically heritable changes in gene expression without alternations in the DNA sequence [1]. Several studies are actively examining the mechanisms of epigenetic regulation, including DNA methylation, histone modification, and RNA-based mechanisms [2]. In mammals, DNA methylation mainly involves the covalent addition of a methyl group at the 5-position of a cytosine followed by a guanine (5′-CpG-3′) [3]. This modification results in gene silencing by direct and/or indirect inhibition of transcription factor-DNA interactions. Mechanisms of DNA methylation are divided into two classes according to the corresponding DNA methyltransferase (DNMT): maintenance methylation carried out by DNMT1 and de novo methylation carried out by DNMT3a and DNMT3b [4,5]. Histone posttranslational modifications include acetylation, methylation, ubiquitination, and phosphorylation [6]. Acetylation of histones is one of the most highly studied processes and is regulated by opposing enzymes: histone acetyl transferases and histone deacetylases (HDACs) [7]. Acetylation of lysine residues by histone acetyl transferases neutralizes the positive charge of histones and consequently decreases the interaction of histones with the negatively charged phosphate group of DNA. Then, the chromatin structure becomes relaxed, allowing easier access of transcription factors to DNA. Conversely, deacetylation of histones by HDACs induces gene silencing [8,9]. Histone methylation occurs mainly on lysine and arginine residues. These residues can be methylated multiple times (lysine: three times; arginine: twice), making the effects on gene regulation complex [1,7]. Diverse classes of RNA also regulate gene expression [10]. For example, small interfering RNAs directed to promoter regions result in transcriptional gene silencing through heterochromatin formation [11].

These epigenetic modifications have emerged as key factors in functions of the central nervous system (CNS) and in the development of common neurodegenerative diseases. So far, several studies into the role of epigenetics in CNS functions (e.g., learning and memory processes, fear memory formation, and drug addiction) have been reported [8,12–15]. Meanwhile, neurodegenerative diseases that may involve epigenetic alterations include Alzheimer's disease (AD), Parkinson's disease, and Huntington's disease [2]. For example, significantly reduced levels of DNA methylation were observed in temporal neocortex neuronal nuclei in an AD monozygotic twin compared to his normal sibling [16]. Understanding the expression of enzymes associated with epigenetics and associated alterations in the human brain will help elucidate the pathological mechanisms of these neurodegenerative diseases and will accelerate development of therapeutic agents targeting epigenetics.

The noninvasive imaging technique, positron emission tomography (PET), has received increasing attention in the past few decades, and now PET can be used to monitor various pathological changes in the living brain with a neurodegenerative disease [17,18]. This review summarizes the recent advances in development of PET ligands for imaging HDAC, as a representative example of an epigenetic enzyme, in the brain. With the aim of radiolabeling compounds with affinity for HDACs, various labeling methods using radioisotopes such as carbon-11 and fluorine-18 were investigated in previous studies. Unfortunately, even though more than 20 HDAC imaging radioligands have been developed since 2006, only one report describing imaging results in humans has been published. Summarizing the current advances and issues with HDAC PET ligands will help with future development of imaging techniques and will enable monitoring of the HDAC state in neurodegenerative diseases.

2. HDACs in the Brain

In humans, 18 HDACs can be classified into five classes based on their characteristics and homology to yeast HDACs: class I (HDAC1, 2, 3, and 8), class IIa (HDAC4, 5, 7, and 9), class IIb (HDAC6 and 10), class III (sirtuins, SIRT1–7), and class IV (HDAC11) [9,19]. Generally, class III sirtuins, which are nicotinamide adenine dinucleotide-dependent deacetylases, are considered as separate from other HDAC classes, which have zinc-dependent catalytic activity. Class I HDACs mainly reside in the nucleus and are ubiquitously expressed. Class II HDACs shuttle between the nucleus and the cytoplasm and deacetylate non-histone proteins [20]. Compared to class I HDACs, class II HDACs show tissue specificity, and consequently, class II HDAC knockout mice tend to show local defects [21]. HDAC11, the only class IV HDAC, contains conserved residues in the catalytic core regions shared by both class I and II HDACs [22].

The distribution and expression of HDACs in normal and neurodegenerative brains of mammals, including humans, have been reported. Broide et al. conducted comprehensive gene expression mapping of the 11 HDAC isoforms in the rat brain using high-resolution in situ hybridization in 2007 [23]. The distribution of HDAC subtypes showed overlapping and distinct patterns, suggesting that HDACs play distinct physiological roles in the brain. Of the 11 isoforms, HDAC3, 4, 5, and 11 demonstrated the highest expression in the brain, particularly in the cortex. In addition, immunohistochemistry (IHC) showed that most HDACs are expressed primarily in neurons, whereas expression in other cell types is isoform-specific or limited in the rat brain. Regarding age-related changes in HDAC expression in rodent brain, global HDAC enzymatic activity in the hippocampus and frontal cortex of 18-month-old rats is higher than that in 3-month-old rats [24]. In disease models, concentrations of HDAC3 and 4 in whole brain hemispheres of 5xFAD mice, an AD model with amyloid-β deposition, are about 1.5 times higher than those in wild-type mice [25]. Expression of class I and II HDACs in the non-human primate (NHP) brain was reported by Yeh et al. in a study evaluating a radioligand for HDAC PET imaging [26]. In this report, they performed qualitative IHC analysis of the expression of 11 HDAC isoforms in brain regions including the nucleus accumbens, hippocampus, cortex, and cerebellum. In 2008, Lucio-Eterovic et al. assessed mRNA expression of HDACs in the human brain with quantitative real-time polymerase chain reaction as part of an investigation into HDAC expression levels in gliomas [27]. In normal brain tissue, expression of class I HDACs, HDAC6, and HDAC7 is relatively low, whereas HDAC4, 5, 9, 10, and 11 show higher expression. Anderson et al. determined HDAC expression in human brain with multiple reaction monitoring mass spectrometry [25]. Concentrations of HDAC1 + 2, 5, and 6 were reported as 1.10, 0.083, and 0.106 pmol/mg tissue protein, respectively, in the frontal cortex of aged controls. In 2016, Wey et al. assessed expression levels of HDAC1, 2, 3, and 6 in human brains diagnosed with no neuropathological abnormalities as part of a clinical study of an HDAC PET ligand [28]. The amounts of HDAC2 and 3 are significantly higher in the superior frontal gyrus relative to the corpus callosum. The expression levels of HDAC2, 3, and 6 are comparable in the superior frontal gyrus (0.12–0.16 pmol/mg total protein), whereas that of HDAC1 is obviously higher than others (1.7 pmol/mg total protein). In brains with neurodegenerative disease, Ding et al. reported that compared with aged controls, the expression of HDAC6 in the cerebral cortex and hippocampus of AD patients is increased by 52% and 91%, respectively [29]. In a study by Anderson et al. in 2015, the concentrations of HDAC1 + 2, 5, and 6 in the frontal cortex of AD patients were reported to be 0.7, 1.5, and 1.3 times those of aged controls [25]. Whitehouse et al. investigated the distribution and intensity of HDAC4, 5, and 6 in FTLD with IHC [30]. HDAC4 and 6 show higher immunoreactivity in the dentate gyrus in FTLD cases, especially in FTLD-tau Picks, compared with controls, and the difference in HDAC6 was more prominent. No changes were observed for HDAC5 between FTLD and controls.

3. HDAC Inhibitors

In the last two decades, many types of inhibitors against zinc-dependent HDACs have been developed for treatment of cancer and neurodegenerative disease [21,31–33]. Typical small molecule HDAC inhibitors can be classified into four classes: hydroxamic acids, alkanoic acids, cyclic peptides, and *ortho*-aminoanilides [19]. Following the discovery of the class I and II HDAC inhibitor, suberoylanilide hydroxamic acid (SAHA, also known as vorinostat) in 1996 [34], hydroxamic acid-based HDAC inhibitors compose the biggest compound library of the four classes. The chemical structure of HDAC inhibitors is generally composed of three motifs: a zinc-binding group that holds the zinc ion of HDAC, a cap group that interacts with the protein surface in the binding pocket, and a linker group that bridges the other two groups. For example, in SAHA, the zinc-binding, cap, and linker groups are *N*-phenylformamide, hydroxamic acid, and hexane groups, respectively (Figure 1). SAHA is the first US Food and Drug Administration (FDA)-approved HDAC inhibitor for use in patients with cutaneous T-cell lymphoma. The anticancer effects of SAHA are thought to be caused by modulation of both gene expression and acetylation of proteins that regulate cell proliferation [35]. Besides SAHA, other HDAC inhibitors (e.g., panobinostat and belinostat) have been approved by the US FDA for the treatment of cutaneous T-cell lymphoma, peripheral T-cell lymphoma, or multiple myeloma [36,37].

HDAC inhibitors have shown promising results in preclinical studies using animal models of neurodegenerative diseases [33,38]. Ricobaraza et al. reported that sodium 4-phenylbutyrate treatment reverses spatial memory deficits in Tg2576 AD model mice without affecting β-amyloid levels [39]. They attributed this effect to both activation of gene transcription by histone deacetylation inhibition and normalization of tau hyperphosphorylation by glycogen synthase kinase 3β inhibition. Gardian et al. showed neuroprotective effects of phenylbutylate treatment in 1-methyl-4-phenyl-1,2,3,6-tetrahydropyridine (MPTP)-induced Parkinson's disease model mice [40]. Following phenylbutylate administration, depletion of dopamine in the striatum and reduction in tyrosine hydroxylase-positive neurons in the substantia nigra are attenuated. In R6/2 Huntington's disease model mice, SAHA treatment improves motor impairment in the mice as measured with the rotarod performance test [41].

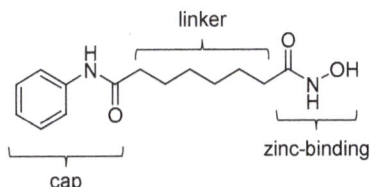

Figure 1. Chemical structure of SAHA. Three groups that constitute a typical HDAC inhibitor are indicated.

Although HDAC inhibitors are expected to be promising therapeutic tools for neurodegenerative diseases as described above, they may have adverse events [42]. HDAC inhibitors have pleiotropic effects on the different types of cells in the CNS because they often inhibit multiple HDACs that deacetylate a wide variety of substrates, including histones, transcription factors, and cytoplasmic proteins [32]. For this reason, isoform-specific HDAC inhibitors are likely to broaden the application of HDAC-targeted therapy [19]. In addition, the in vivo target engagement studies of HDAC inhibitors using HDAC PET imaging techniques described in this review will help estimate the desired and undesired CNS effects of the drugs.

4. Radioligands for HDACs

Radiosynthesis and preclinical characteristics of reported HDAC radioligands are summarized in the following sections. In this review, the radioligands were classified into four groups according to their structural properties: SAHA-based, adamantane-conjugated hydroxamic acid-based, other carboxylic acid- and hydroxamic acid-based, and *ortho*-aminoanilide-based ligands. Radioligands intended to image HDACs not only in the brain but also in tumors are included.

4.1. SAHA-Based Ligands

In 2006, radiosynthesis of the first radiolabeled ligand for imaging of HDAC expression and activity was reported [43]. Aiming to evaluate the HDAC expression and to estimate the effectiveness of SAHA, Mukhopadhyay et al. developed an [18F]-labeled SAHA analogue, 6-([18F]fluoroacetamido)-1-hexanoicanilide ([18F]FAHA, [18F]**1**) (Table 1), which has an [18F]-fluoroacetamido group as a substitute for a hydroxamic acid group. Although the fluoroacetamido group appeared to be a substrate for HDACs because of its similarity to acetyllysine, no characterization of its binding affinity and selectivity for HDACs was reported in the paper. Radiosynthesis of [18F]**1** was achieved by [18F]-fluorination with a bromide precursor (Scheme 1). Using [18F]tetrabutylammonium fluoride ([18F]TBAF), reactions at 80 °C in several solvents such as acetonitrile (MeCN), tetrahydrofuran (THF), and dimethylsulfoxide (DMSO) for 20 min gave decay-corrected radiochemical yields of only 1.0–3.8% with an average of 1.8%, although non-radioactive fluorination with TBAF gave the desired product in >50% yield. The authors presumed that the low radiochemical yield was caused by loss of the product during a high-performance liquid chromatography (HPLC) purification step due to poor solubility of the product in the aqueous HPLC eluent. Meanwhile, reactions at 115 °C in MeCN for 25 min using [18F]KF/Kryptofix 2.2.2 gave [18F]**1** in a decay-corrected radiochemical yield of 11 ± 1.7% after HPLC purification. The radiochemical purity and the molar activity following this method were >99% and >74 GBq/μmol, respectively, at the end of synthesis. Nishii et al. performed in vivo characterization of [18F]**1** in rats using PET [44,45]. The uptake of [18F]**1** in the rat brain increased rapidly after intravenous (i.v.) injection and reached 0.44% injected dose (ID)/g at 5 min post-injection (p.i.). Brain uptake was significantly decreased ($p < 0.01$; *t*-test) by pre-treatment with SAHA [50 mg/kg; intraperitoneal (i.p.) administration] an hour before a PET scan [44]. Tumor uptake of [^{18}F]1 was also assessed using human breast carcinoma xenografts in rats, and as seen in the rat brain, the uptake was inhibited by pre-treatment with SAHA [45]. In 2009, Reid et al. reported a further biological study to evaluate the utility of [^{18}F]1 for PET imaging of HDAC activity [46]. They anticipated that HDAC may cleave [^{18}F]1 and generate a radiometabolite, [^{18}F]fluoroacetate ([^{18}F]FACE) (Figure 2). [^{18}F]FACE crosses the blood-brain barrier (BBB) and is observed as a radiometabolite of PET tracers containing a 2-[^{18}F]fluoroethyl group [47–50]. In vivo generation of [^{18}F]FACE could cause high background radioactivity and complicate interpretation of PET data [48]. Consequently, they attempted to assess the in vivo biodistribution and metabolism of [^{18}F]1 as well as the biodistribution of independently synthesized [^{18}F]FACE [51]. In baboons, almost all [^{18}F]1 was rapidly metabolized within 5 min in plasma, and regional differences in brain radioactivity uptake and kinetics were observed. In contrast, [^{18}F]FACE showed gradual and uniform uptake in the same baboon brain, and the radioactivity peak was much smaller than that of [^{18}F]1. Seven min after administration of [^{18}F]1 in rats, all of the radioactivity in the brain homogenates was in the form of [^{18}F]FACE. Radioactivity accumulation in the bone suggested some in vivo defluorination in rats. Altogether, they concluded that the rapid metabolism and ability of the radiometabolite to penetrate the BBB complicated PET image analysis and that further studies about the interaction between [^{18}F]1 and HDACs were needed. Four years after this study, an international research group performed an [^{18}F]1 PET study in rhesus macaques combining computed tomography (CT) and magnetic resonance imaging (MRI) to quantify the rate of HDAC-mediated accumulation of [^{18}F]1 in the brain using an optimized pharmacokinetics model [26]. The pharmacokinetics model involved two blood plasma input functions for [^{18}F]1 and [^{18}F]FACE and three tissue compartments. In their study, high substrate specificity of **1** for class IIa HDACs was demonstrated with an in vitro assay, which was consistent with a preceding study using trifluoroacetyl-lysine [52]. In addition to the multimodal imaging, they also assessed expression of individual HDACs and the deacetylation status of histones in the brain with IHC, and confirmed their heterogeneity in different brain structures and cell types. Quantitative analysis of PET/CT/MRI showed high accumulation of [^{18}F]1 in the nucleus accumbens and cerebellum in which high expression levels of class IIa HDACs such as HDAC 4 and 5 are observed. These results indicated that [^{18}F]1 accumulation in macaque brains depends on class IIa HDAC expression and activity, and consequently, [^{18}F]1 PET could be used for pharmacodynamics studies of class IIa HDAC inhibitors in the brain. Recently, the utility of [^{18}F]1 for PET imaging of lung cancer was evaluated using small animal PET/CT and a mouse model [53]. HDACs are also expected to be promising therapeutic targets in lung cancer [54]. A/J mice treated with 4-(methylnitrosamino)-1-(3-pyridyl)-1-butanone, a tobacco-specific precarcinogen, were used as lung cancer model mice [55]. In vivo PET imaging showed significant [^{18}F]1 uptake in lung tumors with a >2.0 tumor/nontumor ratio, which was slightly higher than those of [^{18}F]fluorodeoxyglucose and [^{18}F]nifene, an $\alpha4\beta2$ nicotinic receptor radioligand [56].

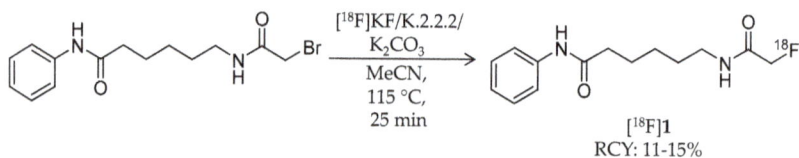

Scheme 1. Radiosynthesis of [^{18}F]1.

Figure 2. Chemical structure of [^{18}F]FACE.

Table 1. Physiochemical properties of radiolabeled SAHA-based HDAC ligands.

Compound	MW	Log D	RCY	Target	IC$_{50}$ for HDACs	In Vivo Properties in Rodents	Reference
[^{18}F]1	265.32	1.39	11–15%	Class IIa	–	Brain uptake: 0.44 and 0.40% ID/g at 5 and 60 min, respectively (rat)	[26,43,44,46]
[^{18}F]2	283.31	–	25%	Class IIa	–	Brain uptake: SUVs were around 1 in various brain regions during the 60-min scan (rat)	[57]
[^{18}F]3	301.30	–	22%	Class IIa	–	Brain uptake: SUVs were around 1 in various brain regions during the 60-min scan (rat)	[57]
[^{18}F]4	309.37	1.01	19%	Non-selective	HDAC1: 56 nM HDAC5: 67 nM HDAC6: 3 nM HDAC7: 85 nM	Brain uptake: 1.0, 0.7, and 0.4% ID/g at 30, 60, and 120 min, respectively Defluorination: 9.6, 11.1, and 13.4% ID/g bone uptake at 30, 60, and 120 min, respectively (tumor-bearing mouse)	[58]
[^{18}F]5	281.32	–	40%	HDAC1–3/6	HDAC1: 9.0 nM HDAC2: 13 nM HDAC3: 24 nM HDAC6: 50 nM	Brain uptake: Very limited in biodistribution study (mouse/tumor-bearing mouse)	[59]
[^{11}C]6	278.34	0.5	23%	–	–	–	[60]
[^{11}C]7	247.33	1.7 [1]	48%	–	–	–	[60]

[1] Clog *P*. MW: molecular weight; RCY: radiochemical yield (decay corrected); SUV: standardized uptake value; ID: injected dose.

In 2015, two novel ^{18}F-labeled successors to [^{18}F]**1** were reported [57]. Bonimi et al. developed 6-(difluoroacetamido)-1-hexanoicanilide (DFAHA, [^{18}F]**2**) and 6-(trifluoroacetamido)-1-hexanoicanilide (TFAHA, [^{18}F]**3**) with improved selectivity and substrate efficiency for class IIa HDACs, which have important roles in the brain development and function (Table 1) [61]. These ligands were rationally designed in accordance with previous reports describing the substrate specificity of HDACs [52,62]. Non-radiolabeled **2** and **3** were synthesized from 6-amino-1-hexanoicanilide by reactions with corresponding di/trifluoroacetic anhydride. Radiosynthesis of [^{18}F]**2** and [^{18}F]**3** was achieved in a similar manner as that of [^{18}F]**1** (Scheme 2). For [^{18}F]**2** radiosynthesis, a bromofluoro precursor in MeCN was added to dried [^{18}F]KF/K.2.2.2 and then stirred at 100 °C for 20 min. HPLC purification gave the desired product with a decay-corrected radiochemical yield of 25%, >95% radiochemical purity, and molar activity of 60–70 GBq/μmol. For [^{18}F]**3** radiosynthesis, the same procedure except ^{18}F-fluorination with additional heating at 110 °C for 25 min was performed to give the desired product with a decay-corrected radiochemical yield of 22%, >95% radiochemical purity, and molar activity of 70–80 GBq/μmol. In vitro comparison of substrate affinity of **1**, **2**, and **3** demonstrated that substitution of a fluorine atom for the hydrogen atom increased k_{cat} values for class IIa HDACs and decreased v_{max} values for class I HDAC8. In dynamic PET/CT imaging, high accumulation of [^{18}F]**2** and [^{18}F]**3** was observed in rat brain regions including the cerebellum, nucleus accumbens, and hippocampus, where class IIa HDACs are highly expressed. The brain radioactivity of [^{18}F]**3** was higher than that of [^{18}F]**2**, and it was significantly decreased by pre-treatment with SAHA (100 mg/kg; i.p. 30 min before [^{18}F]**3** i.v.). These results suggested that [^{18}F]**3** is a more suitable substrate-based radioligand to image the expression and activity of class IIa HDACs in the brain.

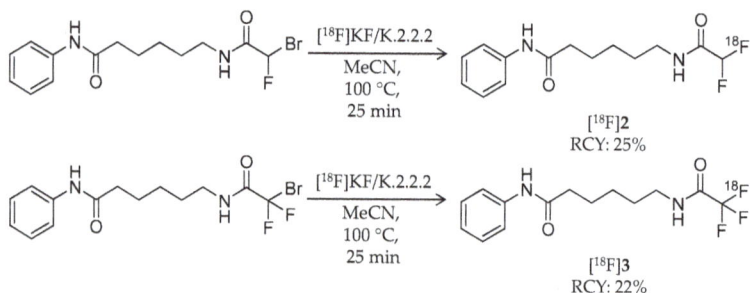

Scheme 2. Radiosynthesis of [^{18}F]**2** and [^{18}F]**3**.

In 2011, Zeglis et al. reported the synthesis and evaluation of a hydroxamic acid-based HDAC radioligand, N^1-(4-(2-fluoroethyl)phenyl)-N^8-hydroxyoctanediamide (FESAHA, [^{18}F]**4**) (Table 1) [58]. Unlike [^{18}F]**1** that could be an HDAC substrate, [^{18}F]**4** was expected to image the HDAC expression itself. [^{18}F]**4** was radiosynthesized from a tosylate precursor with a one-pot, two-step reaction (Scheme 3). The precursor was ^{18}F-fluorinated by a reaction with [^{18}F]TBAF in MeCN at 90 °C for 20 min to obtain an ^{18}F-intermediate, and then the solvent was removed by heating at 90 °C under a slow stream of argon. After the reaction vial was cooled to 0 °C, a solution of hydroxylamine in methanol was added and reacted for 30 min to convert the ethylester group into a hydroxamic acid group. HPLC purification and formulation gave the desired product in 19 ± 9% decay-corrected radiochemical yield and >99% radiochemical purity. The molar activity was about 2.3 GBq/μmol, which was relatively low for an ^{18}F-labeled ligand. To ^{18}F-radiolabel SAHA, they positioned a fluoroethyl group at the para position of the aniline ring. In silico modeling suggested that the modified structure would bind to the active site of an HDAC-like protein [63]. Furthermore, in vitro HDAC inhibition assays confirmed its submicromolar IC_{50} values ranging from 3 nM (HDAC6) to 474 nM (HDAC9) except 1.8 μM for HDAC4. Cell proliferation studies revealed identical cytostatic effects for **4** and SAHA against the prostate cancer cell lines, LNCaP and PC-3, both of which have increased HDAC expression [35,64]. A biodistribution study of [^{18}F]**4** was performed in mice bearing LNCaP xenografts. Tumor uptake was 2.8 ± 0.3% ID/g at 30 min p.i. and decreased with time. Brain uptake was 1.0% ID/g or less over the course of 120 min p.i. Furthermore, significant radioactivity accumulation was observed in the bone (9.6 ± 2.2% ID/g at 30 min p.i.), suggesting substantial in vivo defluorination of [^{18}F]**4**, which was confirmed by small animal PET imaging. Altogether, the radiolabeling position of [^{18}F]**4** has little effect on its HDAC inhibition properties, but as a PET ligand, structural optimization is needed due to its metabolic instability.

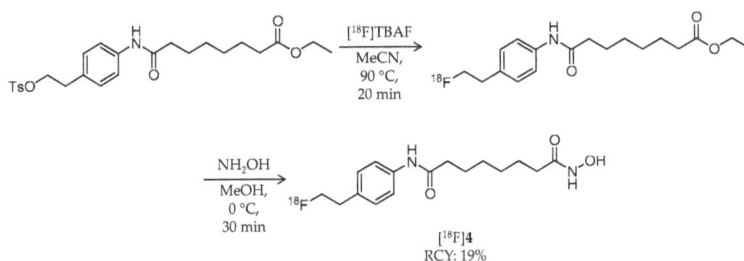

Scheme 3. Radiosynthesis of [^{18}F]**4**.

The same year, Hendricks et al. reported a close analogue of SAHA, *N*-hydroxy-*N'*-(4-fluoro-phenyl)octanediamide (*p*-fluoro SAHA, [^{18}F]**5**), for characterization of SAHA as a cancer drug (Table 1) [59]. Non-radiolabeled *p*-fluoro SAHA was synthesized from 4-fluoroaniline by following reported strategies [65]. For radiosynthesis of [^{18}F]**5**, 1,4-dinitrobenzene was used as a starting reagent (Scheme 4). The first intermediate, [^{18}F]1-fluoro-4-nitrobenzene, was obtained by microwave heating of dinitrobenzene and [^{18}F]KF/K$_2$CO$_3$/K.2.2.2 in DMSO at 120 °C for 5 min. A nitro group of the intermediate was reduced by NaBH$_4$ and Pd/C to obtain [^{18}F]4-fluoroaniline, and then it was isolated by solid phase extraction (SPE). Methyl 8-chloro-8-oxooctanoate was added, and the mixture was stirred at room temperature for 5 min. Finally, 50% hydroxylamine and 1 M NaOH in methanol was added and reacted for 3 min to convert the methyl ester group into a hydroxamic acid group. HPLC purification gave the desired product with 39.5 ± 6.0% decay-corrected radiochemical yield and 97.0 ± 4.7% radiochemical purity. **5** has identical inhibitory effects as SAHA against HDAC1, 2, 3, and 6. They determined the tissue distribution of [^{18}F]**5** in mice. After i.v. injection of the radioligand, the highest radioactivity was observed in the kidney, liver, and blood, whereas brain uptake was the lowest. Bone uptake of radioactivity was not noticeable. They next assessed its accumulation in tumors using a mouse with a human ovarian cancer xenograft and small animal PET/CT. Tumor accumulation of [^{18}F]**5** was observed in a time-dependent manner, and it decreased by 23% on pre-treatment with SAHA.

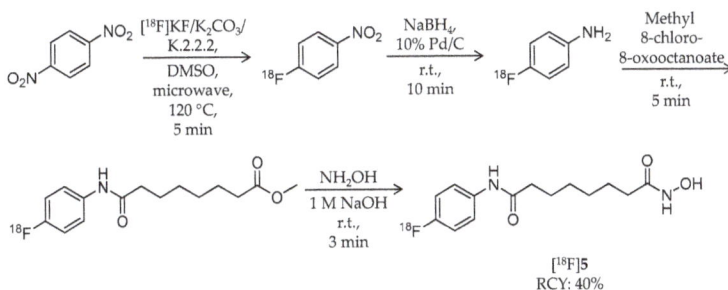

Scheme 4. Radiosynthesis of [^{18}F]**5**.

In 2013, Seo et al. reported radiosynthesis and biological evaluation of two ^{11}C-labeled SAHA-based ligands ([^{11}C]**6** and [^{11}C]**7**) for HDAC imaging in the brain (Table 1) [60]. [^{11}C]**6** was synthesized from a hydroxyl precursor by a two-step reaction of ^{11}C-methylation and conversion of a methylester group into a hydroxamic acid group (Scheme 5). A mixture of the precursor, [^{11}C]CH$_3$I, and sodium hydride in DMSO was heated to 70 °C for 5 min, and then hydroxylamine hydrochloride was added. The authors used potassium cyanide as a base in the second step instead of sodium hydroxide because of the slow reaction rate and by-product formation [66]. HPLC purification gave the desired product in 23 ± 3% decay-corrected radiochemical yield (calculated from [^{11}C]CH$_3$I) and >99% radiochemical purity. The molar activity was 10.4 ± 1.9 GBq/μmol at the time of use. Synthesis of [^{11}C]**7**, an analogue of **1**, was achieved using [2-^{11}C]CH$_3$COCl [67] (Scheme 5). Distillation of [2-^{11}C]CH$_3$COCl into a mixture of the precursor and trimethylamine in THF resulted in the formation of [^{11}C]**7**. HPLC purification gave the desired product in 48 ± 7% decay-corrected radiochemical yield and >99% radiochemical purity. The molar activity was 6.0 ± 1.4 GBq/μmol at the time of use. Using PET, organ uptake and clearance of [^{11}C]**6** and [^{11}C]**7** in baboons were assessed over a 90-min period after i.v. injection. Arterial blood plasma was also collected, and the fraction of intact radioligand was measured with HPLC at different time points during the PET scan. [^{11}C]**6** showed poor brain uptake in baboons, and the distribution pattern was homogeneous in different brain regions. Furthermore, no difference was observed in time-activity curves in peripheral organs between the presence and absence of SAHA pre-treatment, suggesting that no specific binding of [^{11}C]**6** occurs in these organs. Contrary to a prediction by the authors, brain uptake of [^{11}C]**7** was also poor even though the chemical structure of [^{11}C]**7** closely resembles that of [^{18}F]**1**. They confirmed that the unchanged [^{11}C]**7** fraction in plasma at 10 min was higher than [^{18}F]**1** (10% versus 1%), and therefore, the difference was not due to a metabolic difference. Thus, substitution of a hydrogen atom for a fluorine atom affected brain uptake and substrate specificity for HDACs more than expected. In summary, these radioligands are not suitable for imaging HDACs in the brain.

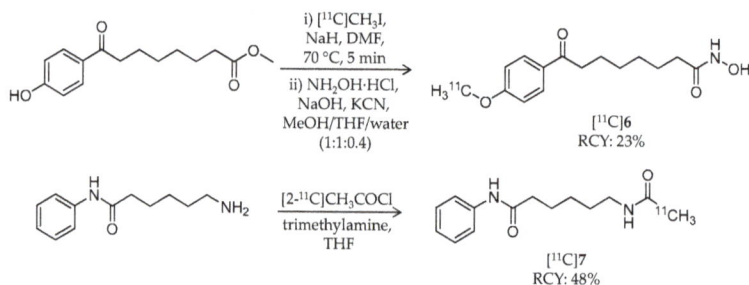

Scheme 5. Radiosynthesis of [^{11}C]**6** and [^{11}C]**7**.

4.2. Adamantane-Conjugated Ligands

In 2014, the first adamantane-conjugated HDAC imaging radioligand was reported by Hooker et al. at Massachusetts General Hospital [68]. The adamantane "lipophilic bullet" is often conjugated to drugs to improve their brain penetrance [69–71], and adamantane-based hydroxamates are highly potent HDAC inhibitors [72]. Based on prior reports, Wang et al. synthesized (E)-3-(4-((((3r,5r,7r)-adamantan-1-ylmethyl)([^{11}C]methyl)amino)methyl)phenyl)-N-hydroxyacrylamide ([^{11}C]Martinostat, [^{11}C]**8**) (Scheme 6) as a radioligand for quantification of the density of HDACs in the CNS and major peripheral organs (Table 2). The radiosynthesis of [^{11}C]**8** was achieved by ^{11}C-methylation of a desmethylated precursor. [^{11}C]CH$_3$I was trapped in a solution of the precursor in DMSO, and the solution was heated to 110 °C for 4 min. HPLC purification gave the desired product in 3–5% non-decay-corrected radiochemical yield (calculated from trapped [^{11}C]CH$_3$I) and ≥95% radiochemical purity with a molar activity of 37 ± 7.4 GBq/μmol. The log *D* value of [^{11}C]**8** was

2.03. In vitro inhibitory activities of **8** against HDACs were measured using recombinant human enzymes. Low IC_{50} values were observed for HDAC1 (0.3 nM), 2 (2.0 nM), 3 (0.6 nM), and 6 (4.1 nM), moderate values were observed for HDAC4 (1970 nM) and 5 (352 nM), and high values were observed for HDAC7 (>20,000), **8** (>15,000), and 9 (>15,000). An in vitro radioligand displacement assay with four zinc-dependent enzymes and 80 additional non-HDAC and Zn-dependent CNS targets demonstrated that the dopamine transporter was a potential off-target binding site of [^{11}C]**8**; however, binding of 2-β-carbomethoxy-3β-(4-fluorophenyl)-[N-^{11}C-methyl]tropane, a dopamine transporter radioligand [73], in rat brains was not inhibited by **8**. In vitro autoradiography (ARG) confirmed specific binding of [^{11}C]**8** in rat brain sections, and the binding was completely inhibited in the presence of excess SAHA. The authors assessed brain uptake of [^{11}C]**8** with PET imaging using both rats and baboons. In rat brains, radioactivity accumulation was observed after i.v. injection of [^{11}C]**8**, and accumulation decreased following pre-treatment of unlabeled **8** in a dose-dependent manner. Furthermore, brain uptake of [^{11}C]**8** was not altered by pre-treatment with the P-glycoprotein (P-gp) inhibitor cyclosporine A (25 mg/kg, 30 min before administration of radioligand), suggesting that [^{11}C]**8** is not a substrate of P-gp. Metabolism of [^{11}C]**8** in rat brains was not noticeable. To analyze baboon PET data for [^{11}C]**8**, a two-tissue compartmental model was used with a metabolite-corrected plasma time-activity curve. Plasma radioactivity was rapidly cleared after i.v. injection of [^{11}C]**8**, although about 40% unchanged fraction still remained in plasma 30 min post-injection. As in rats, high brain uptake of [^{11}C]**8** was observed in baboons, and uptake decreased following pre-treatment with unlabeled **8** (Figure 3). Following these promising results, the same group reported in vivo target engagement studies of a subset of HDAC inhibitors (5 hydroxamates, 4 *ortho*-aminoanilides, and 1 short-chain fatty acid) using [^{11}C]**8** PET imaging in the rat brain [74]. Small molecule HDAC inhibitors modulate CNSdisease-related behaviors in rodents [13,75]; however, direct evidence of target engagement in the brain has not been demonstrated. In this study, they found that i.v. pre-treatment with hydroxamates containing heterocyclic capping groups [76] 3–10 min before [^{11}C]**8** administration resulted in 20–40% blockage of [^{11}C]**8** binding in the brain. Meanwhile, i.p. pre-treatment with *ortho*-aminoanilides [75,77] resulted in limited blockage of [^{11}C]**8** binding. To test whether adamantane could improve brain penetrance of *ortho*-aminoanilides like hydroxamates, they synthesized an adamantane-conjugated *ortho*-aminoanilide named CN147. This novel compound showed 25% blockage of [^{11}C]**8** binding, suggesting that it enhanced HDAC engagement in the brain. Furthermore, an antidepressant-like effect of CN147 was confirmed with a rat behavioral test. These results suggested the utility of [^{11}C]**8** PET as a tool for investigating in vivo target engagement of HDAC inhibitors. Kinetic analysis with arterial plasma sampling of [^{11}C]**8** in different brain regions was also reported [78].

Besides [^{11}C]**8**, Strebl et al., from the same group at Massachusetts General Hospital, developed ^{18}F-labeled adamantane- or cyclohexane-conjugated ligands in 2016 [79]. The three novel compounds have a similar structure to [^{11}C]**8** and a fluorine atom bound directly to a benzene ring ([^{18}F]**9–11**) (Scheme 7, Table 2). [^{18}F]**9** and [^{18}F]**10** were synthesized using the same precursor, 3-(4-formyl-2-nitrostyryl)-5,5-dimethyl-1,4,2-dioxazole. A mixture of the precursor, Cs^{18}F, and DMSO was heated to 110 °C for 5 min followed by transfer into another vial containing the corresponding amine and stirring at room temperature for 5 min. Then a saturated solution of sodium borohydride in isopropanol was added. After stirring at 60 °C for 10 min, the mixture was purified by semi-preparative HPLC. The isolated intermediate was passed through a C18 SPE cartridge and eluted with isopropanol. Camphor sulfonic acid was added, and the mixture was heated to 110 °C for 30 min. HPLC purification gave the desired products in 7% and 0.9% decay-corrected radiochemical yield for [^{18}F]**9** and [^{18}F]**10**, respectively. [^{18}F]**11** was synthesized using methyl 4-formyl-3-nitrocinnamate as the precursor. A mixture of the precursor, [^{18}F]KF/K$_2$CO$_3$/K.2.2.2, and DMSO was heated to 110 °C for 5 min. After SPE with a C18 cartridge, (1-adamantylmethyl)-*N*-methylamine was introduced. The mixture was reacted with NaB(CN)H$_3$ and then purified with HPLC. Finally, a methyl ester group was converted to a hydroxamic acid group using hydroxylamine and sodium hydroxide.

Table 2. Physiochemical properties of radiolabeled adamantane-conjugated HDAC ligands.

Compound	MW	Log D	RCY	Target	IC$_{50}$ for HDACs	In Vivo Properties in Rodents	Reference
[11C]8	353.49	2.03	3–5% [2]	Class I/IIb	HDAC1: 0.3 nM HDAC2: 2.0 nM HDAC3: 0.6 nM HDAC6: 4.1 nM	Brain uptake: Around 0.5% ID/cc during the 60-min scan Metabolism: Radiometabolites in the brain were limited (rat)	[28,68,74,78]
[18F]9	305.38	3.07 [1]	7%	Class I/IIb	HDAC1: 1.6 nM HDAC2: 14 nM HDAC3: 0.5 nM HDAC6: 12 nM	Brain uptake: Rapid brain uptake and limited washout were observed Specific binding in the brain was confirmed in a blocking study (rat)	[79]
[18F]10	357.46	3.77 [1]	0.9%	Class I/IIb	HDAC1: 0.8 nM HDAC2: 6.4 nM HDAC3: 0.5 nM HDAC6: 9.5 nM	Brain uptake: Rapid brain uptake and limited washout were observed Specific binding in the brain was confirmed in a blocking study (rat)	[79]
[18F]11	371.49	4.17 [1]	0.3%	Class I/IIb	HDAC1: 0.8 nM HDAC2: 7.0 nM HDAC3: 0.8 nM HDAC6: 12 nM	Brain uptake: Rapid brain uptake and limited washout were observed Specific binding in the brain was confirmed in a blocking study (rat)	[79]
[18F]12	345.45	–	8.1% [2]	HDAC6	HDAC6: 60 nM Others: $\geq$1 µM	Brain uptake: Rapid brain uptake and limited washout were observed Specific binding in the brain was confirmed in a blocking study (rat)	[80]

[1] Clog *P*. [2] Non-decay corrected. MW: molecular weight; RCY: radiochemical yield (decay corrected).

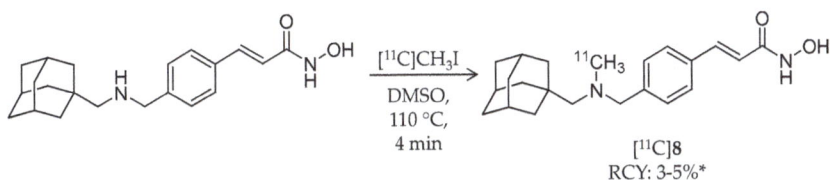

Scheme 6. Radiosynthesis of [^{11}C]8. * Non-decay corrected.

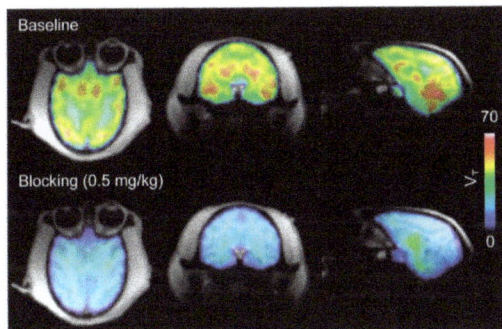

Figure 3. The total volume of distribution (V_T) images from [^{11}C]8 PET in baboon brains. Robust brain uptake (**top**) of [^{11}C]8 was decreased by pre-treatment with unlabeled **8** (**bottom**). (Reproduced from Wang et al. J Med Chem; published by The American Chemical Society, 2014 [68]).

HPLC purification gave the desired product in 0.3% decay-corrected radiochemical yield. All three compounds showed high affinity for HDAC1, 2, 3, and 6 (IC$_{50}$ values: <20 nM). Brain kinetics of the radioligands was assessed with PET imaging in rats and baboons. In rats, all radioligands showed rapid brain uptake after injection, followed by a gradual decrease in radioactivity over the 2-h scan. By pre-treatment with unlabeled **8**, washout rates of the radioligands from the brain were increased, and brain radioactivity decreased to about 50% of peak within 2 h. Meanwhile, PET imaging in baboons revealed differences between the radioligands in brain kinetics. A brain standardized uptake value (SUV) from 30 to 60 min of [^{18}F]**9**, the cyclohexyl compound, was only 0.57, whereas those of [^{18}F]**10** and [^{18}F]**11** reached 1.22 and 1.80, respectively, suggested that the adamantane group enhanced the BBB permeability. On the other hand, the brain uptake of [^{18}F]**11** was slightly lower than that of [^{11}C]**8** although they are close analogues. Furthermore, the regional SUV of ^{18}F-labeled ligands in baboon brains were compared to that of [^{11}C]**8**.

Although all three radioligands showed significant correlation with [^{11}C]**8**, correlation between [^{18}F]**11** and [^{11}C]**8** was the highest. Taken together, [^{18}F]**11** exhibited comparable properties to [^{11}C]**8**, such as HDAC binding and brain uptake, although further optimization of radiosynthesis is needed.

Very recently, Strebl et al. reported synthesis and evaluation of an ^{18}F-labeled hydroxamic acid-based radioligand, [^{18}F]Bavarostat ([^{18}F]**12**, Scheme 8), with selectivity for HDAC6 (Table 2) [80]. HDAC6 is receiving increased attention as a therapeutic target for the neurodegenerative diseases [81]. [^{18}F]**12** has a similar structure as [^{18}F]**11** but without a vinyl group in the linker. Close approach of a bulky moiety to the zinc-binding group is one of the factors that achieve the selectivity [82]. Radiolabeling of [^{18}F]**12** was achieved by a recently reported ^{18}F-deoxyfluorination method [83–85]. First, a mixture of 4-((((adamantan-1-yl)methyl)(methyl)amino)methyl)-3-hydroxybenzoate, CpRu(cod)Cl, N,N-bis(2,6-diisopropyl)phenyl-2-chloroimidazolium chloride, and ethanol was heated to 85 °C for 30 min to form an η^6 coordinated ruthenium–phenol complex. The resulting solution was passed through an ^{18}F anion-bearing anion exchange cartridge, which was flushed with MeCN and DMSO.

The elution was heated to 130 °C for 30 min, and then a hydroxylamine solution was added to convert a methyl ester group to a hydroxamic acid group. HPLC purification gave the desired product in 8.1% non-decay-corrected radiochemical yield with a molar activity of 148 GBq/μmol. Radiochemical purity was not stated in the published article. Regarding selectivity, IC$_{50}$ values of **12**, determined using recombinant enzymes, were 0.06 μM for HDAC6 and >1 μM for other HDACs. **12** also showed selective deacetylation inhibition of α-tubulin, a substrate of HDAC6, in human induced pluripotent stem cell-derived neural progenitor cells. In vitro ARG using rat brain sections demonstrated specific binding of [^{18}F]**12**, and the binding was blocked by the HDAC6 selective inhibitor, tubastatin A [82]. Brain kinetics of [^{18}F]**12** was assessed with PET imaging in a baboon. The brain uptake of [^{18}F]**12** reached an SUV of 3 immediately after injection and gradually decreased over a 120-min period. Pre-treatment with 1.0 mg/kg unlabeled **12** resulted in a substantial decrease in baboon brain uptake of [^{18}F]**12**. Preliminary analysis showed a good correlation between SUV and total distribution volume (V_T). In summary, a promising HDAC6 PET radioligand with BBB permeability was successfully developed, and further biological evaluation and optimization of radiosynthesis are awaited for a clinical study.

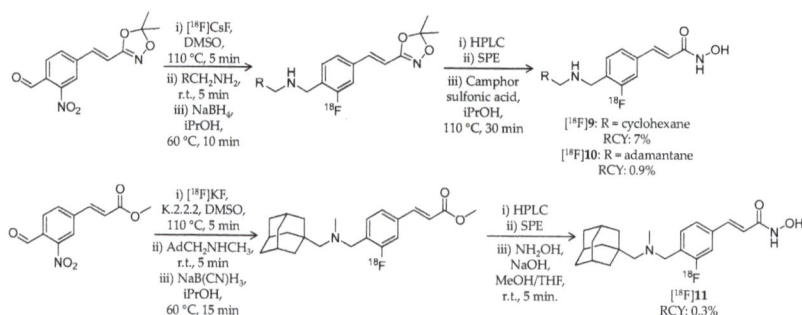

Scheme 7. Radiosynthesis of [^{18}F]**9–11**.

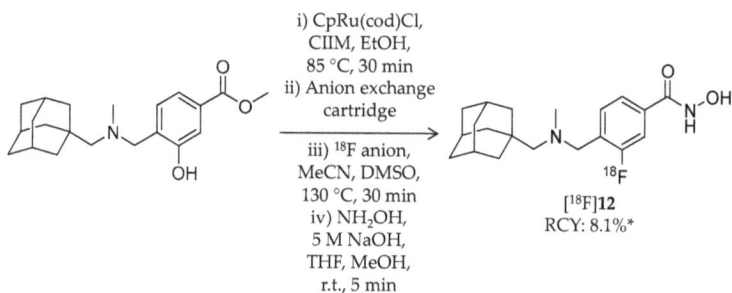

Scheme 8. Radiosynthesis of [^{18}F]**12**. * Non-decay corrected.

4.3. Other Carboxylic Acid- and Hydroxamic Acid-Based Ligands

Pharmacokinetic evaluation of three ^{11}C-labeled alkanoic acid-based HDAC inhibitors, n-butyric acid, 4-phenylbutylic acid, and valproic acid, was reported by Kim et al. in 2013 (Table 3) [86]. Generally, HDAC inhibitory activities of these alkanoic acids (IC$_{50}$ in the μM range) are weaker than those of hydroxamic acid- or benzamide-based inhibitors [87]. Although these radioligands have been used to treat several diseases including CNS diseases, information about their distribution and pharmacokinetics in the brain was limited. The authors introduced carbon-11 at the carbonyl carbon of carboxylic acid by reaction of ^{11}C-carbon dioxide and the respective Grignard reagents (Scheme 9).

Briefly, [¹¹C]CO$_2$ was passed through a THF solution of the precursor. The reaction was quenched by water followed by sequential addition of 6 N hydrochloric acid and 6 N sodium hydroxide. For the synthesis of [¹¹C]**15**, the reaction was performed with non-radioactive CO$_2$ as a carrier with heating to 45 °C for 2 min. The products were purified with HPLC. Radiochemical yield (decay corrected at the end of cyclotron bombardment) and molar activity were 31–50% and 7.4–37 GBq/µmol for [¹¹C]**13**, 40–55% and 7.4–30 GBq/µmol for [¹¹C]**14**, and 38–50% for [¹¹C]**15**, respectively (the molar activity of [¹¹C]**15** was not mentioned). Radiochemical purities of all radioligands were greater than 98%. Whole-body pharmacokinetics of these radioligands was determined with PET imaging in baboons. Of the three radioligands, [¹¹C]**15** was the most stable in baboon plasma (>90% intact ligand over 90 min). PET images demonstrated very low uptake of these radioligands in the baboon brain of less than 0.006% ID/cc, presumably due to their poor BBB penetrance. Interestingly, [¹¹C]**15** showed relatively high heart uptake consistent with the cardiac side effects of **15** [88]. These findings obtained by the imaging studies can help in understanding therapeutic profiles of the inhibitors.

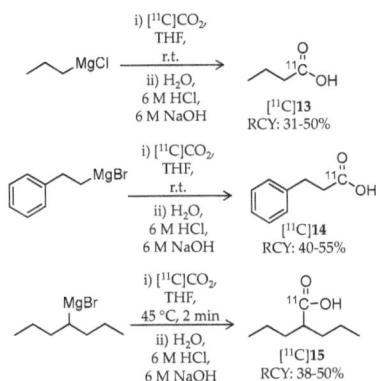

Scheme 9. Radiosynthesis of [¹¹C]**13–15**.

In 2014, Wang et al. reported the synthesis and evaluation of radiolabeled hydroxamic acid-based ligands for HDAC imaging [89]. Aiming to investigate the brain kinetics of hydroxamates, they selected three known HDAC inhibitors and modified their structures for ¹¹C-labeling (Scheme 10, Table 3) [90–92].

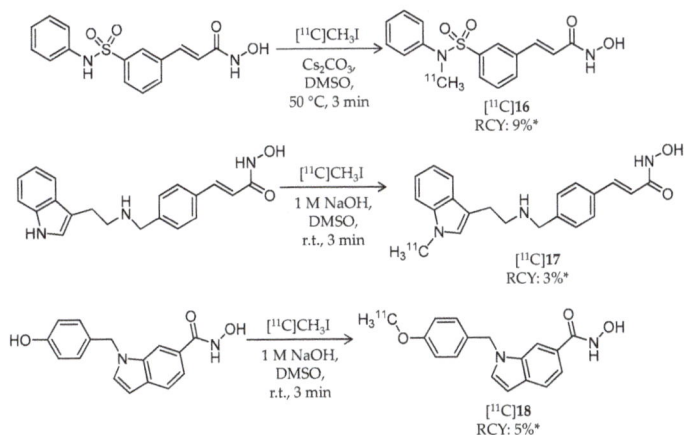

Scheme 10. Radiosynthesis of [¹¹C]**16–18**. * Non-decay corrected.

Table 3. Physiochemical properties of radiolabeled carboxylic acid- and hydroxamic acid-based HDAC ligands.

Compound	MW	Log D	RCY	Target	IC$_{50}$ for HDACs	In Vivo Properties in Rodents	Reference
[11C]13	87.11	1.02	31–50%	Class I	HDAC1: 16 µM HDAC2: 12 µM HDAC3: 9 µM HDAC8: 15 µM	–	[86,87]
[11C]14	149.18	−0.20	40–55%	Class I	HDAC1: 64 µM HDAC2: 65 µM HDAC3: 260 µM HDAC8: 93 µM	–	[86,87]
[11C]15	143.21	0.26	38–50%	Class I	HDAC1: 39 µM HDAC2: 62 µM HDAC3: 161 µM HDAC8: 1103 µM	–	[86,87]
[11C]16	331.37	2.03[1]	9%[3]	HDAC1–3/6	HDAC1: 5 nM HDAC2: 32 nM HDAC3: 3.4 nM HDAC6: 5 nM	Brain uptake: Around 0.1% ID/cc immediately after i.v. injection (rat)	[89]
[11C]17	348.43	2.66[1]	3%[3]	HDAC1–3/6	HDAC1: 0.2 nM HDAC2: 1.2 nM HDAC3: 0.4 nM HDAC6: 1.6 nM	Brain uptake: Around 0.15% ID/cc immediately after i.v. injection (rat)	[89]
[11C]18	295.33	2.61[1]	5%[3]	HDAC8	HDAC8: 18 nM	Brain uptake: Around 0.2% ID/cc immediately after i.v. injection (rat)	[89]
[11C]19	334.41	1.33[2]	8%	HDAC6	HDAC1: 5.2 µM HDAC6: 1.4 nM	Brain uptake: Radioactivity peaked in forebrain at 0.44 SUV and in cerebellum at 0.48 SUV (rat)	[93,94]
[11C]20	334.41	1.33[2]	16%	HDAC6	HDAC6: 4 nM Others: ≥1.3 µM	–	[82,95]
[64Cu]21	1009.02	–	≥98%	Class I/IIb/III	Class I, IIb, III: 94 nM	Brain uptake: Biodistribution study was performed but brain uptake was not assessed (tumor-bearing mouse)	[96]

[1] Log *p*. [2] clog *D*. [3] Non-decay corrected. MW: molecular weight; RCY: radiochemical yield (decay corrected).

Radiosynthesis of [^{11}C]**16–18** was achieved by a reaction of precursors and [^{11}C]CH$_3$I in DMSO in the presence of base. Radiochemical yield (non-decay corrected, calculated from trapped [^{11}C]CH$_3$I) and molar activity were 9% and 29.6 ± 7.4 GBq/μmol for [^{11}C]**16**, 3% and 33.3 ± 3.7 GBq/μmol for [^{11}C]**17**, and 5% and 25.9 ± 7.4 GBq/μmol for [^{11}C]**18**, respectively. Chemical and radiochemical purities were ≥95% for all radioligands. **16** and **17** showed low IC$_{50}$ values against class I HDACs (HDAC1–3) and HDAC6, whereas **18** exhibited high HDAC8 selectivity with an IC$_{50}$ value of 18.3 nM. The brain kinetics of all three radioligands was evaluated with PET imaging in rodents and NHPs. In rats, the radioligands exhibited poor brain uptake with less than 0.25% ID/cc over a 30-min p.i. period, and brain uptake was not changed by pre-treatment with 2 mg/kg unlabeled compound. Brain uptake in a baboon was also very limited. Altogether, these imaging results indicated that these radioligands are not suitable for in vivo HDAC imaging in the brain.

Lu et al. reported development of an HDAC6-selective hydroxamic acid-based radioligand [93]. [^{11}C]**19** is a ^{11}C-labeled analogue of tubastatin A, a selective HDAC6 inhibitor with a tetrahydro-γ-carboline structure (IC$_{50}$: 15 nM for HDAC6; 1640 nM for HDAC1) (Scheme 11, Table 3) [82,94]. ^{11}C-Methylation at the piperidine nitrogen atom was achieved using a desmethyl precursor and [^{11}C]CH$_3$I. [^{11}C]CH$_3$I was trapped in a solution of the precursor and KOH in DMSO, and the mixture was heated to 80 °C for 4 min. HPLC purification gave the desired product in 7.6% radiochemical yield from [^{11}C]CO$_2$ and >99% radiochemical purity. The molar activity was 96.2 GBq/μmol. IC$_{50}$ values of **16** were 1.40 nM and 5180 nM for HDAC6 and HDAC1, respectively [94]. PET imaging with a rhesus monkey demonstrated limited radioactivity in the forebrain (SUV = 0.18) and cerebellum (SUV = 0.38). Furthermore, in rats pre-treated with a P-gp inhibitor 20 min before radioligand injection, [^{11}C]**19** uptake in the forebrain (SUV = 0.44) and cerebellum (SUV = 0.48) was also low. The authors assumed this low BBB permeability was due to low lipophilicity of the compound (cLog*D* = 1.33).

Recently, the same group explored ^{11}C-labeling of tubastatin A at the carbonyl carbon of hydroxamic acid using [^{11}C]carbon monoxide [95]. [^{11}C]CO has the potential to radiolabel a wide variety of carbonyl-containing molecules [97], and this labeling method could be applied to other HDAC ligands with a hydroxamic acid group. Initially, the authors attempted to synthesize [^{11}C]**20** in a one-step palladium-mediated reaction using the iodinated precursor, hydroxylamine, and [^{11}C]CO; however, the decay-corrected radiochemical yield of [^{11}C]**20** was less than 10% (calculated from [^{11}C]CO$_2$), and a carboxylic acid derivative was obtained as a major product. Then they tested a two-step reaction of ^{11}C-methyl ester formation followed by conversion to a hydroxamate. The first step was achieved with a moderate radiochemical yield of 18.5 ± 6.2%, but the second step with hydroxylamine hydrochloride did not work. Through trial and error, they decided to investigate the use of a *p*-nitrophenyl ester, with greater susceptibility toward aminolysis, as a labeled intermediate (Scheme 11).

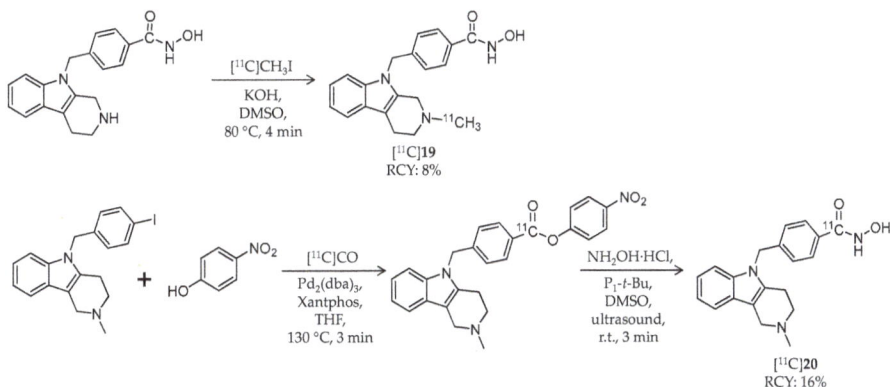

Scheme 11. Radiosynthesis of [^{11}C]**19** and [^{11}C]**20**.

Using Pd$_2$(dba)$_3$ and Xantphos, [^{11}C]CO insertion was successfully obtained with a radiochemical yield of 54.6 ± 8.0%. Furthermore, after investigation into the reaction condition, the second step with a phosphazene base gave the desired product in 16.1 ± 5.6% decay-corrected radiochemical yield with 8.2 GBq/μmol molar activity. Aiming to evaluate the versatility of this labeling method, three simple aryl iodides (e.g., 1-chloro-4-iodobenzene or 4-iodoanisole) were assessed. However, although corresponding *p*-nitrophenyl esters were obtained in good radiochemical yields, the second step reaction gave carboxylic acid derivatives. Taken together, [^{11}C]**20** was labeled at the carbonyl carbon using [^{11}C]CO, but application of the labeling method for the other hydroxamates required further optimization.

In addition to ^{11}C- and ^{18}F-labeled HDAC radioligands, radio-metal-labeled ligands have also been developed. In 2013, Meng et al. reported synthesis and biological evaluation of a ^{64}Cu-labeled hydroxamic acid-based radioligand (Table 3) [96]. In this study, a chelator, 1,4,7,10-tetraazacyclododecane-1,4,7,10-tetraacetic acid, was conjugated to an HDAC inhibitor, CUDC-101 [98,99], with an aliphatic linker by Huisgen cycloaddition. [^{64}Cu]**21** was labeled following a reaction of the precursor and [^{64}Cu]CuCl$_2$ in ammonium acetate buffer at 60 °C for 1 h (Scheme 12). HPLC purification gave the desired product in ≥98% radiochemical yield and radiochemical purity. The molar activity was estimated to 2.4–2.9 GBq/μmol. The IC$_{50}$ value of **21** against class I and II HDACs was 94.47 ± 19.92 nM. An in vitro binding assay using MDA-MB-231, a breast cancer cell line with high HDAC expression [100], demonstrated dose-dependent inhibition of [^{64}Cu]**21** cell uptake in the presence of CUDC-101. Subsequently, biodistribution of [^{64}Cu]**21** in mice bearing an MD-MBA-231 xenograft was evaluated. PET imaging showed tumor uptake of radioactivity (1.20 ± 0.21, 2.16 ± 0.08, and 2.36 ± 0.31% ID/g at 2, 6, and 24 h p.i., respectively), and the uptake was reduced by half by coinjection of 20 mg/kg CUDC-101. The mice were sacrificed immediately after the PET scan and tissue radioactivity was measured. The biodistribution results demonstrated tumor/muscle and tumor/blood uptake ratios of 9.61 ± 1.54 and 4.44 ± 0.88, respectively. The brain uptake is not described in the paper probably because radioligands with a bulky motif such as metal chelators are generally not expected to cross the BBB. These results suggested the feasibility of HDAC imaging with radiometal-labeled ligands in cancer.

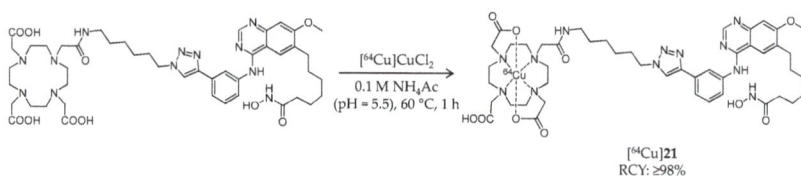

Scheme 12. Radiosynthesis of [^{64}Cu]**21**.

4.4. Ortho-Aminoanilide-Based Ligands

In 2010, Hooker et al. reported synthesis and evaluation of the first benzamide-based HDAC radioligand, [^{11}C]MS-275 ([^{11}C]**22**, Scheme 13), and characterized its pharmacokinetics in the brain (Table 4) [101]. **22** (IC$_{50}$: ~300 nM for HDAC1; ~8 μM for HDAC3 [102]) is a potent, long-lasting, brain region-selective HDAC inhibitor that shows a dose-dependent increase in histone 3 acetylation in the rat brain [103]; however, no direct evidence was obtained for BBB permeability of **22**. They synthesized [^{11}C]**22** by direct incorporation of [^{11}C]CO$_2$ into the carbamate carbon [104]. [^{11}C]CO$_2$ was trapped in a mixture of 3-picolyl chloride hydrochloride, 4-(aminomethyl)-*N*-(2-aminophenyl)-benzamide dihydrochloride, 1,8-diazabicyclo[5.4.0]undec-7-ene (DBU), and DMF, which was then heated to 75 °C for 7 min. HPLC purification gave the desired product in 25% decay-corrected radiochemical yield and >98% radiochemical purity. The molar activity was 100–229 GBq/μmol. Using [^{11}C]**22**, they performed PET imaging studies in baboon and rat brains. In baboons, no radioactivity uptake was observed in the brain after i.v. injection of [^{11}C]**22**. Pre-treatment with the P-gp substrate, verapamil (0.5 mg/kg, i.v., 5 min before [^{11}C]**22** injection), had no influence on brain pharmacokinetics of [^{11}C]**22**, suggesting that P-gp was not involved in [^{11}C]**22** exclusion from the brain. Arterial plasma analysis demonstrated moderate metabolic stability of [^{11}C]**22**; greater than 50% of radioactivity was derived from intact [^{11}C]**22** at 90 min p.i. A further PET study in rats also confirmed poor brain uptake of [^{11}C]**22**, and brain uptake was not altered by pre-treatment with unlabeled **19**. These results suggested that MS-275 may be not suitable for the treatment of neurodegenerative diseases targeting HDACs, but it is also likely to have minimal unwanted CNS effects in treatment of peripheral cancer.

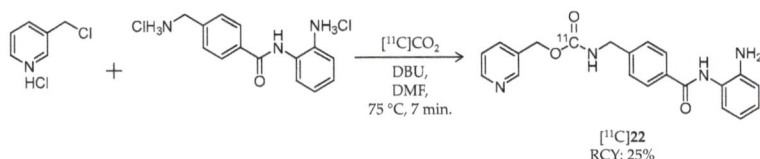

Scheme 13. Radiosynthesis of [^{11}C]**22**.

Table 4. Physiochemical properties of radiolabeled benzamide-based HDAC ligands.

Compound	MW	Log *D*	RCY	Target	IC$_{50}$ for HDACs	In Vivo Properties in Rodents	Reference
[^{11}C]**22**	375.42	1.8	25%	Class I	HDAC1: 60 nM HDAC2: 153 nM	Brain uptake: <0.10% ID/cm^3 after 3 min Metabolism: 80% of radioactivity in the brain was unchanged [^{11}C]**22** (rat)	[101,105]
[^{11}C]**23**	268.30	1.0	10–15%	Class I	HDAC1: 45 nM HDAC2: 31 nM HDAC3: 20 nM	–	[105]
[^{11}C]**24**	344.45	2.1	40%	Class I	HDAC1: 10 nM HDAC2: 20 nM	–	[105]

MW: molecular weight; RCY: radiochemical yield (decay corrected).

Seo et al. conducted PET imaging-guided systematic development of BBB-permeable HDAC inhibitors (Table 4) [105]. They repeated radiosynthesis, performed PET imaging in the baboon brain, and incorporated the imaging data into the compound design. For this study, several ^{11}C-labeled benzamides were radiosynthesized using [^{11}C]methyl iodide, [^{11}C]methyl triflate, and [^{11}C]acetyl chloride. Radiosynthesis of two representative benzamides, [^{11}C]**23** and [^{11}C]**24**, is shown in Scheme 14. **23** is the known HDAC inhibitor, CI-994, with class I HDAC selectivity (dissociation constant: 0.055 µM for HDAC1; 0.255 µM for HDAC2; 0.024 µM for HDAC3 [106]). For [^{11}C]**23** radiosynthesis, [1-^{11}C]CH$_3$COCl, prepared from [^{11}C]CO$_2$, was added to a mixture of a precursor, 4-*N,N*-dimethylaminopyridine, and THF and heated to 60 °C for 5 min. After HPLC purification, the fraction containing the ^{11}C-labeled intermediate was mixed with trifluoroacetic acid, followed by removal of the solvent *in vacuo*. The desired product was obtained, and the decay-corrected radiochemical yield and radiochemical purity were 10–15% and >95%, respectively. The molar activity was 200 MBq/µmol. For [^{11}C]**24** radiosynthesis, [^{11}C]CH$_3$OTf was trapped in the precursor containing DMSO, and the mixture was heated to 50 °C for 3 min. HPLC purification gave the desired product in 40% decay-corrected radiochemical yield and 99% radiochemical purity. The molar activity was 185–666 GBq/µmol. IC$_{50}$ values of **24** were 0.01 µM and 0.02 µM against HDAC1 and HDAC2, respectively. Initially, the authors assessed the BBB permeability of some compounds including [^{11}C]**23**, but brain uptake in baboon was very limited. After image-guided structural optimization, *N,N*-dimethylamino derivatives such as [^{11}C]**24** demonstrated improved brain uptake (~0.015% ID/cc at 5 min p.i.). [^{11}C]**24** showed the highest area under the curve ratio of the brain versus plasma of 7.5. Furthermore, pre-treatment with unlabeled **24** (1 mg/kg) resulted in a decrease in V_T in the range of 13–25% compared with baseline in the cerebellum, thalamus, and temporal cortex. In this study, the polar surface area of less than 65 appeared to be a critical property for BBB permeability, and a key element that improved brain uptake of the benzamides was a benzylic amine.

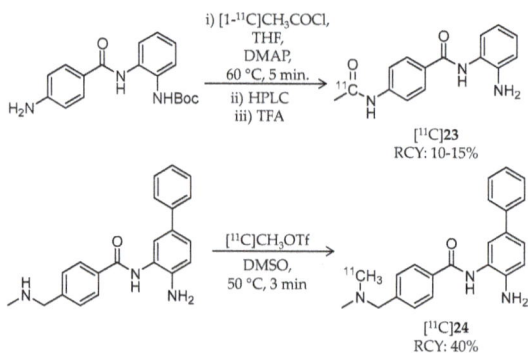

Scheme 14. Radiosynthesis of [^{11}C]**23** and [^{11}C]**24**.

4.5. First-in-Human PET Study

Findings in HDAC radioligand studies with NHPs are summarized in Table 5. Unfortunately, almost all HDAC radioligands showed poor BBB penetrance in PET studies with NHPs, probably due to their low lipophilicity or metabolic instability.

Table 5. Brain PET imaging studies of HDAC radioligands in NHPs.

Compound	Target	Findings in PET Studies in NHPs	Reference
[^{18}F]**1**	Class IIa	Radioactivity accumulation was brain region specific in rhesus macaques (~0.03% ID/g) By 30 min p.i., almost all [^{18}F]**1** was metabolized to [^{18}F]FACE Radioactivity in the brain was decreased by pre-treatment with SAHA in a dose-dependent manner	[26,46]
[^{11}C]**6** [^{11}C]**7**	–	Brain uptake of both ligands was very low in baboons (~0.004% ID/cc) The unchanged fraction of both ligands in baboon plasma was less than 20% at 30 min p.i.	[60]
[^{11}C]**8**	Class I/IIb	Regional V_T (90-min scan) in the baboon brain ranged from 29.9 to 54.4 mL/cm^3 Parent fraction in plasma decreased gradually (50% at 30 min p.i. and 40% at 60 min p.i.) The mean V_T in the brain decreased by 82.3 ± 5.5% with a 1-mg/kg blocking dose	[28,68,78]
[^{18}F]**9** [^{18}F]**10** [^{18}F]**11**	Class I/IIb	Whole brain SUV$_{30-60 min}$ of [^{18}F]**9**, [^{18}F]**10**, and [^{18}F]**11** in baboons were 0.57, 1.2, and 1.8, respectively (2.3 for [^{11}C]**8**) [^{18}F]**11** showed the highest correlation in regional brain distribution with [^{11}C]**8**	[79]
[^{18}F]**12**	HDAC6	Excellent brain uptake (SUV ≈ 3 around 30 min p.i.) was observed in baboons Nonspecific binding in the brain determined with 1 mg/kg unlabeled 12 was low (<1 SUV)	[80]
[^{11}C]**13** [^{11}C]**14** [^{11}C]**15**	Class I	In the baboon brain, uptake of the three ligands was low (~0.006% ID/cc)	[86]
[^{11}C]**16** [^{11}C]**17**	HDAC1–3/6	Brain uptake in baboons was very low over the 80-min scan time	[89]
[^{11}C]**18**	HDAC8	Brain uptake in baboons was very low over the 80-min scan time	[89]
[^{11}C]**22**	Class I	Brain uptake in baboons was very low (<0.001% ID/cc) over the 90-min scan time Approximately 60% of the plasma radioactivity was unchanged ligand at 40 min p.i.	[101]
[^{11}C]**23** [^{11}C]**24**	Class I	Total V_T values of [^{11}C]**23** and [^{11}C]**24** in the baboon brain were 0.41 and 12 mL/cm^3, respectively The degree of V_T reduction of [^{11}C]**24** by unlabeled 24 or SAHA (1 mg/kg) ranged from 8–24% in various brain regions	[105]

Incidentally, in vivo P-gp blocking studies in rodents or NHPs were performed for some radioligands but showed no difference in brain uptake compared to baseline, suggesting that they are not P-gp substrates [68,93,101]. [^{18}F]**1**, the first HDAC-targeting radioligand, showed a significant accumulation of radioactivity in the rhesus macaque brain, and substrate specificity analysis and immunohistochemical analysis for HDAC isoforms confirmed that these results reflect the level of class IIa HDACs in the brain [26]. However, quantitative analysis required consideration of the influence of brain uptake of a major radiometabolite, [^{18}F]FACE. After this report in 2013, we could not find publications on clinical studies of [^{18}F]1. Adamantane-conjugated hydroxamic acid-based HDAC radioligands tended to show preferable brain uptake in PET studies with NHPs, and especially, the study of kinetic analysis of [^{11}C]8 is well advanced [68,79,80]. An in vivo blocking experiment using [^{11}C]8 PET is also expected to be a tool to estimate brain penetrance of HDAC inhibitors for CNS diseases [74]. For ^{18}F-labeled adamantane-conjugated ligands, further preclinical studies including metabolism analysis and brain kinetics analysis in NHPs are awaited. Following the promising preclinical studies, Wey and Gilbert et al. reported the first-in-human evaluation of [^{11}C]8 PET in 2016 [28]. They conducted [^{11}C]8 PET imaging on eight healthy volunteers (four females and four males; 28.6 ± 7.6 years old). [^{11}C]8 showed rapid brain uptake after injection followed by a slight increase in brain radioactivity during 90 min p.i. Cortical SUV from 60 to 90 min was twice that of white matter, and consequently, white matter was selected as a reference region to determine SUV$_{60-90\,min}$ ratios (SUVRs) (Figure 4). The lowest SUVRs were observed in the hippocampus and amygdala (around 1.5), and the highest SUVRs were observed in the putamen and cerebellum (around 2). V_T values determined using the metabolite-corrected arterial plasma as input fraction were stable beyond 50 min p.i., whereas information about the degree of unchanged radioligand in plasma was not described in the paper. The regional SUV$_{60-90\,min}$ correlated well with the V_T values ($r = 0.98$, $p < 0.0001$); therefore, SUV$_{60-90\,min}$ may be useful for quantification of [^{11}C]8 in future studies without arterial blood sampling. Test-retest scans (3 h apart) in three subjects showed less than 3% variability in SUV$_{60-90\,min}$. Western blotting analysis of postmortem brain tissues, which were obtained independently from the imaging study, demonstrated that amounts of HDAC2 and 3 in the superior frontal gyrus (gray matter) were significantly higher than those in the corpus callosum (white matter), whereas differences in HDAC1 and 6 amounts were not significant between the superior frontal gyrus and corpus callosum. In further biochemical profiling using a thermal shift assay, [^{11}C]8 was confirmed to selectively bind to HDAC1, 2, and 3. Although more research is needed to evaluate age-related and disease-related changes in HDAC expression, this study provided a critical foundation for quantification of epigenetic activity in the human brain.

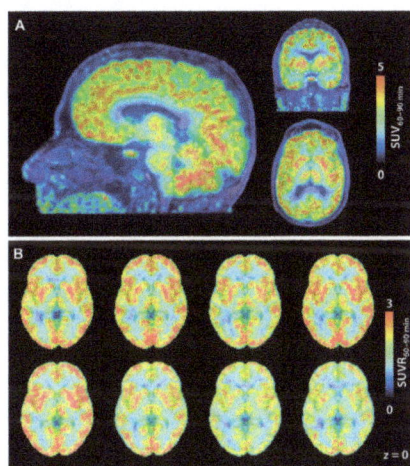

Figure 4. [^{11}C]8 PET images in a first-in-human study. (**A**) Brain SUV images averaged from 60 to 90 min p.i. of [^{11}C]8 (174 MBq). PET images are overlaid on MRI; (**B**) [^{11}C]8 SUVR images of eight individual subjects. SUV$_{60-90\,min}$ was normalized to white matter as a reference. (Reproduced with permission from Wey et al. Sci Transl Med; The American Association for the Advancement of Science, 2016 [28]).

5. Conclusions

Epigenetic abnormalities in the brain have been attracting more attention as therapeutic targets for neurodegenerative diseases. In this regard, non-invasive imaging techniques that can quantify the expression and/or activity of epigenetics-related enzymes in the living brain may deepen our understanding of the role of epigenetics in neurodegeneration and also help in development of new treatments; nevertheless, no clinically applicable technique currently exists. Although several types of epigenetic enzymes are expressed, development of PET imaging ligands for HDACs is particularly advanced at present. In this review, we summarized the radiosynthesis and preclinical characteristics of dozens of potential HDAC PET ligands developed in this decade to identify the achievements and challenges in the field.

Brain uptake of just under 20 HDAC radioligands has already been assessed in NHPs with PET (Table 5); however, few of these radioligands readily enter the brain. Some radioligands showed low brain uptake in NHPs, although they had a preferable molecular weight and lipophilicity to cross the BBB by passive transport [107]; therefore, investigation into the background of the poor BBB permeability could help optimize radioligands for HDAC imaging in the brain. Besides the radioligands with binding affinity for multiple HDACs, the development of isoform-specific ligands is also awaited because disease-related changes in HDAC expression may be isoform specific. In this regard, evaluation of in vivo selectivity of radioligands (e.g., in vivo blocking assays using selective HDAC inhibitors and comparison of ex vivo ARG and immunohistochemical staining in rodent brain sections) is important in addition to in vitro HDAC inhibition assays. Moreover, PET imaging studies with disease model animals could reveal the time course of pathological changes and cognitive and behavioral manifestations by performing multiple longitudinal scans in the same subject. At the moment, the adamantane-conjugated radioligand, [^{11}C]8, seems to be the most promising radioligand to image HDACs in the human brain. Clinical studies using [^{11}C]8 to assess age-related and disease-related changes in brain HDACs could provide valuable information about the roles of HDACs in neurodegenerative diseases.

Acknowledgments: This work was supported by the Japan Society for the Promotion of Science KAKENHI Grant No. 16H07486.

Author Contributions: Tetsuro Tago and Jun Toyohara wrote and revised the manuscript.

Conflicts of Interest: The authors declare no conflict of interest.

References

1. Landgrave-Gomez, J.; Mercado-Gomez, O.; Guevara-Guzman, R. Epigenetic mechanisms in neurological and neurodegenerative diseases. *Front. Cell. Neurosci.* **2015**, *9*, 58. [PubMed]
2. Lovrečić, L.; Maver, A.; Zadel, M.; Peterlin, B. The Role of Epigenetics in Neurodegenerative Diseases. In *Neurodegenerative Diseases*; InTech: London, UK, 2013; pp. 345–365. ISBN 978-953-51-1088-0.
3. Klose, R.J.; Bird, A.P. Genomic DNA methylation: The mark and its mediators. *Trends Biochem. Sci.* **2006**, *31*, 89–97. [CrossRef] [PubMed]
4. Jeltsch, A. On the Enzymatic Properties of Dnmt1: Specificity, Processivity, Mechanism of Linear Diffusion and Allosteric Regulation of the Enzyme. *Epigenetics* **2014**, *1*, 63–66. [CrossRef]
5. Turek-Plewa, J.; Jagodzinski, P.P. The role of mammalian DNA methyltransferases in the regulation of gene expression. *Cell. Mol. Biol. Lett.* **2005**, *10*, 631–647. [PubMed]
6. Kouzarides, T. Chromatin modifications and their function. *Cell* **2007**, *128*, 693–705. [CrossRef] [PubMed]
7. Strahl, B.D.; Allis, C.D. The language of covalent histone modifications. *Nature* **2000**, *403*, 41–45. [CrossRef] [PubMed]
8. Barrett, R.M.; Wood, M.A. Beyond transcription factors: The role of chromatin modifying enzymes in regulating transcription required for memory. *Learn. Mem.* **2008**, *15*, 460–467. [CrossRef] [PubMed]

9. De Ruijter, A.J.; van Gennip, A.H.; Caron, H.N.; Kemp, S.; van Kuilenburg, A.B. Histone deacetylases (HDACs): Characterization of the classical HDAC family. *Biochem. J.* **2003**, *370*, 737–749. [CrossRef] [PubMed]

10. Holoch, D.; Moazed, D. RNA-mediated epigenetic regulation of gene expression. *Nat. Rev. Genet.* **2015**, *16*, 71–84. [CrossRef] [PubMed]

11. Volpe, T.A.; Kidner, C.; Hall, I.M.; Teng, G.; Grewal, S.I.; Martienssen, R.A. Regulation of heterochromatic silencing and histone H3 lysine-9 methylation by RNAi. *Science* **2002**, *297*, 1833–1837. [CrossRef] [PubMed]

12. Lubin, F.D.; Roth, T.L.; Sweatt, J.D. Epigenetic regulation of BDNF gene transcription in the consolidation of fear memory. *J. Neurosci.* **2008**, *28*, 10576–10586. [CrossRef] [PubMed]

13. Graff, J.; Joseph, N.F.; Horn, M.E.; Samiei, A.; Meng, J.; Seo, J.; Rei, D.; Bero, A.W.; Phan, T.X.; Wagner, F.; et al. Epigenetic priming of memory updating during reconsolidation to attenuate remote fear memories. *Cell* **2014**, *156*, 261–276. [CrossRef] [PubMed]

14. Renthal, W.; Nestler, E.J. Epigenetic mechanisms in drug addiction. *Trends Mol. Med.* **2008**, *14*, 341–350. [CrossRef] [PubMed]

15. Jiang, Y.; Langley, B.; Lubin, F.D.; Renthal, W.; Wood, M.A.; Yasui, D.H.; Kumar, A.; Nestler, E.J.; Akbarian, S.; Beckel-Mitchener, A.C. Epigenetics in the nervous system. *J. Neurosci.* **2008**, *28*, 11753–11759. [CrossRef] [PubMed]

16. Mastroeni, D.; McKee, A.; Grover, A.; Rogers, J.; Coleman, P.D. Epigenetic differences in cortical neurons from a pair of monozygotic twins discordant for Alzheimer's disease. *PLoS ONE* **2009**, *4*, e6617. [CrossRef] [PubMed]

17. Politis, M. Neuroimaging in Parkinson disease: From research setting to clinical practice. *Nat. Rev. Neurol.* **2014**, *10*, 708–722. [CrossRef] [PubMed]

18. Holland, J.P.; Liang, S.H.; Rotstein, B.H.; Collier, T.L.; Stephenson, N.A.; Greguric, I.; Vasdev, N. Alternative approaches for PET radiotracer development in Alzheimer's disease: Imaging beyond plaque. *J. Label. Comp. Radiopharm.* **2014**, *57*, 323–331. [CrossRef] [PubMed]

19. Wagner, F.F.; Wesmall yi, U.M.; Lewis, M.C.; Holson, E.B. Small molecule inhibitors of zinc-dependent histone deacetylases. *Neurotherapeutics* **2013**, *10*, 589–604. [CrossRef] [PubMed]

20. Yao, Y.L.; Yang, W.M. Beyond histone and deacetylase: An overview of cytoplasmic histone deacetylases and their nonhistone substrates. *J. Biomed. Biotechnol.* **2011**, *2011*, 146493. [CrossRef] [PubMed]

21. Dokmanovic, M.; Clarke, C.; Marks, P.A. Histone deacetylase inhibitors: Overview and perspectives. *Mol. Cancer Res.* **2007**, *5*, 981–989. [CrossRef] [PubMed]

22. Gao, L.; Cueto, M.A.; Asselbergs, F.; Atadja, P. Cloning and functional characterization of HDAC11, a novel member of the human histone deacetylase family. *J. Biol. Chem.* **2002**, *277*, 25748–25755. [CrossRef] [PubMed]

23. Broide, R.S.; Redwine, J.M.; Aftahi, N.; Young, W.; Bloom, F.E.; Winrow, C.J. Distribution of histone deacetylases 1–11 in the rat brain. *J. Mol. Neurosci.* **2007**, *31*, 47–58. [CrossRef] [PubMed]

24. Anna, G.D.S.S.; Elsner, V.R.; Moyses, F.; Cechinel, R.L.; Lovatel, G.A.; Siqueira, I.R. Histone deacetylase activity is altered in brain areas from aged rats. *Neurosci. Lett.* **2013**, *556*, 152–154. [CrossRef] [PubMed]

25. Anderson, K.W.; Chen, J.; Wang, M.; Mast, N.; Pikuleva, I.A.; Turko, I.V. Quantification of histone deacetylase isoforms in human frontal cortex, human retina, and mouse brain. *PLoS ONE* **2015**, *10*, e0126592. [CrossRef] [PubMed]

26. Yeh, H.H.; Tian, M.; Hinz, R.; Young, D.; Shavrin, A.; Mukhapadhyay, U.; Flores, L.G.; Balatoni, J.; Soghomonyan, S.; Jeong, H.J.; et al. Imaging epigenetic regulation by histone deacetylases in the brain using PET/MRI with ^{18}F-FAHA. *Neuroimage* **2013**, *64*, 630–639. [CrossRef] [PubMed]

27. Lucio-Eterovic, A.K.; Cortez, M.A.; Valera, E.T.; Motta, F.J.; Queiroz, R.G.; Machado, H.R.; Carlotti, C.G., Jr.; Neder, L.; Scrideli, C.A.; Tone, L.G. Differential expression of 12 histone deacetylase (HDAC) genes in astrocytomas and normal brain tissue: Class II and IV are hypoexpressed in glioblastomas. *BMC Cancer* **2008**, *8*, 243. [CrossRef] [PubMed]

28. Wey, H.Y.; Gilbert, T.M.; Zurcher, N.R.; She, A.; Bhanot, A.; Taillon, B.D.; Schroeder, F.A.; Wang, C.; Haggarty, S.J.; Hooker, J.M. Insights into neuroepigenetics through human histone deacetylase PET imaging. *Sci. Transl. Med.* **2016**, *8*, 351ra106. [CrossRef] [PubMed]

29. Ding, H.; Dolan, P.J.; Johnson, G.V. Histone deacetylase 6 interacts with the microtubule-associated protein tau. *J. Neurochem.* **2008**, *106*, 2119–2130. [CrossRef] [PubMed]

30. Whitehouse, A.; Doherty, K.; Yeh, H.H.; Robinson, A.C.; Rollinson, S.; Pickering-Brown, S.; Snowden, J.; Thompson, J.C.; Davidson, Y.S.; Mann, D.M. Histone deacetylases (HDACs) in frontotemporal lobar degeneration. *Neuropathol. Appl. Neurobiol.* **2015**, *41*, 245–257. [CrossRef] [PubMed]

31. Marks, P.A.; Breslow, R. Dimethyl sulfoxide to vorinostat: Development of this histone deacetylase inhibitor as an anticancer drug. *Nat. Biotechnol.* **2007**, *25*, 84–90. [CrossRef] [PubMed]

32. Dietz, K.C.; Casaccia, P. HDAC inhibitors and neurodegeneration: At the edge between protection and damage. *Pharmacol. Res.* **2010**, *62*, 11–17. [CrossRef] [PubMed]

33. Chuang, D.M.; Leng, Y.; Marinova, Z.; Kim, H.J.; Chiu, C.T. Multiple roles of HDAC inhibition in neurodegenerative conditions. *Trends Neurosci.* **2009**, *32*, 591–601. [CrossRef] [PubMed]

34. Richon, V.M.; Webb, Y.; Merger, R.; Sheppard, T.; Jursic, B.; Ngo, L.; Civoli, F.; Breslow, R.; Rifkind, R.A.; Marks, P.A. Second generation hybrid polar compounds are potent inducers of transformed cell differentiation. *Proc. Natl. Acad. Sci. USA* **1996**, *93*, 5705–5708. [CrossRef] [PubMed]

35. Marks, P.A. Discovery and development of SAHA as an anticancer agent. *Oncogene* **2007**, *26*, 1351–1356. [CrossRef] [PubMed]

36. Mottamal, M.; Zheng, S.; Huang, T.L.; Wang, G. Histone deacetylase inhibitors in clinical studies as templates for new anticancer agents. *Molecules* **2015**, *20*, 3898–3941. [CrossRef] [PubMed]

37. Laubach, J.P.; Moreau, P.; San-Miguel, J.F.; Richardson, P.G. Panobinostat for the Treatment of Multiple Myeloma. *Clin. Cancer Res.* **2015**, *21*, 4767–4773. [CrossRef] [PubMed]

38. Didonna, A.; Opal, P. The promise and perils of HDAC inhibitors in neurodegeneration. *Ann. Clin. Transl. Neurol.* **2015**, *2*, 79–101. [CrossRef] [PubMed]

39. Ricobaraza, A.; Cuadrado-Tejedor, M.; Perez-Mediavilla, A.; Frechilla, D.; Del Rio, J.; Garcia-Osta, A. Phenylbutyrate ameliorates cognitive deficit and reduces tau pathology in an Alzheimer's disease mouse model. *Neuropsychopharmacology* **2009**, *34*, 1721–1732. [CrossRef] [PubMed]

40. Gardian, G.; Yang, L.; Cleren, C.; Calingasan, N.Y.; Klivenyi, P.; Beal, M.F. Neuroprotective effects of phenylbutyrate against MPTP neurotoxicity. *Neuromol. Med.* **2004**, *5*, 235–241. [CrossRef]

41. Hockly, E.; Richon, V.M.; Woodman, B.; Smith, D.L.; Zhou, X.; Rosa, E.; Sathasivam, K.; Ghazi-Noori, S.; Mahal, A.; Lowden, P.A.; et al. Suberoylanilide hydroxamic acid, a histone deacetylase inhibitor, ameliorates motor deficits in a mouse model of Huntington's disease. *Proc. Natl. Acad. Sci. USA* **2003**, *100*, 2041–2046. [CrossRef] [PubMed]

42. Subramanian, S.; Bates, S.E.; Wright, J.J.; Espinoza-Delgado, I.; Piekarz, R.L. Clinical Toxicities of Histone Deacetylase Inhibitors. *Pharmaceuticals* **2010**, *3*, 2751–2767. [CrossRef] [PubMed]

43. Mukhopadhyay, U.; Tong, W.P.; Gelovani, J.G.; Alauddin, M.M. Radiosynthesis of 6-([^{18}F]fluoroacetamido)-1-hexanoicanilide ([^{18}F]FAHA) for PET imaging of histone deacetylase (HDAC). *J. Label. Compd. Radiopharm.* **2006**, *49*, 997–1006. [CrossRef]

44. Nishii, R.; Mukhopadhyay, U.; Yeh, H.; Soghomonyan, S.; Volgin, A.; Alauddin, M.; Tong, W.; Gelovani, J. PET imaging of histone deacetylase activity in a rat brain using 6-([^{18}F]-fluoroacetamide)-1-hexanoicanilide ([^{18}F]-FAHA). *J. Nucl. Med.* **2007**, *48* (Suppl. 2), 336.

45. Nishii, R.; Mukhopadhyay, U.; Yeh, H.; Soghomonyan, S.; Volgin, A.; Alauddin, M.; Tong, W.; Gelovani, J. Non-invasive imaging of histone deacetylase activity in human breast carcinoma xenografts in rats using positron emission tomography (PET) with [^{18}F]-FAHA. *J. Nucl. Med.* **2007**, *48* (Suppl. 2), 34.

46. Reid, A.E.; Hooker, J.; Shumay, E.; Logan, J.; Shea, C.; Kim, S.W.; Collins, S.; Xu, Y.; Volkow, N.; Fowler, J.S. Evaluation of 6-([^{18}F]fluoroacetamido)-1-hexanoicanilide for PET imaging of histone deacetylase in the baboon brain. *Nucl. Med. Biol.* **2009**, *36*, 247–258. [CrossRef] [PubMed]

47. Lear, J.L.; Ackermann, R.F. Evaluation of radiolabeled acetate and fluoroacetate as potential tracers of cerebral oxidative metabolism. *Metab. Brain Dis.* **1990**, *5*, 45–56. [CrossRef] [PubMed]

48. Pan, J.; Pourghiasian, M.; Hundal, N.; Lau, J.; Benard, F.; Dedhar, S.; Lin, K.S. 2-[^{18}F]fluoroethanol and 3-[^{18}F]fluoropropanol: Facile preparation, biodistribution in mice, and their application as nucleophiles in the synthesis of [^{18}F]fluoroalkyl aryl ester and ether PET tracers. *Nucl. Med. Biol.* **2013**, *40*, 850–857. [CrossRef] [PubMed]

49. Luurtsema, G.; Schuit, R.C.; Takkenkamp, K.; Lubberink, M.; Hendrikse, N.H.; Windhorst, A.D.; Molthoff, C.F.; Tolboom, N.; van Berckel, B.N.; Lammertsma, A.A. Peripheral metabolism of [^{18}F]FDDNP and cerebral uptake of its labelled metabolites. *Nucl. Med. Biol.* **2008**, *35*, 869–874. [CrossRef] [PubMed]

50. Zoghbi, S.S.; Shetty, H.U.; Ichise, M.; Fujita, M.; Imaizumi, M.; Liow, J.S.; Shah, J.; Musachio, J.L.; Pike, V.W.; Innis, R.B. PET imaging of the dopamine transporter with ^{18}F-FECNT: A polar radiometabolite confounds brain radioligand measurements. *J. Nucl. Med.* **2006**, *47*, 520–527. [PubMed]

51. Ponde, D.E.; Dence, C.S.; Oyama, N.; Kim, J.; Tai, Y.C.; Laforest, R.; Siegel, B.A.; Welch, M.J. ^{18}F-fluoroacetate: A potential acetate analog for prostate tumor imaging—In vivo evaluation of ^{18}F-fluoroacetate versus ^{11}C-acetate. *J. Nucl. Med.* **2007**, *48*, 420–428. [PubMed]

52. Lahm, A.; Paolini, C.; Pallaoro, M.; Nardi, M.C.; Jones, P.; Neddermann, P.; Sambucini, S.; Bottomley, M.J.; Lo Surdo, P.; Carfi, A.; et al. Unraveling the hidden catalytic activity of vertebrate class IIa histone deacetylases. *Proc. Natl. Acad. Sci. USA* **2007**, *104*, 17335–17340. [CrossRef] [PubMed]

53. Tang, W.; Kuruvilla, S.A.; Galitovskiy, V.; Pan, M.L.; Grando, S.A.; Mukherjee, J. Targeting histone deacetylase in lung cancer for early diagnosis: ^{18}F-FAHA PET/CT imaging of NNK-treated A/J mice model. *Am. J. Nucl. Med. Mol. Imaging* **2014**, *4*, 324–332. [PubMed]

54. Neal, J.W.; Sequist, L.V. Exciting new targets in lung cancer therapy: ALK, IGF-1R, HDAC, and Hh. *Curr. Treat. Opt. Oncol.* **2010**, *11*, 36–44. [CrossRef] [PubMed]

55. Gordon, W.; Galitovskiy, V.; Edwards, R.; Andersen, B.; Grando, S.A. The tobacco carcinogen nitrosamine induces a differential gene expression response in tumour susceptible A/J and resistant C_3H mouse lungs. *Eur. J. Cancer* **2013**, *49*, 725–733. [CrossRef] [PubMed]

56. Galitovskiy, V.; Kuruvilla, S.A.; Sevriokov, E.; Corches, A.; Pan, M.L.; Kalantari-Dehaghi, M.; Chernyavsky, A.I.; Mukherjee, J.; Grando, S.A. Development of novel approach to diagnostic imaging of lung cancer with ^{18}F-Nifene PET/CT using A/J mice treated with NNK. *J. Cancer Res. Ther.* **2013**, *1*, 128–137.

57. Bonomi, R.; Mukhopadhyay, U.; Shavrin, A.; Yeh, H.H.; Majhi, A.; Dewage, S.W.; Najjar, A.; Lu, X.; Cisneros, G.A.; Tong, W.P.; et al. Novel Histone Deacetylase Class IIa Selective Substrate Radiotracers for PET Imaging of Epigenetic Regulation in the Brain. *PLoS ONE* **2015**, *10*, e0133512. [CrossRef] [PubMed]

58. Zeglis, B.M.; Pillarsetty, N.; Divilov, V.; Blasberg, R.A.; Lewis, J.S. The synthesis and evaluation of N^1-(4-(2-[^{18}F]-fluoroethyl)phenyl)-N^8-hydroxyoctanediamide ([^{18}F]-FESAHA), a PET radiotracer designed for the delineation of histone deacetylase expression in cancer. *Nucl. Med. Biol.* **2011**, *38*, 683–696. [CrossRef] [PubMed]

59. Hendricks, J.A.; Keliher, E.J.; Marinelli, B.; Reiner, T.; Weissleder, R.; Mazitschek, R. In vivo PET imaging of histone deacetylases by ^{18}F-suberoylanilide hydroxamic acid (^{18}F-SAHA). *J. Med. Chem.* **2011**, *54*, 5576–5582. [CrossRef] [PubMed]

60. Seo, Y.J.; Muench, L.; Reid, A.; Chen, J.; Kang, Y.; Hooker, J.M.; Volkow, N.D.; Fowler, J.S.; Kim, S.W. Radionuclide labeling and evaluation of candidate radioligands for PET imaging of histone deacetylase in the brain. *Bioorg. Med. Chem. Lett.* **2013**, *23*, 6700–6705. [CrossRef] [PubMed]

61. Majdzadeh, N.; Morrison, B.E.; D'Mello, S.R. Class IIA HDACs in the regulation of neurodegeneration. *Front. Biosci.* **2008**, *13*, 1072–1082. [CrossRef] [PubMed]

62. Riester, D.; Hildmann, C.; Grunewald, S.; Beckers, T.; Schwienhorst, A. Factors affecting the substrate specificity of histone deacetylases. *Biochem. Biophys. Res. Commun.* **2007**, *357*, 439–445. [CrossRef] [PubMed]

63. Finnin, M.S.; Donigian, J.R.; Cohen, A.; Richon, V.M.; Rifkind, R.A.; Marks, P.A.; Breslow, R.; Pavletich, N.P. Structures of a histone deacetylase homologue bound to the TSA and SAHA inhibitors. *Nature* **1999**, *401*, 188–193. [CrossRef] [PubMed]

64. Wang, L.; Zou, X.; Berger, A.D.; Twiss, C.; Peng, Y.; Li, Y.; Chiu, J.; Guo, H.; Satagopan, J.; Wilton, A.; et al. Increased expression of histone deacetylaces (HDACs) and inhibition of prostate cancer growth and invasion by HDAC inhibitor SAHA. *Am. J. Transl. Res.* **2009**, *1*, 62–71. [PubMed]

65. Gediya, L.K.; Chopra, P.; Purushottamachar, P.; Maheshwari, N.; Njar, V.C. A new simple and high-yield synthesis of suberoylanilide hydroxamic acid and its inhibitory effect alone or in combination with retinoids on proliferation of human prostate cancer cells. *J. Med. Chem.* **2005**, *48*, 5047–5051. [CrossRef] [PubMed]

66. Ho, C.Y.; Strobel, E.; Ralbovsky, J.; Galemmo, R.A., Jr. Improved solution- and solid-phase preparation of hydroxamic acids from esters. *J. Org. Chem.* **2005**, *70*, 4873–4875. [CrossRef] [PubMed]

67. Hooker, J.M.; Xu, Y.; Schiffer, W.; Shea, C.; Carter, P.; Fowler, J.S. Pharmacokinetics of the potent hallucinogen, salvinorin A in primates parallels the rapid onset and short duration of effects in humans. *Neuroimage* **2008**, *41*, 1044–1050. [CrossRef] [PubMed]

68. Wang, C.; Schroeder, F.A.; Wey, H.Y.; Borra, R.; Wagner, F.F.; Reis, S.; Kim, S.W.; Holson, E.B.; Haggarty, S.J.; Hooker, J.M. In vivo imaging of histone deacetylases (HDACs) in the central nervous system and major peripheral organs. *J. Med. Chem.* **2014**, *57*, 7999–8009. [CrossRef] [PubMed]

69. Banister, S.D.; Wilkinson, S.M.; Longworth, M.; Stuart, J.; Apetz, N.; English, K.; Brooker, L.; Goebel, C.; Hibbs, D.E.; Glass, M.; et al. The synthesis and pharmacological evaluation of adamantane-derived indoles: Cannabimimetic drugs of abuse. *ACS Chem. Neurosci.* **2013**, *4*, 1081–1092. [CrossRef] [PubMed]

70. Tsuzuki, N.; Hama, T.; Kawada, M.; Hasui, A.; Konishi, R.; Shiwa, S.; Ochi, Y.; Futaki, S.; Kitagawa, K. Adamantane as a brain-directed drug carrier for poorly absorbed drug. 2. AZT derivatives conjugated with the 1-adamantane moiety. *J. Pharm. Sci.* **1994**, *83*, 481–484. [CrossRef] [PubMed]

71. Wanka, L.; Iqbal, K.; Schreiner, P.R. The lipophilic bullet hits the targets: Medicinal chemistry of adamantane derivatives. *Chem. Rev.* **2013**, *113*, 3516–3604. [CrossRef] [PubMed]

72. Gopalan, B.; Ponpandian, T.; Kachhadia, V.; Bharathimohan, K.; Vignesh, R.; Sivasudar, V.; Narayanan, S.; Mandar, B.; Praveen, R.; Saranya, N.; et al. Discovery of adamantane based highly potent HDAC inhibitors. *Bioorg. Med. Chem. Lett.* **2013**, *23*, 2532–2537. [CrossRef] [PubMed]

73. Kawamura, K.; Oda, K.; Ishiwata, K. Age-related changes of the [^{11}C]CFT binding to the striatal dopamine transporters in the Fischer 344 rats: A PET study. *Ann. Nucl. Med.* **2003**, *17*, 249–253. [CrossRef] [PubMed]

74. Schroeder, F.A.; Wang, C.; Van de Bittner, G.C.; Neelamegam, R.; Takakura, W.R.; Karunakaran, A.; Wey, H.Y.; Reis, S.A.; Gale, J.; Zhang, Y.L.; et al. PET imaging demonstrates histone deacetylase target engagement and clarifies brain penetrance of known and novel small molecule inhibitors in rat. *ACS Chem. Neurosci.* **2014**, *5*, 1055–1062. [CrossRef] [PubMed]

75. Malvaez, M.; McQuown, S.C.; Rogge, G.A.; Astarabadi, M.; Jacques, V.; Carreiro, S.; Rusche, J.R.; Wood, M.A. HDAC3-selective inhibitor enhances extinction of cocaine-seeking behavior in a persistent manner. *Proc. Natl. Acad. Sci. USA* **2013**, *110*, 2647–2652. [CrossRef] [PubMed]

76. Binaschi, M.; Boldetti, A.; Gianni, M.; Maggi, C.A.; Gensini, M.; Bigioni, M.; Parlani, M.; Giolitti, A.; Fratelli, M.; Valli, C.; et al. Antiproliferative and differentiating activities of a novel series of histone deacetylase inhibitors. *ACS Med. Chem. Lett.* **2010**, *1*, 411–415. [CrossRef] [PubMed]

77. Schroeder, F.A.; Lewis, M.C.; Fass, D.M.; Wagner, F.F.; Zhang, Y.L.; Hennig, K.M.; Gale, J.; Zhao, W.N.; Reis, S.; Barker, D.D.; et al. A selective HDAC 1/2 inhibitor modulates chromatin and gene expression in brain and alters mouse behavior in two mood-related tests. *PLoS ONE* **2013**, *8*, e71323. [CrossRef] [PubMed]

78. Wey, H.Y.; Wang, C.; Schroeder, F.A.; Logan, J.; Price, J.C.; Hooker, J.M. Kinetic Analysis and Quantification of [^{11}C]Martinostat for In Vivo HDAC Imaging of the Brain. *ACS Chem. Neurosci.* **2015**, *6*, 708–715. [CrossRef] [PubMed]

79. Strebl, M.G.; Wang, C.; Schroeder, F.A.; Placzek, M.S.; Wey, H.Y.; Van de Bittner, G.C.; Neelamegam, R.; Hooker, J.M. Development of a Fluorinated Class-I HDAC Radiotracer Reveals Key Chemical Determinants of Brain Penetrance. *ACS Chem. Neurosci.* **2016**, *7*, 528–533. [CrossRef] [PubMed]

80. Strebl, M.G.; Campbell, A.J.; Zhao, W.N.; Schroeder, F.A.; Riley, M.M.; Chindavong, P.S.; Morin, T.M.; Haggarty, S.J.; Wagner, F.F.; Ritter, T.; et al. HDAC6 Brain Mapping with [^{18}F]Bavarostat Enabled by a Ru-Mediated Deoxyfluorination. *ACS Cent. Sci.* **2017**, *3*, 1006–1014. [CrossRef] [PubMed]

81. Simoes-Pires, C.; Zwick, V.; Nurisso, A.; Schenker, E.; Carrupt, P.A.; Cuendet, M. HDAC6 as a target for neurodegenerative diseases: What makes it different from the other HDACs? *Mol. Neurodegener.* **2013**, *8*, 7. [CrossRef] [PubMed]

82. Butler, K.V.; Kalin, J.; Brochier, C.; Vistoli, G.; Langley, B.; Kozikowski, A.P. Rational design and simple chemistry yield a superior, neuroprotective HDAC6 inhibitor, tubastatin A. *J. Am. Chem. Soc.* **2010**, *132*, 10842–10846. [CrossRef] [PubMed]

83. Beyzavi, M.H.; Mandal, D.; Strebl, M.G.; Neumann, C.N.; D'Amato, E.M.; Chen, J.; Hooker, J.M.; Ritter, T. ^{18}F-Deoxyfluorination of Phenols via Ru pi-Complexes. *ACS Cent. Sci.* **2017**, *3*, 944–948. [CrossRef] [PubMed]

84. Neumann, C.N.; Hooker, J.M.; Ritter, T. Concerted nucleophilic aromatic substitution with ^{19}F$^-$ and ^{18}F$^-$. *Nature* **2016**, *534*, 369–373. [CrossRef] [PubMed]

85. Fujimoto, T.; Ritter, T. PhenoFluorMix: Practical chemoselective deoxyfluorination of phenols. *Org. Lett.* **2015**, *17*, 544–547. [CrossRef] [PubMed]

86. Kim, S.W.; Hooker, J.M.; Otto, N.; Win, K.; Muench, L.; Shea, C.; Carter, P.; King, P.; Reid, A.E.; Volkow, N.D.; et al. Whole-body pharmacokinetics of HDAC inhibitor drugs, butyric acid, valproic acid and 4-phenylbutyric acid measured with carbon-11 labeled analogs by PET. *Nucl. Med. Biol.* **2013**, *40*, 912–918. [CrossRef] [PubMed]

87. Fass, D.M.; Shah, R.; Ghosh, B.; Hennig, K.; Norton, S.; Zhao, W.N.; Reis, S.A.; Klein, P.S.; Mazitschek, R.; Maglathlin, R.L.; et al. Short-Chain HDAC Inhibitors Differentially Affect Vertebrate Development and Neuronal Chromatin. *ACS Med. Chem. Lett.* **2010**, *2*, 39–42. [CrossRef] [PubMed]

88. Sodhi, P.; Poddar, B.; Parmar, V. Fatal cardiac malformation in fetal valproate syndrome. *Indian J. Pediatr.* **2001**, *68*, 989–990. [CrossRef] [PubMed]

89. Wang, C.; Eessalu, T.E.; Barth, V.N.; Mitch, C.H.; Wagner, F.F.; Hong, Y.; Neelamegam, R.; Schroeder, F.A.; Holson, E.B.; Haggarty, S.J.; et al. Design, synthesis, and evaluation of hydroxamic acid-based molecular probes for in vivo imaging of histone deacetylase (HDAC) in brain. *Am. J. Nucl. Med. Mol. Imaging* **2013**, *4*, 29–38. [PubMed]

90. Plumb, J.A.; Finn, P.W.; Williams, R.J.; Bandara, M.J.; Romero, M.R.; Watkins, C.J.; La Thangue, N.B.; Brown, R. Pharmacodynamic response and inhibition of growth of human tumor xenografts by the novel histone deacetylase inhibitor PXD101. *Mol. Cancer Ther.* **2003**, *2*, 721–728. [PubMed]

91. Giles, F.; Fischer, T.; Cortes, J.; Garcia-Manero, G.; Beck, J.; Ravandi, F.; Masson, E.; Rae, P.; Laird, G.; Sharma, S.; et al. A phase I study of intravenous LBH589, a novel cinnamic hydroxamic acid analogue histone deacetylase inhibitor, in patients with refractory hematologic malignancies. *Clin. Cancer Res.* **2006**, *12*, 4628–4635. [CrossRef] [PubMed]

92. Balasubramanian, S.; Ramos, J.; Luo, W.; Sirisawad, M.; Verner, E.; Buggy, J.J. A novel histone deacetylase 8 (HDAC8)-specific inhibitor PCI-34051 induces apoptosis in T-cell lymphomas. *Leukemia* **2008**, *22*, 1026–1034. [CrossRef] [PubMed]

93. Lu, S.; Zhang, L.; Kalin, J.; Liow, J.S.; Gladding, R.L.; Innis, R.B.; Kozikowski, A.P.; Pike, V.W. Synthesis and evaluation of [methyl-^{11}C]KB631—A candidate radioligand for histone deacetylase isozyme 6 (HDAC6). *J. Label. Comp. Radiopharm.* **2013**, *56*, S319.

94. Kalin, J.H.; Butler, K.V.; Akimova, T.; Hancock, W.W.; Kozikowski, A.P. Second-generation histone deacetylase 6 inhibitors enhance the immunosuppressive effects of Foxp3+ T-regulatory cells. *J. Med. Chem.* **2012**, *55*, 639–651. [CrossRef] [PubMed]

95. Lu, S.; Zhang, Y.; Kalin, J.H.; Cai, L.; Kozikowski, A.P.; Pike, V.W. Exploration of the labeling of [^{11}C]tubastatin A at the hydroxamic acid site with [^{11}C]carbon monoxide. *J. Label. Comp. Radiopharm.* **2016**, *59*, 9–13. [CrossRef] [PubMed]

96. Meng, Q.; Li, F.; Jiang, S.; Li, Z. Novel ^{64}Cu-Labeled CUDC-101 for In Vivo PET Imaging of Histone Deacetylases. *ACS Med. Chem. Lett.* **2013**, *4*, 858–862. [CrossRef] [PubMed]

97. Miller, P.W.; Long, N.J.; Vilar, R.; Gee, A.D. Synthesis of ^{11}C, ^{18}F, ^{15}O, and ^{13}N radiolabels for positron emission tomography. *Angew. Chem. Int. Ed. Engl.* **2008**, *47*, 8998–9033. [CrossRef] [PubMed]

98. Lai, C.J.; Bao, R.; Tao, X.; Wang, J.; Atoyan, R.; Qu, H.; Wang, D.G.; Yin, L.; Samson, M.; Forrester, J.; et al. CUDC-101, a multitargeted inhibitor of histone deacetylase, epidermal growth factor receptor, and human epidermal growth factor receptor 2, exerts potent anticancer activity. *Cancer Res.* **2010**, *70*, 3647–3656. [CrossRef] [PubMed]

99. Cai, X.; Zhai, H.X.; Wang, J.; Forrester, J.; Qu, H.; Yin, L.; Lai, C.J.; Bao, R.; Qian, C. Discovery of 7-(4-(3-ethynylphenylamino)-7-methoxyquinazolin-6-yloxy)-N-hydroxyheptanamide (CUDc-101) as a potent multi-acting HDAC, EGFR, and HER2 inhibitor for the treatment of cancer. *J. Med. Chem.* **2010**, *53*, 2000–2009. [CrossRef] [PubMed]

100. Feng, W.; Lu, Z.; Luo, R.Z.; Zhang, X.; Seto, E.; Liao, W.S.; Yu, Y. Multiple histone deacetylases repress tumor suppressor gene ARHI in breast cancer. *Int. J. Cancer* **2007**, *120*, 1664–1668. [CrossRef] [PubMed]

101. Hooker, J.M.; Kim, S.W.; Alexoff, D.; Xu, Y.; Shea, C.; Reid, A.; Volkow, N.; Fowler, J.S. Histone deacetylase inhibitor, MS-275, exhibits poor brain penetration: PK studies of [^{11}C]MS-275 using Positron Emission Tomography. *ACS Chem. Neurosci.* **2010**, *1*, 65–73. [CrossRef] [PubMed]

102. Hu, E.; Dul, E.; Sung, C.M.; Chen, Z.; Kirkpatrick, R.; Zhang, G.F.; Johanson, K.; Liu, R.; Lago, A.; Hofmann, G.; et al. Identification of novel isoform-selective inhibitors within class I histone deacetylases. *J. Pharmacol. Exp. Ther.* **2003**, *307*, 720–728. [CrossRef] [PubMed]

103. Simonini, M.V.; Camargo, L.M.; Dong, E.; Maloku, E.; Veldic, M.; Costa, E.; Guidotti, A. The benzamide MS-275 is a potent, long-lasting brain region-selective inhibitor of histone deacetylases. *Proc. Natl. Acad. Sci. USA* **2006**, *103*, 1587–1592. [CrossRef] [PubMed]

104. Hooker, J.M.; Reibel, A.T.; Hill, S.M.; Schueller, M.J.; Fowler, J.S. One-pot, direct incorporation of [^{11}C]CO$_2$ into carbamates. *Angew. Chem. Int. Ed. Engl.* **2009**, *48*, 3482–3485. [CrossRef] [PubMed]

105. Seo, Y.J.; Kang, Y.; Muench, L.; Reid, A.; Caesar, S.; Jean, L.; Wagner, F.; Holson, E.; Haggarty, S.J.; Weiss, P.; et al. Image-guided synthesis reveals potent blood-brain barrier permeable histone deacetylase inhibitors. *ACS Chem. Neurosci.* **2014**, *5*, 588–596. [CrossRef] [PubMed]

106. Wang, Y.; Zhang, Y.L.; Hennig, K.; Gale, J.P.; Hong, Y.; Cha, A.; Riley, M.; Wagner, F.; Haggarty, S.J.; Holson, E.; et al. Class I HDAC imaging using [^{3}H]CI-994 autoradiography. *Epigenetics* **2013**, *8*, 756–764. [CrossRef] [PubMed]

107. Pike, V.W. PET radiotracers: Crossing the blood-brain barrier and surviving metabolism. *Trends Pharmacol. Sci.* **2009**, *30*, 431–440. [CrossRef] [PubMed]

molecules

MDPI

Article

Investigation of an ^{18}F-labelled Imidazopyridotriazine for Molecular Imaging of Cyclic Nucleotide Phosphodiesterase 2A

Susann Schröder [1,*], Barbara Wenzel [1], Winnie Deuther-Conrad [1], Rodrigo Teodoro [1], Mathias Kranz [1], Matthias Scheunemann [1], Ute Egerland [2], Norbert Höfgen [2], Detlef Briel [3], Jörg Steinbach [1] and Peter Brust [1]

[1] Department of Neuroradiopharmaceuticals, Institute of Radiopharmaceutical Cancer Research, Helmholtz-Zentrum Dresden-Rossendorf, Leipzig 04318, Germany; b.wenzel@hzdr.de (B.W.); w.deuther-conrad@hzdr.de (W.D.-C.); r.teodoro@hzdr.de (R.T.); m.kranz@hzdr.de (M.K.); m.scheunemann@hzdr.de (M.S.); j.steinbach@hzdr.de (J.S.); p.brust@hzdr.de (P.B.)
[2] BioCrea GmbH, Radebeul 01445, Germany; ute.egerland@outlook.de (U.E.); norbert.hoefgen@dynabind.com (N.H.)
[3] Pharmaceutical/Medicinal Chemistry, Institute of Pharmacy, Faculty of Medicine, Leipzig University, Leipzig 04103, Germany; briel@uni-leipzig.de
* Correspondence: s.schroeder@hzdr.de; Tel.: +49-341-234-179-4631

Received: 19 January 2018; Accepted: 23 February 2018; Published: 2 March 2018

Abstract: Specific radioligands for in vivo visualization and quantification of cyclic nucleotide phosphodiesterase 2A (PDE2A) by positron emission tomography (PET) are increasingly gaining interest in brain research. Herein we describe the synthesis, the ^{18}F-labelling as well as the biological evaluation of our latest PDE2A (radio-)ligand 9-(5-Butoxy-2-fluorophenyl)-2-(2-([^{18}F])fluoroethoxy)-7-methylimidazo[5,1-*c*]pyrido[2,3-*e*][1,2,4]triazine (([^{18}F])**TA5**). It is the most potent PDE2A ligand out of our series of imidazopyridotriazine-based derivatives so far (IC$_{50}$ hPDE2A = 3.0 nM; IC$_{50}$ hPDE10A > 1000 nM). Radiolabelling was performed in a one-step procedure starting from the corresponding tosylate precursor. In vitro autoradiography on rat and pig brain slices displayed a homogenous and non-specific binding of the radioligand. Investigation of stability in vivo by reversed-phase HPLC (RP-HPLC) and micellar liquid chromatography (MLC) analyses of plasma and brain samples obtained from mice revealed a high fraction of one main radiometabolite. Hence, we concluded that [^{18}F]**TA5** is not appropriate for molecular imaging of PDE2A neither in vitro nor in vivo. Our ongoing work is focusing on further structurally modified compounds with enhanced metabolic stability.

Keywords: Phosphodiesterase 2A (PDE2A); secondary messengers; PDE2A radioligands; positron emission tomography (PET); neuroimaging; metabolic stability; micellar liquid chromatography (MLC)

1. Introduction

The dual-substrate specific enzyme cyclic nucleotide phosphodiesterase 2A (PDE2A) degrades the secondary messengers cyclic adenosine monophosphate (cAMP) as well as cyclic guanosine monophosphate (cGMP) and thus, considerably affects the signaling cascades of these cyclic nucleotides by altering their intracellular levels [1–3]. The PDE2A protein is mainly expressed in the brain and predominantly in structures of the limbic system such as cortex, hippocampus, striatum, substantia nigra, globus pallidus, habenulae, bulbus olfactorius, tuberculum olfactorium, and amygdala [4,5]. This specific localization indicates a regulatory role of PDE2A in important neuronal processes associated to learning, memory and emotion [2,3]. Therefore, PDE2A is suggested to be involved in the pathophysiology of neurodegenerative and neuropsychiatric disorders like Alzheimer´s disease and depression [2,3,5,6].

Pharmacological inhibition of PDE2A activity has been proven to enhance neuronal plasticity due to increased intracellular levels of cAMP and cGMP [2,7–11]. This effect is considered as a highly promising approach in drug development regarding treatment of related neurological diseases [2,6–9,12]. However, the complex relationship between PDE2A activity and pathological changes in the brain is not entirely understood so far [2].

Accordingly, specific radioligands for in vivo imaging and quantification of PDE2A in the brain by positron emission tomography (PET) have been gaining importance during the last years [13]. Besides the lack of brain-penetrating radiometabolites, the most significant criterion for an appropriate PDE2A radioligand is a high selectivity versus the PDE10A protein due to the comparable distribution pattern of both enzymes in the brain [14].

The first two PDE2A radioligands have been published in 2013 by Janssen Pharmaceutica NV (Beerse, Belgium), [18F]B-23 [6,15], and Pfizer Inc., (New York, NY, USA) [18F]PF-05270430 [6,16] (Figure 1). In biodistribution and microPET imaging studies in rats, [18F]B-23 showed a high uptake in the striatum [15]. However, due to the low PDE2A/PDE10A selectivity of this radioligand (IC$_{50}$ hPDE2A = 1 nM; IC$_{50}$ rPDE10A = 11 nM) and the detection of radiometabolites in the brain (at 2 min post injection (p.i.): 4%; at 10 min p.i.: 18% of total activity) [15], [18F]B-23 is not recommended to be suitable for molecular imaging of the PDE2A protein. The highly potent PDE2A radioligand [18F]PF-05270430 (IC$_{50}$ hPDE2A = 0.5 nM; IC$_{50}$ hPDE10A > 3000 nM) has been evaluated preclinically in monkeys [16] and already in a clinical PET study in humans [17,18]. The promising results stated so far, such as PDE2A-specific accumulation with highest uptake in putamen, caudate and nucleus accumbens, a good metabolic stability (intact radioligand at 120 min p.i. in plasma: 40% of total activity) and a favorable kinetic profile [18], point out that [18F]PF-05270430 is an appropriate radioligand for PET imaging of PDE2A in the human brain.

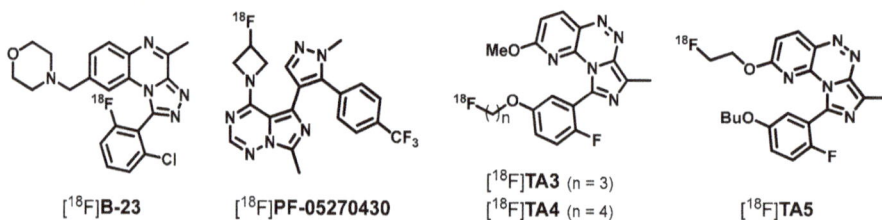

Figure 1. PDE2A radioligands developed by Janssen ([18F]B-23 [6,15]), Pfizer ([18F]PF-05270430 [6,16]), and our group ([18F]TA3–5 [13,19,20]).

Recently, the development of three further PDE2A radioligands, [18F]TA3, [18F]TA4 and [18F]TA5 (TA stands for Triazine) (Figure 1), has been reported by our group [13,19–21].

For ([18F])TA3 and ([18F])TA4, the optimized (radio-)syntheses, the in vitro characterization as well as the biological evaluation in mice have been described previously [19]. Briefly, these two radioligands are suitable for imaging of the PDE2A protein in vitro as demonstrated by the region-specific and displaceable binding in autoradiographic studies on rat brain slices. However, [18F]TA3 and [18F]TA4 undergo a fast metabolic degradation in mice with a high fraction of polar radiometabolites in the brain (at 30 min p.i.: >70% of total activity). It is supposed that these radiometabolites are formed by cytochrome P450 (CYP450) enzyme-induced cleavage of the 18F-bearing alkoxyphenyl side chains resulting in the corresponding brain-penetrating 18F-alkyl alcohols, aldehydes or carboxylic acids [22,23]. Consequently, [18F]TA3 and [18F]TA4 are not applicable for PET neuroimaging of PDE2A [19].

It should be noted that radiotracers bearing a [18F]fluoroalkoxyphenyl group do not per se undergo a metabolic O-dealkylation. For example, the radioligand [18F]FET (O-(2-[18F]fluoroethyl)-L-tyrosine) for PET imaging of brain tumors [24–26] as well as the O-(2-[18F]fluoromethyl) and the O-(2-[^{18}F]fluoropropyl) derivatives ([^{18}F]**FMT** and [^{18}F]**FPT**) demonstrated high in vivo stability (intact [^{18}F]**FET**, [^{18}F]**FMT**, and [^{18}F]**FPT** at 60 min p.i. in mouse plasma: > 90% of total activity [27]; intact [^{18}F]**FET** at 60 min p.i. in human plasma: >90% [24,28]). Furthermore, the PDE10A radioligand [^{18}F]**MNI-659** (2-(2-(3-(4-(2-[^{18}F]Fluoroethoxy)phenyl)-7-methyl-4-oxo-3,4-dihydrochinazolin-2-yl) ethyl)-4-isopropoxyisoindolin-1,3-di-on) [29–31] showed a high and region-specific accumulation in the human brain [29,31] although it is of moderate metabolic stability (intact [^{18}F]**MNI-659** at 120 min p.i. in human plasma: 20% of total activity [29]). Thus, it is suggested that no brain-penetrating radiometabolites are formed indicating [^{18}F]**MNI-659** does also not undergo a CYP450-induced cleavage of the [^{18}F]fluoroethoxyphenyl side chain. For those reasons, we did not exclude the fluoroalkoxy moiety in order to develop PDE2A ligands with enhanced in vivo stability compared to [^{18}F]**TA3** and [^{18}F]**TA4**. Instead, we intended to reach this purpose by changing the position of the fluoroalkoxy group from the phenolic side chain in **TA3** and **TA4** to the pyridinyl moiety. Regarding that, it has been described for diacylethylenediamine derivatives as diacylglycerol aycltransferase-1 inhibitors that replacement of an ethoxybenzoyl group with a less lipophilic 2-ethoxypyridinyl or 2-(2,2,2-trifluoroethoxy)pyridinyl moiety revealed a significant increasing in vitro metabolic stability in mouse, rat and human hepatic microsomal preparations [32].

Finally, our efforts led to the novel PDE2A (radio-)ligand ([^{18}F])**TA5** (Figure 1) as shortly mentioned in former publications [13,20]. Herein, we report on the synthetic route, the ^{18}F-labelling, and the in vitro and in vivo characterization of ([^{18}F])**TA5** in detail.

2. Results and Discussion

2.1. Organic Syntheses and Inhibitory Potency

The synthesis of our selected lead compound **TA1** comprises five steps [7], which have already been optimized [19]. Starting from **TA1** we established appropriate O-dealkylation procedures to selectively split the alkoxy groups either (a) at the 5′-phenol moiety [19] or (b) at the 2-pyridine function (Scheme 1).

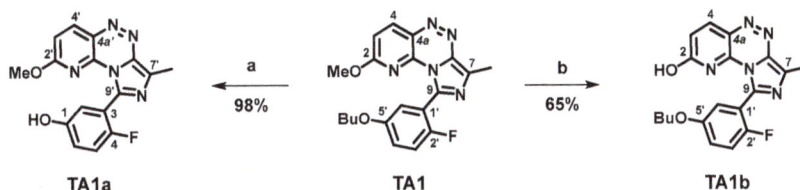

Scheme 1. Syntheses of the 1-phenol **TA1a** [19] and the 2-pyridinol **TA1b**. Reagents and Conditions: (a) 3.05 eq. BBr$_3$ (1 M in CH$_2$Cl$_2$), CH$_2$Cl$_2$, ≤5 °C, 2 h; (b) 1 eq. K$_2$CO$_3$·1.5H$_2$O, 1.5 eq. NH(CH$_3$)$_2$, DMSO/H$_2$O (2.5:1, v/v), 100 °C, 5 h and room temperature (RT) overnight.

The usage of boron tribromide for the cleavage of aromatic alkylethers is a very common and well-known method [33–37]. However and to the best of our knowledge, the regioselective O-dealkylation of an phenolic ether with boron tribromide in the presence of a 2-methoxy pyridine has been described only once before [38]. This approach led to the easy accessibility of our 5′-fluoroalkoxy derivatives (**TA2–4**) as well as the corresponding tosylate precursors for ^{18}F-labelling [19]. Notably, the 2-methoxy function in **TA1** generally showed a very high stability against acidic reagents (e.g., up to 10 eq. BBr$_3$ [19] or conc. HCl under reflux). Due to these experiences, we supposed that the cleavage of the 2-methoxy group requires basic conditions instead of an acidic strategy. This assumption was confirmed by the reaction of **TA1** with dimethylamine as nitrogenous nucleophile in the presence of potassium carbonate and the formation of the desired 2-pyridinol **TA1b** (Scheme 1). Remarkably, similar demethylation reactions at 2-methoxy pyridines have been reported by using sodium thiolates [39,40] but with the herein described method, one could forego the need of a sulfurous nucleophile that brings some advantages in the laboratory work.

Finally, the novel 2-fluoroethoxy derivative **TA5** [13,20] was successfully synthesized with 63% yield using the 2-pyridinol **TA1b** and fluoroethyl iodide as fluoroalkylating agent (Scheme 2).

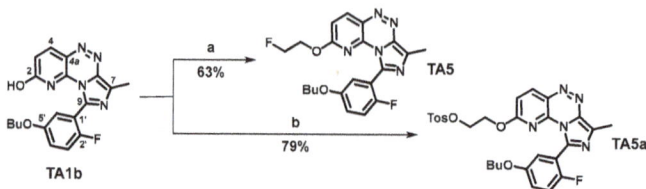

Scheme 2. Syntheses of the novel 2-fluoroethoxy PDE2A ligand **TA5** and the tosylate precursor **TA5a**. Reagents and Conditions: (a) 1.5 eq. F-$(CH_2)_2$-I, 3 eq. $K_2CO_3 \cdot 1.5H_2O$, MeCN, 70–80 °C, 5 h and room temperature (RT) overnight; (b) 2 eq. TosO-$(CH_2)_2$-OTos, 4 eq. $K_2CO_3 \cdot 1.5H_2O$, MeCN, 60–70 °C, 5 h and RT overnight.

The 2-pyridinol **TA1b** exists in equilibrium with its cyclic amide as 2-pyridone or 2-lactam that is suggested to be the more stable form in both the solid state and in solution [41–43]. It has been described that reactions of metal salts of 2-pyridones with alkyl halides often result in *N*- and *O*-alkylated product mixtures depending on the nature of the metal cation, the substitution pattern at the pyridone ring, the molecular structure of the alkyl halide and the solvent used [44–46]. In the herein reported synthesis, only the formation of the preferred *O*-fluoroethylated compound **TA5** was observed, as confirmed by two-dimensional NMR spectroscopy (see Supplementary Materials). Therefore, it is assumed that the nitrogen of the 2-pyridone in **TA5** is sterically hindered against the electrophilic attack of the fluoroethyl iodide by the adjacent imidazotriazine moiety.

Evaluation of the novel 2-fluoroethoxy derivative **TA5** in an enzyme assay [7] resulted in a slightly higher affinity towards the human PDE2A protein as compared with the lead compound **TA1** (Table 1). Above all, **TA5** showed a considerably increased PDE2A/PDE10A selectivity compared to **TA1** as well as the former developed 5′-fluoroalkoxy derivatives **TA2–4** and thus, **TA5** is the most potent PDE2A ligand out of this series so far [13,19,20].

Table 1. IC_{50} values of the novel 2-fluoroethoxy derivative **TA5** for the inhibition of human PDE2A and human PDE10A compared to our already published data for the lead compound **TA1** and the PDE2A ligands **TA2–4** [13,19,20].

Ligand	IC_{50} hPDE2A	IC_{50} hPDE10A	Selectivity Ratio PDE10A/PDE2A
TA5 (2-fluoroethoxy)	3.0 nM	>1000 nM	>330
TA1 (lead)	4.5 nM	670 nM	149
TA2 (5′-fluoroethoxy)	10.4 nM	77 nM	7
TA3 (5′-fluoropropoxy)	11.4 nM	318 nM	28
TA4 (5′-fluorobutoxy)	7.3 nM	913 nM	125

This tendency points out that the substitution patterns at both the 5′-phenol and the 2-pyridine position in **TA1** play a decisive role for the selectivity versus PDE10A, which is needed for PET neuroimaging of PDE2A due to the similar distribution of both enzymes in the brain [14]. The herein observed strong impact of the chain length of the 5′-fluoroalkoxy group in **TA2–4** on the PDE2A affinity and mainly the PDE2A/PDE10A selectivity is in accordance to the effect seen for the previously reported imidazotriazine-based compounds where the selectivity versus PDE10A increases in the following order: 5′-methoxy < 5′-ethoxy < 5′-propoxy < 5′-butoxy with PDE10A/PDE2A selectivity ratios of 3, 4, 40, and >120, respectively [7]. Notably, in this series the 2-pyridine was permanently substituted by a methoxy function [7].

Accordingly, **TA5** was selected for ^{18}F-labelling and the corresponding tosylate precursor **TA5a** was synthesized by reaction of 2-pyridinol **TA1b** with ethane-1,2-diyl bis(4-methylbenzenesulfonate) in 79% yield (see Scheme 2).

2.2. Radiosynthesis, In Vitro Stability and Lipophilicity

Based on our experiences in the radiosyntheses of [^{18}F]**TA3** and [^{18}F]**TA4** [19], the optimized parameters for the one-step ^{18}F-labelling procedure have been adopted to generate the novel PDE2A radioligand [^{18}F]**TA5** (Scheme 3). Nucleophilic substitution of the tosylate group of the precursor **TA5a** with the anhydrous K$^+$/[^{18}F]F$^-$/K$_{222}$-carbonate complex in acetonitrile resulted in [^{18}F]**TA5** with a high radiochemical yield of 65.3 ± 2.1% (n = 3; based on radio-TLC analysis of the crude product). Stability of the precursor under the reaction conditions over 20 min was proven by HPLC and no ^{18}F-labelled by-product was detected.

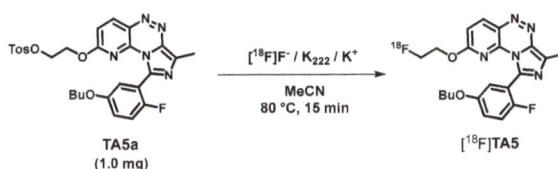

Scheme 3. One-step nucleophilic ^{18}F-labelling procedure to generate the novel PDE2A radioligand [^{18}F]**TA5**.

Isolation of [^{18}F]**TA5** was performed by semi-preparative HPLC (t_R = 34–38 min, see Figure 2) followed by purification and concentration via solid-phase extraction on a pre-conditioned reversed-phase (RP) cartridge and elution with absolute ethanol. After evaporation of the solvent at 70 °C, the radioligand was finally formulated in sterile isotonic saline with a maximum ethanol content of 10% (v/v) for better solubility. The identity of [^{18}F]**TA5** was confirmed by analytical HPLC using an aliquot of the final product spiked with the non-radioactive reference compound **TA5** (Figure 2).

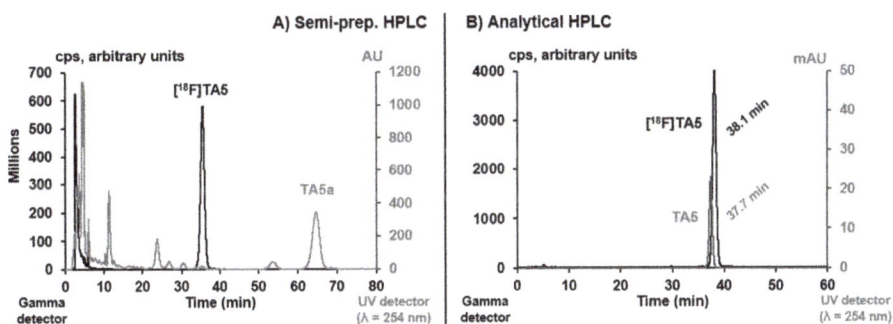

Figure 2. (**A**) Semi-preparative HPLC profile of the crude reaction mixture for isolation of [^{18}F]**TA5** (column: Reprosil-Pur C18-AQ, 250 × 10 mm, particle size: 10 μm; eluent: 50% MeCN/20 mM NH$_4$OAc$_{aq.}$; flow: 5 mL/min); (**B**) analytical HPLC profile of the formulated radioligand [^{18}F]**TA5** spiked with the non-radioactive reference compound **TA5** (column: Reprosil-Pur C18-AQ, 250 × 4.6 mm, particle size: 5 μm; eluent: 52% MeCN/20 mM NH$_4$OAc$_{aq.}$; flow: 1 mL/min).

The novel PDE2A radioligand [^{18}F]**TA5** was synthesized with an overall radiochemical yield of 44.8 ± 5.7% (n = 3), a molar activity of 47.0 ± 7.6 GBq/μmol (n = 3, end of synthesis (EOS)) and a high radiochemical purity of ≥99%.

In vitro stability of [^{18}F]**TA5** was proven in phosphate-buffered saline (PBS, pH 7.4), *n*-octanol and pig plasma. Samples of each medium were analyzed by radio-TLC and radio-HPLC after 1 h incubation at 37 °C and no degradation or defluorination of the radioligand has been observed.

The distribution coefficient of [^{18}F]**TA5** was determined by partitioning between *n*-octanol and phosphate-buffered saline (PBS, pH 7.4) at ambient temperature using the conventional shake-flask method. The obtained logD value of 2.52 ± 0.23 (n = 4) indicates a lipophilicity of [^{18}F]**TA5** which should allow moderate passive diffusion at the blood-brain barrier. However, we observed a strong discrepancy between the experimentally determined logD value and the calculated distribution coefficients (ChemBioDraw Ultra 12.0 (CambridgeSoft Corporation, Cambridge, MA, USA): clogP = 5.26; ACD/Labs 12.0: $clogD_{7.4}$ = 4.78). The experimentally determined higher hydrophilicity of [^{18}F]**TA5** could be a result of solvation effects associated with formation of hydrogen bonds or ionization of the radioligand in the buffered aqueous system. These effects may be underestimated in the software-based determination and thus, the calculated values of lipophilicity are often higher than those obtained experimentally [47].

Remarkably, this has also been observed for [^{18}F]**TA4** while for [^{18}F]**TA3** the logD values obtained from the shake-flask method correlated well with the calculated data. Compared to [^{18}F]**TA5**, the experimentally determined logD values of [^{18}F]**TA3** (3.57 [19]) and [^{18}F]**TA4** (2.99 [19]) are higher indicating that [^{18}F]**TA5** is the least lipophilic derivative in this series. In contrast, the calculated data show the reverse tendency. Regarding the molecular structures of the herein discussed PDE2A ligands, we expect that the lipophilicity increases in the order **TA3** < **TA4** < **TA5** due to the additional methylene groups in **TA4** and **TA5**, respectively, which corresponds with the calculated tendency. This assumption is supported by HPLC where **TA3** elutes at the shortest retention time followed by **TA4** and **TA5** (gradient and isocratic mode). A probable conclusion could be: the more lipophilic a compound is, the higher is the difference between the logD values obtained from the shake-flask method and the calculated data for lipophilicity. Nevertheless, this postulation needs to be confirmed.

2.3. Biological Evaluation—In Vitro Autoradiography and In Vivo Metabolism

For autoradiographic studies, sagittal slices of rat and pig brain were incubated with [^{18}F]**TA5**. Non-specific binding was assessed via co-incubation with an access of either **TA1** or **TA5**. However, the activity pattern on both, rat and pig brain sections, showed a homogenous and non-displaceable distribution which indicates insufficient specificity of [^{18}F]**TA5** under in vitro conditions. This is in contrast to the demonstrated suitability of [^{18}F]**TA3** and [^{18}F]**TA4** for in vitro imaging of PDE2A [19]. Compared to these radioligands, the lipophilicity of [^{18}F]**TA5** is suggested to be higher as indicated by the elution order in the HPLC analyses and the calculated distribution coefficients. This would result in an increased plasma protein binding leading to a reduced availability of free [^{18}F]**TA5**. However, up to now we have no reasonable explanation for the observed high non-specific binding of [^{18}F]**TA5** in vitro. Notably, with availability of [^{18}F]**TA5** the in vitro and in vivo investigations have been performed in parallel.

Encouraged by our experiences in the metabolism studies with [^{18}F]**TA3** and [^{18}F]**TA4** [19] regarding reliable qualitative and quantitative data, radiometabolites of [^{18}F]**TA5** and parent compound were analyzed by (i) reversed-phase HPLC (RP-HPLC) after conventional extraction and (ii) micellar liquid chromatography (MLC [48]). Blood plasma and brain homogenate samples were obtained from CD-1 mice at 30 min post injection of ~70 MBq of the radioligand. In both, plasma and brain, only 7–10% of total activity were representing non-metabolized [^{18}F]**TA5** (see Figure 3). A high fraction of a main radiometabolite [^{18}F]**M1** (~90%) was detected in RP-HPLC and MLC eluting at a very short retention time of 3–4 min indicating a high hydrophilicity.

Plasma sample

A) **RP-HPLC**

B) **MLC**

RP-HPLC (after extraction)				
	Plasma		Brain	
	Amount (%)	t_R (min)	Amount (%)	t_R (min)
[18F]TA5	8	36.4	7	36.5
[18F]M1	89	3.3	93	3.3
[18F]M2	3	22.1	-	

MLC (directly injected)				
	Plasma		Brain	
	Amount (%)	t_R (min)	Amount (%)	t_R (min)
[18F]TA5	7	48.6	10	48.6
[18F]M1	88	3.8	90	3.9
[18F]M2	5	5.3	-	

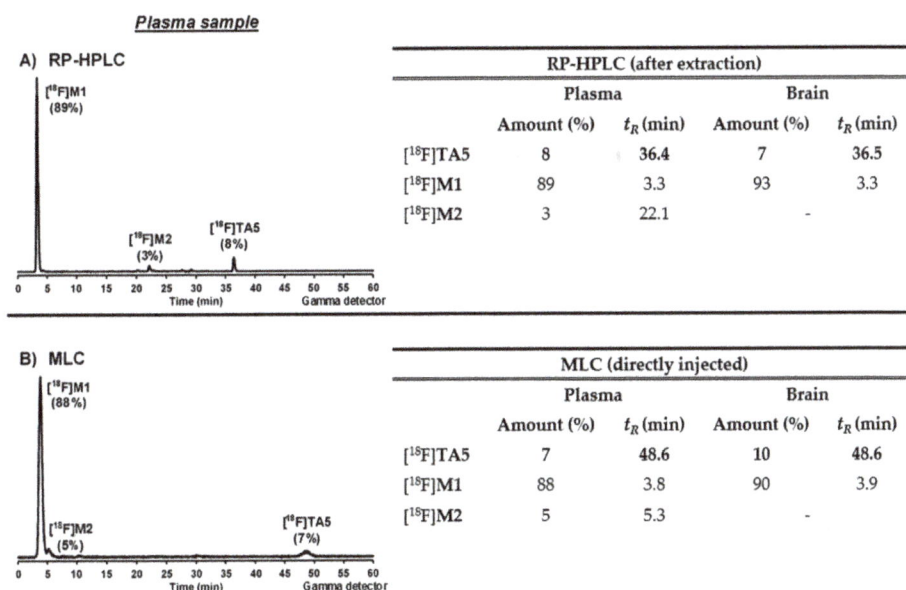

Figure 3. Representative in vivo metabolism study of mouse plasma and brain samples at 30 min p.i. of [18F]TA5 (~70 MBq): (**A**) RP-HPLC chromatogram of extracted plasma sample (column: Reprosil-Pur C18-AQ, 250 x 4.6 mm, particle size: 5 μm; gradient: 10–90–10% MeCN/20 mM NH$_4$OAc$_{aq.}$; flow: 1 mL/min); (**B**) MLC chromatogram of directly injected plasma sample (column: Reprosil-Pur C18-AQ, 250 x 4.6 mm, particle size: 10 μm; gradient: 3-30-3% 1-PrOH/100 mM SDS$_{aq.}$, 10 mM Na$_2$HPO$_{4aq.}$; flow: 1 mL/min); Tables: percentages and retention times of intact [18F]TA5 and radiometabolites from RP-HPLC analysis after extraction and MLC analysis of directly injected samples.

The MLC method was recently established in our group for rapid analysis of radiometabolites [19,49]. Briefly, plasma and homogenized brain samples were dissolved in aqueous sodium dodecyl sulphate (SDS), as an important part of the MLC eluent, and injected directly into the MLC system. There is no further work-up needed and thus, there is no loss of activity prior analysis of the samples. Hence, it is possible to quantify the real composition of total activity in the injected biological material. In contrast, the twofold extraction procedure is work-intensive and time-consuming and most important, the extractability of highly polar radiometabolites from denatured proteins might be low resulting in only a partial recovery of total activity. With the herein applied protocol we previously observed that for [18F]fluoride [19]. Consequently, the amount of intact radioligand would be overestimated if radiometabolites with an ionic character are formed leading to misinterpretation of the data.

The percentages of intact [18F]TA5 achieved from the RP-HPLC analysis of the extracted samples fit well with those from samples directly analyzed by MLC (see Figure 3). This is not surprising, because with the conventional extraction procedure high recoveries of ≥93% of total activity were observed. The chromatograms obtained with both methods differ only regarding the elution profile. While in the RP-HPLC radiometabolite [18F]M2 elutes after 22 min, in the MLC it elutes already after 5 min (Figure 3). This might be a result of the various retention mechanisms in these two systems. Therefore, analysis of biological samples with RP-HPLC as well as with MLC is beneficial to reliably characterize the metabolic profile of a newly developed radioligand.

The highly polar main radiometabolite [18F]M1 was detected in both, plasma and brain samples, pointing out that [18F]M1 may cross the blood-brain barrier. Regarding the identity of [18F]M1, it is presumed that no defluorination or formation of any ionic radiometabolite of [^{18}F]**TA5** occurs due to the fact that ionic compounds are not completely extractable from biological material with the method used here. Accordingly, [^{18}F]**M1** is suggested to be 2-[^{18}F]fluoroethanol resulting from a cytochrome P450 enzyme-induced metabolic degradation of the ^{18}F-fluoroethoxy side chain in [^{18}F]**TA5**. This assumption is further supported by the fact that 2-[^{18}F]fluoroethanol and the oxidized 2-[^{18}F]fluoroacetaldehyde or 2-[^{18}F]fluoroacetate are able to enter the brain [50–53]. To clarify whether there is an in vivo defluorination of [^{18}F]**TA5**, PET imaging or biodistribution investigations could provide more information regarding accumulation of activity in the bones. However, due to the low stability of [^{18}F]**TA5** in mice and formation of brain-penetrating radiometabolites we decided to abstain from further in vivo studies with this radioligand. In conclusion, the rate of metabolic degradation could not be reduced by moving the [^{18}F]fluoroalkoxy group from the phenolic position in [^{18}F]**TA3** and [^{18}F]**TA4** to the pyridinyl moiety in [^{18}F]**TA5**.

Finally, with the novel highly potent PDE2A ligand **TA5** we have shown that changes in the substitution pattern at the pyridinyl moiety of our imidazopyridotriazine lead compound **TA1** can significantly increase the PDE2A/PDE10A selectivity. Starting from the 2-pyridone **TA1b** it is possible to introduce different substituents at the 2-pyridinyl position with or without an alkoxy linker. Thus, our current work is focused on further structurally modified derivatives with feasibly enhanced in vivo stability at the 2-pyridinyl side chain, for example with branched fluoroalkoxy or fluoroalkyl groups as well as cyclic and aromatic fluorine-bearing functions [54–56].

3. Materials and Methods

3.1. General Information

Chemicals were purchased from standard commercial sources in analytical grade and were used without further purification. Radio-/TLCs were performed on pre-coated silica gel plates (Alugram® Xtra SIL G/UV$_{254}$; Polygram® SIL G/UV$_{254}$, Roth, Karlsruhe, Germany). The compounds were localized at 254 nm (UV lamp) and/or by staining with aqueous KMnO$_4$ solution or ninhydrin solution. Radio-TLC was recorded using a bioimaging analyzer system (BAS-1800 II, Fuji Photo Film, Co. Ltd., Tokyo, Japan) and images were evaluated with Aida 2.31 software (raytest Isotopenmessgeräte GmbH, Straubenhardt, Germany). Column chromatography was conducted on silica gel (0.06–0.20 mm, Roth). HPLC separations were performed on JASCO systems equipped with UV detectors from JASCO and activity detectors from raytest Isotopenmessgeräte GmbH (GABI Star, Straubenhardt, Germany).

Semi-preparative HPLC conditions were: Column: Reprosil-Pur C18-AQ, 250 × 10 mm, particle size: 10 μm; eluent: 50% MeCN/20 mM NH$_4$OAc$_{aq.}$; flow: 5 mL/min; ambient temperature; UV detection at 254 nm.

Analytical HPLC conditions were: Column: Reprosil-Pur C18-AQ, 250 × 4.6 mm, particle size: 5 μm; gradient: 0–10 min: 10% MeCN, 10–35 min: 10% → 90% MeCN, 35–45 min: 90% MeCN, 45–50 min: 90% → 10% MeCN, 50–60 min: 10% MeCN/20 mM NH$_4$OAc$_{aq.}$; isocratic: 52% MeCN/20 mM NH$_4$OAc$_{aq.}$; flow: 1 mL/min; ambient temperature; UV detection at 254 nm. Molar activity was determined on the base of a calibration curve (0.2–20 μg **TA5**) carried out under isocratic HPLC conditions (52% MeCN/20 mM NH$_4$OAc$_{aq.}$) using chromatograms obtained at 270 nm as the maximum of UV absorbance.

MLC conditions were: Column: Reprosil-Pur C18-AQ, 250 × 4.6 mm, particle size: 10 μm; gradient: 0–15 min: 3% 1-PrOH, 15–40 min: 3% → 30% 1-PrOH; 40–49 min: 30% 1-PrOH, 49–50 min: 30% → 3% 1-PrOH; 50–60 min: 3% 1-PrOH/100 mM SDS, 10 mM Na$_2$HPO$_4$; flow: 1 mL/min; ambient temperature; UV detection at 254 nm. Notably, a pre-column with 10 mm length was used and frequently exchanged to expand the lifetime of the RP-column.

NMR spectra (^{1}H, ^{13}C, ^{19}F) were recorded on Mercury 300/Mercury 400 (Varian, Palo Alto, CA, USA) or Fourier 300/Avance DRX 400 Bruker (Billerica, MA, USA) instruments. The hydrogenated residue of deuteriated solvents and/or tetramethylsilane (TMS) were used as internal standards for ^{1}H-NMR (CDCl$_3$, δ_H = 7.26; DMSO-d_6, δ_H = 2.50) and ^{13}C-NMR (CDCl$_3$, δ_C = 77.2; DMSO-d_6, δ_C = 39.5). The chemical shifts (δ) are reported in ppm (s, singlet; d, doublet; t, triplet; q, quartet; p, pentet (quintet); h, hexett (sextet); m, multiplet) and the related coupling constants (J) are reported in Hz. High resolution mass spectra (ESI +/−) were recorded on an Impact II^{TM} instrument (Bruker Daltonics).

No-carrier-added (n.c.a.) [^{18}F]fluoride ($t_{1/2}$ = 109.8 min) was produced via the [^{18}O(p,n)^{18}F] nuclear reaction by irradiation of [^{18}O]H$_2$O (Hyox 18 enriched water, Rotem Industries Ltd, Arava, Israel) on a Cyclone$^®$18/9 (iba RadioPharma Solutions, Louvain-la-Neuve, Belgium) with fixed energy proton beam using Nirta$^®$ [^{18}F]fluoride XL target.

3.2. Organic Syntheses

The optimized syntheses of the lead compound **TA1** and the 1-phenol intermediate **TA1a** are published previously [19]. All final compounds described in this manuscript meet the purity requirements determined by HPLC, NMR and HR-MS.

3.2.1. 9-(5-Butoxy-2-fluorophenyl)-7-methylimidazo[5,1-c]pyrido[2,3-e][1,2,4]triazin-2-ol (**TA1b**)

Compound **TA1** (0.50 g, 1 eq.) was dissolved in dimethyl sulfoxide and water (14 mL, 2.5:1, v/v) followed by addition of K$_2$CO$_3$·1.5H$_2$O (0.22 g, 1 eq.) and dimethylamine (250 µL, 1.5 eq.). The reaction mixture was stirred at 100 °C for 5 h and at ambient temperature overnight. After evaporation of the solvent, the residue was dissolved in CH$_2$Cl$_2$ (10 mL) and washed once with aq. saturated solutions of NaHCO$_3$ and NaCl, and water (5 mL each). The aqueous phase was extracted with CH$_2$Cl$_2$ (5 mL). The combined organic phases were dried over Na$_2$SO$_4$ and filtered. Evaporation of the solvent and subsequent purification by column chromatography (EtOAc/CH$_2$Cl$_2$, 1:4 to 100% EtOAc, v/v) afforded a yellow solid of **TA1b** (0.31 g, 65%). ^{1}H-NMR (400 MHz, CDCl$_3$): δ_H = 0.88 (t, J = 7.4, 3H, O(CH$_2$)$_3$CH$_3$); 1.31 (h-like, J = 7.3, 2H, O(CH$_2$)$_2$CH$_2$CH$_3$); 1.52–1.63 (m, 2H, OCH$_2$CH$_2$CH$_2$CH$_3$); 2.81 (s, 3H, 7-C-CH$_3$); 3.66 (t, J = 6.6, 2H, OCH$_2$(CH$_2$)$_2$CH$_3$); 6.35 (dt, J = 9.0, 3.6, 1H$_{Ar}$, 4′-H); 6.64 (t, J = 9.0, 1H$_{Ar}$, 3′-H); 6.95 (d, J = 8.8, 1H$_{Ar}$, 3-H); 7.07 (dd, J = 5.6, 3.1, 1H$_{Ar}$, 6′-H); 8.54 (d, J = 8.8, 1H$_{Ar}$, 4-H); 11.08 (br s, 1H, 2-C-OH). ^{13}C-NMR (75 MHz, CDCl$_3$): δ_C = 12.4 (s, 1C$_{prim}$, 7-C-CH$_3$); 13.9 (s, 1C$_{prim}$, O(CH$_2$)$_3$CH$_3$); 19.2 (s, 1C$_{sec}$, O(CH$_2$)$_2$CH$_2$CH$_3$); 31.2 (s, 1C$_{sec}$, OCH$_2$CH$_2$CH$_2$CH$_3$); 68.4 (s, 1C$_{sec}$, OCH$_2$(CH$_2$)$_2$CH$_3$); 112.1 (s, 1C$_{ArH}$, 3-C); 115.6 (d, J = 23.2, 1C$_{ArH}$, 3′-C); 115.9 (d, J = 2.3, 1C$_{ArH}$, 6′-C); 117.2 (d, J = 8.0, 1C$_{ArH}$, 4′-C); 119.7 (d, J = 16.3, 1C$_{Ar}$, 1′-C); 128.2 (s, 1C$_{Ar}$, 4a-C); 132.6 (s, 1C$_{Ar}$, 9-C); 134.9 (s, 1C$_{Ar}$, 10a-C); 137.6 (s, 1C$_{Ar}$, 7-C); 139.2 (s, 1C$_{Ar}$, 6a-C); 140.9 (s, 1C$_{ArH}$, 4-C); 154.8 (d, J = 1.9, 1C$_{Ar}$, 5′-C); 155.1 (d, overlap, J = 243.2, 1C$_{Ar}$, 2′-C); 164.9 (s, 1C$_{Ar}$, 2-C). ^{19}F-NMR (377 MHz, CDCl$_3$): δ_F = −119.95 (p, J = 5.0, 1F$_{Ar}$, 2′-C-F).

3.2.2. 9-(5-Butoxy-2-fluorophenyl)-2-(2-fluoroethoxy)-7-methylimidazo[5,1-c]pyrido[2,3-e]-[1,2,4]triazine (**TA5**)

To a solution of compound **TA1b** (0.31 g, 1 eq.) in MeCN (25 mL), K$_2$CO$_3$·1.5H$_2$O (0.42 g, 3 eq.) and 1-fluoro-2-iodoethane (104 µL, 1.5 eq.) were added. The yellow suspension was stirred at 70–80 °C for 5 h and at ambient temperature overnight. The mixture was filtered and the solvent was evaporated. The residue was dissolved in CH$_2$Cl$_2$ (10 mL), washed with water (5 mL) and then with citric acid (5 mL, 25%). The aqueous phase was extracted with CH$_2$Cl$_2$ (2 mL). The combined organic phases were dried over Na$_2$SO$_4$ and filtered. After evaporation of the solvent the crude product was purified by column chromatography (EtOAc/CH$_2$Cl$_2$, 1:4 to 100% EtOAc, v/v) yielding a yellow solid of **TA5** (0.22 g, 63%). ^{1}H-NMR (400 MHz, DMSO-d_6): δ_H = 0.90 (t, J = 7.4, 3H, O(CH$_2$)$_3$CH$_3$); 1.41 (h-like, J = 7.4, 2H, O(CH$_2$)$_2$CH$_2$CH$_3$); 1.67 (p-like, J = 6.6, 2H, OCH$_2$CH$_2$CH$_2$CH$_3$); 2.80 (s, 3H, 7C-CH$_3$); 3.98 (t, distorted, J = 6.5, 3H, OCH$_2$(CH$_2$)$_2$CH$_3$, OCHH′-CH$_2$F); 4.04 (t, poorly resolved, partly overlapped A part of AA′BB′X, J = 3.9, 1H, OCHH′-CH$_2$F); 4.47 (dt, poorly resolved, J = 47.7, 4.0, B part of AA′BB′X, 2H, OCH$_2$-CH$_2$F); 7.08–7.16 (m, 1H$_{Ar}$, 4′-H); 7.17–7.25 (m, 2H$_{Ar}$, 4′-H, 3-H); 7.31 (t, J = 9.2,

$1H_{Ar}$, 3'-H); 8.71 (d, J = 8.8, $1H_{Ar}$, 4-H). ^{13}C-NMR (75 MHz, DMSO-d_6): δ_C = 12.4 (s, $1C_{prim}$, 7-C-$\underline{C}H_3$); 13.6 (s, $1C_{prim}$, O(CH$_2$)$_3\underline{C}H_3$); 18.7 (s, $1C_{sec}$, O(CH$_2$)$_2\underline{C}H_2$CH$_3$); 30.7 (s, $1C_{sec}$, OCH$_2\underline{C}H_2$CH$_2$CH$_3$); 66.1 (d, J = 19.1, $1C_{sec}$, O$\underline{C}H_2$CH$_2$F); 68.0 (s, $1C_{sec}$, O$\underline{C}H_2$(CH$_2$)$_2$CH$_3$); 81.1 (d, J = 167.0, $1C_{sec}$, OCH$_2\underline{C}H_2$F); 112.2 (s, $1C_{ArH}$, 3-C); 116.0 (d, J = 22.9, $1C_{ArH}$, 3'-C); 117.4 (d, J = 8.4, $1C_{ArH}$, 4'-C); 117.5 (d, J = 1.7, $1C_{ArH}$, 6'-C); 120.3 (d, J = 16.6, $1C_{Ar}$, 1'-C); 128.0 (s, $1C_{Ar}$, 4a-C); 131.6 (s, $1C_{Ar}$, 9-C); 133.4 (s, $1C_{Ar}$, 10a-C); 136.7 (s, $1C_{Ar}$, 7-C); 138.9 (s, $1C_{Ar}$, 6a-C); 141.1 (s, $1C_{ArH}$, 4-C); 154.3 (d, J = 2.2, $1C_{Ar}$, 5'-C); 154.6 (d, overlapping, J = 239.9, $1C_{Ar}$, 2'-C); 162.9 (s, $1C_{Ar}$, 2-C). ^{19}F-NMR (282 MHz, DMSO-d_6): δ_F = −122.53 (dt, J = 9.5, 4.7, $1F_{Ar}$, 2'-F); −223.50 (tt, J = 47.8, 29.7, 1F, O(CH$_2$)$_2$F). HR-MS (ESI) m/z: calcd. for [C$_{21}$H$_{22}$F$_2$N$_5$O$_2$]$^+$ = 414.1736; found = 414.1739 [M + H]$^+$.

3.2.3. 2-((9-(5-Butoxy-2-fluorophenyl)-7-methylimidazo[5,1-c]pyrido[2,3-e][1,2,4]-triazin-2-yl)oxy)ethyl-4-methylbenzenesulfonate (TA5a)

Compound **TA1b** (0.85 g, 1 eq.) was dissolved in MeCN (20 mL) and K$_2$CO$_3$·1.5 H$_2$O (0.15 g, 4 eq.) and ethane-1,2-diyl bis(4-methylbenzenesulfonate) (0.17 g, 2 eq.; synthesized according to the literature [19,57]) were added. After stirring at 60–70 °C for 5 h and at ambient temperature overnight, the yellow suspension was filtered followed by evaporation of the solvent. The residue was dissolved in CH$_2$Cl$_2$ (10 mL), washed once with water and citric acid (25%, 5 mL each) and then the aqueous phase was extracted with CH$_2$Cl$_2$ (2 mL). The combined organic phases were dried over Na$_2$SO$_4$, filtered and the solvent was evaporated. Purification by column chromatography (EtOAc/CH$_2$Cl$_2$, 1:5, *v/v*) afforded a yellow solid of **TA5a** (0.10 g, 79%). ^{1}H-NMR (400 MHz, DMSO-d_6): δ_H = 0.92 (t, J = 7.4, 3H, O(CH$_2$)$_3\underline{C}H_3$); 1.37–1.48 (m, 2H, O(CH$_2$)$_2\underline{C}H_2$CH$_3$); 1.63–1.77 (m, 2H, OCH$_2\underline{C}H_2$CH$_2$CH$_3$); 2.15 (s, 3H, 4''-C-CH$_3$); 2.83 (s, 3H, 7-C-CH$_3$); 3.87–3.92 (m, 2H, OCH$_2\underline{C}H_2$OTs); 3.97–4.03 (m, 4H, O$\underline{C}H_2$CH$_2$OTs, O$\underline{C}H_2$(CH$_2$)$_2$CH$_3$); 7.03 (d, J = 8.8, $1H_{Ar}$, 3-H); 7.10 (ddd, J = 9.0, 4.0, 3.1, $1H_{Ar}$, 4'-H); 7.15 (dd, J = 8.6, 0.8, $2H_{Ar}$, 3''-H, 5''-H); 7.19–7.29 (m, $2H_{Ar}$, 3'-H, 6'-H); 7.56 (d, J = 8.3, $2H_{Ar}$, 2''-H, 6''-H); 8.69 (d, J = 8.8, $1H_{Ar}$, 4-H). ^{13}C-NMR (75 MHz, DMSO-d_6): δ_C = 12.4 (s, $1C_{prim}$, 7-C-$\underline{C}H_3$); 13.7 (s, $1C_{prim}$, O(CH$_2$)$_3\underline{C}H_3$); 18.7 (s, $1C_{sec}$, O(CH$_2$)$_2\underline{C}H_2$CH$_3$); 20.8 (s, $1C_{prim}$, 4''-C-$\underline{C}H_3$); 30.7 (s, $1C_{sec}$, OCH$_2\underline{C}H_2$CH$_2$CH$_3$); 64.5 (s, $1C_{sec}$, O$\underline{C}H_2$CH$_2$OTs); 67.3 (s, $1C_{sec}$, OCH$_2\underline{C}H_2$OTs); 68.1 (s, $1C_{sec}$, O$\underline{C}H_2$(CH$_2$)$_2$CH$_3$); 112.3 (s, $1C_{ArH}$, 3-C); 116.0 (d, J = 22.9, $1C_{ArH}$, 3'-C); 117.4 (d, J = 8.1, $1C_{ArH}$, 4'-C); 117.5 (d, J = 1.9, $1C_{ArH}$, 6'-C); 120.1 (d, J = 16.6, $1C_{Ar}$, 1'-C); 127.5 (s, $2C_{ArH}$, 2''-C, 6''-C); 127.9 (s, $1C_{Ar}$, 4a-C); 129.7 (s, $2C_{ArH}$, 3''-C, 5''-C); 131.5 (s, $1C_{Ar}$, 9-C); 131.6 (s, $1C_{Ar}$, 1''-C); 133.1 (s, $1C_{Ar}$, 10a-C); 136.9 (s, $1C_{Ar}$, 7-C); 138.9 (s, $1C_{Ar}$, 6a-C); 140.9 (s, $1C_{ArH}$, 4-C); 144.7 (s, $1C_{Ar}$, 4''-C); 154.45 (d, J = 1.9, $1C_{Ar}$, 5'-C); 154.49 (d, overlap, J = 240.2, $1C_{Ar}$, 2'-C); 162.5 (s, $1C_{Ar}$, 2-C). ^{19}F-NMR (282 MHz, DMSO-d_6): δ_F = −122.41 (dt, J = 9.5, 4.9, $1F_{Ar}$, 2'-F). HR-MS (ESI) m/z: calcd. for [C$_{28}$H$_{29}$FN$_5$O$_5$S]$^+$ = 566.1867; found = 566.1868 [M + H]$^+$.

3.3. In Vitro Affinity Assay

The inhibitory potencies of **TA1–5** for human recombinant PDE2A and PDE10A proteins were determined by BioCrea GmbH (Radebeul, Germany) [7].

3.4. Radiochemistry

The aqueous solution of no-carrier-added [^{18}F]fluoride (1–2 GBq) was trapped on a Chromafix® 30 PS-HCO$_3^-$ cartridge (MACHEREY-NAGEL GmbH & Co. KG, Düren, Germany). The activity was eluted with 300 μL of an aqueous K$_2$CO$_3$-solution (1.78 mg, 12.9 μmol) into a 4 mL V-vial containing Kryptofix 2.2.2 (K$_{222}$, 11.2 mg, 29.7 μmol) in 1 mL MeCN. The K$^+$/[^{18}F]F$^-$/K$_{222}$-carbonate complex was azeotropically dried under vacuum and nitrogen flow within 7–10 min using a Discover PETwave Microwave CEM® (75 W, 50–60 °C, power cycling mode). Two aliquots of MeCN (2 × 1.0 mL) were added during the drying procedure and the final complex was dissolved in 500 μL MeCN ready for radiolabelling.

The aliphatic radiolabelling of the tosylate **TA5a** (1 mg in 500 μL MeCN) was performed under conventional heating at 80 °C for 15 min. Aliquots of the reaction mixture were analyzed by radio-TLC (EtOAc/DCM, 1:1, v/v) to determine the radiochemical yield of the crude product. After dilution with water (1:1, v/v), the crude reaction mixture was applied to an isocratic semi-preparative HPLC (see General Information) for isolation of the desired radioligand [^{18}F]**TA5** (t_R = 34–38 min). The collected fractions were diluted with water (total volume: 40 mL), passed through a Sep-Pak® C18 Plus light cartridge (Waters, Milford, MA, USA; pre-conditioned with 20 mL of absolute EtOH and 60 mL water), and eluted with 0.75 mL of absolute EtOH. Evaporation of the solvent at 70 °C under a gentle nitrogen stream and subsequent formulation of the radioligand in sterile isotonic saline containing 10% EtOH (v/v) afforded a [^{18}F]**TA5**-solution usable for biological investigations.

The identity of the radioligand was proved by analytical radio-HPLC (see General Information) of samples of [^{18}F]**TA5** spiked with the non-radioactive reference compound **TA5** using a gradient and an isocratic mode.

3.5. Investigation of In Vitro Stability and Lipophilicity (logD₇.₄)

In vitro stability of [^{18}F]**TA5** was studied by incubation in phosphate-buffered saline (PBS, pH 7.4), *n*-octanol and pig plasma at 37 °C for 60 min (~5 MBq of the radioligand added to 500 µL of each medium). Samples were taken at 15, 30 and 60 min and analyzed by radio-TLC and radio-HPLC (see General Information).

The lipophilicity of [^{18}F]**TA5** was examined by partitioning between *n*-octanol and phosphate-buffered saline (PBS, pH 7.4) at ambient temperature using the conventional shake-flask method. The radioligand (10 µL, ~1 MBq) was added to a tube containing the *n*-octanol/PBS-mixture (6 mL, 1:1, v/v, fourfold determination). The tubes were shaken for 20 min using a mechanical shaker (HS250 basic, IKA Labortechnik GmbH & Co. KG, Staufen, Germany) followed by centrifugation (5000 rpm for 5 min) and separation of the phases. Aliquots were taken from the organic and the aqueous phase (1 mL each) and activity was measured with an automated gamma counter (1480 WIZARD, Fa. Perkin Elmer, Waltham, MA, USA). The distribution coefficient (D) was calculated as [activity (cpm/mL) in *n*-octanol]/[activity (cpm/mL) in PBS, pH 7.4] stated as the decade logarithm (logD₇.₄).

3.6. Animal Studies

All animal procedures were approved by the Animal Care and Use Committee of Saxony (TVV 08/13).

3.6.1. In Vitro Autoradiographic Studies

Cryosections of brains obtained from juvenile female German landrace pigs (10–13 kg) and female SPRD rats (10–12 weeks old) were thawed, dried in a stream of cold air, and preincubated for 10 min with incubation buffer (50 mM TRIS-HCl, pH 7.4, 120 mM NaCl, 5 mM KCl, 2 mM $CaCl_2$, 5 mM $MgCl_2$) at ambient temperature.

Brain sections were incubated with ~1 MBq/mL of [^{18}F]**TA5** in incubation buffer for 60 min at ambient temperature. Afterwards sections were washed twice with 50 mM TRIS-HCl (pH 7.4) for 2 min at 4 °C, dipped briefly in ice-cold deionized water, dried in a stream of cold air and exposed for 60 min to an ^{18}F-sensitive image plate that was analyzed afterwards using an image plate scanner (HD-CR 35; Duerr NDT GmbH, Bietigheim Bissingen, Germany). Non-specific binding of the radioligand was determined by co-incubation with 10 µM **TA1** or **TA5**.

3.6.2. In Vivo Metabolism Studies

The radioligand [^{18}F]**TA5** (~70 MBq in 150 µL isotonic saline) was injected in female CD-1 mice (10–12 weeks old) via the tail vein. Brain and blood samples were obtained at 30 min p.i., plasma separated by centrifugation (14,000× *g*, 1 min), and brain homogenized in ~1 mL isotonic saline on ice (10 strokes of a PTFE plunge at 1000 rpm in a borosilicate glass cylinder; Potter S Homogenizer, B. Braun Melsungen AG, Melsungen, Germany).

Conventional Extraction Method

The twofold extractions were performed as a double determination. Plasma (2 × 50 µL) and brain samples (2 × 250 µL) were added each to an ice-cold acetone/water mixture (4:1, *v/v*; plasma or brain sample/organic solvent, 1:4, *v/v*). The samples were vortexed for 1 min, incubated on ice for 10 min (first extraction) or 5 min (second extraction) and centrifuged at 10,000 rpm for 5 min. Supernatants were collected and the precipitates were re-dissolved in ice-cold acetone/water (100 µL) for the second extraction. Activity of aliquots from supernatants of each extraction step and of the precipitates was quantified using an automated gamma counter (1480 WIZARD, Fa. Perkin Elmer, Waltham, MA, USA). The combined supernatants from both extractions were concentrated at 70 °C under nitrogen stream and analyzed by radio-HPLC (see General Information). Notably, [^{18}F]**TA5** was quantitatively extracted from the biological material as proven by in vitro incubation of the radioligand in pig plasma and subsequent extraction applying the same protocol as for the in vivo metabolism studies.

Micellar Liquid Chromatography (MLC)

Plasma (20–50 µL) was dissolved in 100–300 µL of 200 mM aqueous SDS and injected directly into the MLC system (2000 µL sample loop; see General Information). Homogenized brain material (100–200 µL) was dissolved in 500 µL of 200 mM aqueous SDS, stirred at 75 °C for 5 min and after cooling to ambient temperature injected into the MLC system.

4. Conclusions

The novel PDE2A radioligand [^{18}F]**TA5** proved to be not suitable for molecular imaging of PDE2A protein in the brain due to (i) non-specific binding in vitro and (ii) formation of a high fraction of brain-penetrating radiometabolites in vivo. Nevertheless, **TA5** represents the most potent PDE2A ligand out of our imidazopyridotriazine-based derivatives so far. Besides, further structural modification at the side chains of the tricyclic framework is needed to possibly improve the in vitro binding properties and the in vivo stability of upcoming compounds. Notably, the developed synthetic strategies for the selective *O*-dealkylation at either the 5´-alkoxyphenyl moiety or the 2-alkoxypyridinyl function of the lead compound **TA1** are of highly importance for our ongoing work.

Supplementary Materials: The supplementary materials are available online at www.mdpi.com/1420-3049/23/3/556/s1. General Information; NMR data of compounds **TA1** (incl. HR-MS) and **TA1a**; Figures S1–S6: NMR spectra of compound **TA1** (^{1}H, ^{13}C, ^{19}F, COSY, HSQC, HMBC); Figure S7: ^{1}H-NMR spectrum of compound **TA1a**; Figures S8–S10: NMR spectra of compound **TA1b** (^{1}H, ^{13}C-APT, ^{19}F); Figures S11–S16: NMR spectra of compound **TA5** (^{1}H, ^{13}C-APT, ^{19}F, COSY, HSQC, HMBC); Figures S17–S22: NMR spectra of compound **TA5a** (^{1}H, ^{13}C-APT, ^{19}F, COSY, HSQC, HMBC).

Acknowledgments: The Deutsche Forschungsgemeinschaft (DFG) is acknowledged for financial support (Project No. SCHE 1825/3-1). We thank the staff of the Institute of Analytical Chemistry, Department of Chemistry and Mineralogy of the University of Leipzig, for recording and processing the NMR and HR-MS spectra, Karsten Franke, Helmholtz-Zentrum Dresden-Rossendorf (HZDR), for providing [^{18}F]fluoride as well as Tina Spalholz (HZDR) for technical assistance.

Author Contributions: Susann Schröder, Barbara Wenzel, Matthias Scheunemann, Detlef Briel and Jörg Steinbach designed and performed organic syntheses; Susann Schröder, Barbara Wenzel, Rodrigo Teodoro, Jörg Steinbach and Matthias Scheunemann designed and performed radiosyntheses; Susann Schröder, Barbara Wenzel, Rodrigo Teodoro, Ute Egerland, Norbert Höfgen, Winnie Deuther-Conrad and Peter Brust designed and performed in vitro and in vivo studies; Mathias Kranz, Winnie Deuther-Conrad and Peter Brust designed and performed PET/MR studies with [^{18}F]**TA3**; Susann Schröder, Barbara Wenzel, Winnie Deuther-Conrad, Mathias Kranz and Peter Brust analyzed the data; Susann Schröder, Barbara Wenzel, Rodrigo Teodoro, Winnie Deuther-Conrad, Mathias Kranz, Matthias Scheunemann, Jörg Steinbach and Peter Brust wrote the paper. All authors read and approved the final manuscript.

Conflicts of Interest: The authors declare no conflict of interest.

References

1. Keravis, T.; Lugnier, C. Cyclic nucleotide phosphodiesterase (PDE) isozymes as targets of the intracellular signalling network: Benefits of PDE inhibitors in various diseases and perspectives for future therapeutic developments. *Brit. J. Pharmacol.* **2012**, *165*, 1288–1305. [CrossRef] [PubMed]

2. Zhang, C.; Yu, Y.; Ruan, L.; Wang, C.; Pan, J.; Klabnik, J.; Lueptow, L.; Zhang, H.-T.; O'Donnell, J.M.; Xu, Y. The roles of phosphodiesterase 2 in the central nervous and peripheral systems. *Curr. Pharm. Design* **2015**, *21*, 274–290. [CrossRef]

3. Trabanco, A.A.; Buijnsters, P.; Rombouts, F.J. Towards selective phosphodiesterase 2A (PDE2A) inhibitors: A patent review (2010–present). *Expert Opin. Ther. Pat.* **2016**, *26*, 933–946. [CrossRef] [PubMed]

4. Stephenson, D.T.; Coskran, T.M.; Wilhelms, M.B.; Adamowicz, W.O.; O'Donnell, M.M.; Muravnick, K.B.; Menniti, F.S.; Kleiman, R.J.; Morton, D. Immunohistochemical localization of phosphodiesterase 2A in multiple mammalian species. *J. Histochem. Cytochem.* **2009**, *57*, 933–949. [CrossRef] [PubMed]

5. Stephenson, D.T.; Coskran, T.M.; Kelly, M.P.; Kleiman, R.J.; Morton, D.; O'Neill, S.M.; Schmidt, C.J.; Weinberg, R.J.; Menniti, F.S. The distribution of phosphodiesterase 2A in the rat brain. *Neuroscience* **2012**, *226*, 145–155. [CrossRef] [PubMed]

6. Gomez, L.; Breitenbucher, J.G. PDE2 inhibition: Potential for the treatment of cognitive disorders. *Bioorg. Med. Chem. Lett.* **2013**, *23*, 6522–6527. [CrossRef] [PubMed]

7. Stange, H.; Langen, B.; Egerland, U.; Hoefgen, N.; Priebs, M.; Malamas, M.S.; Erdel, J.J.; Ni, Y. Triazine Derivatives as Inhibitors of Phosphodiesterases. Patent No. WO 2010/054253 A1, 14 May 2010.

8. Masood, A.; Huang, Y.; Hajjhussein, H.; Xiao, L.; Li, H.; Wang, W.; Hamza, A.; Zhan, C.-G.; O'Donnell, J.M. Anxiolytic effects of phosphodiesterase-2 inhibitors associated with increased cGMP signaling. *J. Pharmacol. Exp. Ther.* **2009**, *331*, 690–699. [CrossRef] [PubMed]

9. Fernández-Fernández, D.; Rosenbrock, H.; Kroker, K.S. Inhibition of PDE2A, but not PDE9A, modulates presynaptic short-term plasticity measured by paired-pulse facilitation in the Ca1 region of the hippocampus. *Synapse* **2015**, *69*, 484–496. [CrossRef] [PubMed]

10. Boess, F.G.; Hendrix, M.; van der Staay, F.J.; Erb, C.; Schreiber, R.; van Staveren, W.; de Vente, J.; Prickaerts, J.; Blokland, A.; Koenig, G. Inhibition of phosphodiesterase 2 increases neuronal cGMP, synaptic plasticity and memory performance. *Neuropharmacology* **2004**, *47*, 1081–1092. [CrossRef] [PubMed]

11. Reneerkens, O.A.H.; Rutten, K.; Bollen, E.; Hage, T.; Blokland, A.; Steinbusch, H.W.M.; Prickaerts, J. Inhibition of phoshodiesterase type 2 or type 10 reverses object memory deficits induced by scopolamine or MK-801. *Behav. Brain Res.* **2013**, *236*, 16–22. [CrossRef] [PubMed]

12. Bales, K.; Plath, N.; Svenstrup, N.; Menniti, F. Phosphodiesterase inhibition to target the synaptic dysfunction in Alzheimer's disease. In *Neurodegenerative Diseases*; Dominguez, C., Ed.; Springer: Berlin/Heidelberg, Germany, 2010; Volume 6, pp. 57–90.

13. Schröder, S.; Wenzel, B.; Deuther-Conrad, W.; Scheunemann, M.; Brust, P. Novel radioligands for cyclic nucleotide phosphodiesterase imaging with positron emission tomography: An update on developments since 2012. *Molecules* **2016**, *21*, 650. [CrossRef]

14. Lakics, V.; Karran, E.H.; Boess, F.G. Quantitative comparison of phosphodiesterase mRNA distribution in human brain and peripheral tissues. *Neuropharmacology* **2010**, *59*, 367–374. [CrossRef] [PubMed]

15. Andrés, J.I.; Rombouts, F.J.R.; Trabanco, A.A.; Vanhoof, G.C.P.; De Angelis, M.; Buijnsters, P.J.J.A.; Guillemont, J.E.G.; Bormans, G.M.R.; Celen, S.J.L. 1-Aryl-4-methyl-[1,2,4]triazolo[4,3-*a*]quinoxaline derivatives. Patent No. WO 2013/000924 A1, 3 January 2013.

16. Zhang, L.; Villalobos, A.; Beck, E.M.; Bocan, T.; Chappie, T.A.; Chen, L.; Grimwood, S.; Heck, S.D.; Helal, C.J.; Hou, X.; et al. Design and selection parameters to accelerate the discovery of novel central nervous system positron emission tomography (PET) ligands and their application in the development of a novel phosphodiesterase 2A PET ligand. *J. Med. Chem.* **2013**, *56*, 4568–4579. [CrossRef] [PubMed]

17. Naganawa, M.; Nabulsi, N.; Waterhouse, R.; Lin, S.-F.; Zhang, L.; Cass, T.; Ropchan, J.; McCarthy, T.; Huang, Y.; Carson, R. Human PET studies with [18F]PF-05270430, a PET radiotracer for imaging phosphodiesterase-2A. *J. Nucl. Med.* **2013**, *54*, 201.

18. Naganawa, M.; Waterhouse, R.N.; Nabulsi, N.; Lin, S.F.; Labaree, D.; Ropchan, J.; Tarabar, S.; DeMartinis, N.; Ogden, A.; Banerjee, A.; et al. First-in-human assessment of the novel PDE2A PET radiotracer 18F-PF-05270430. *J. Nucl. Med.* **2016**, *57*, 1388–1395. [CrossRef] [PubMed]

19. Schröder, S.; Wenzel, B.; Deuther-Conrad, W.; Teodoro, R.; Egerland, U.; Kranz, M.; Scheunemann, M.; Höfgen, N.; Steinbach, J.; Brust, P. Synthesis, ^{18}F-radiolabelling and biological characterization of novel fluoroalkylated triazine derivatives for in vivo imaging of phosphodiesterase 2A in brain via positron emission tomography. *Molecules* **2015**, *20*, 9591–9615. [CrossRef] [PubMed]

20. Schröder, S.; Wenzel, B.; Deuther-Conrad, W.; Teodoro, R.; Egerland, U.; Kranz, M.; Fischer, S.; Höfgen, N.; Steinbach, J.; Brust, P. Novel ^{18}F-labelled triazine derivatives for PET imaging of phosphodiesterase 2A. *J. Labelled Compd. Rad.* **2015**, *58*, S221.

21. Schröder, S.; Wenzel, B.; Kranz, M.; Egerland, U.; Teodoro, R.; Deuther-Conrad, W.; Fischer, S.; Höfgen, N.; Steinbach, J.; Brust, P. Development, synthesis and F-18 labelling of a fluoroalkylated triazine derivative for PET imaging of phosphodiesterase 2A. *Eur. J. Nucl. Med. Mol. Imaging* **2014**, *41*, S197.

22. Zoghbi, S.S.; Shetty, H.U.; Ichise, M.; Fujita, M.; Imaizumi, M.; Liow, J.-S.; Shah, J.; Musachio, J.L.; Pike, V.W.; Innis, R.B. PET imaging of the dopamine transporter with ^{18}F]FECNT: A polar radiometabolite confounds brain radioligand measurements. *J. Nucl. Med.* **2006**, *47*, 520–527. [PubMed]

23. Evens, N.; Vandeputte, C.; Muccioli, G.G.; Lambert, D.M.; Baekelandt, V.; Verbruggen, A.M.; Debyser, Z.; Van Laere, K.; Bormans, G.M. Synthesis, in vitro and in vivo evaluation of fluorine-18 labelled FE-GW405833 as a PET tracer for type 2 cannabinoid receptor imaging. *Bioorg. Med. Chem.* **2011**, *19*, 4499–4505. [CrossRef] [PubMed]

24. Langen, K.-J.; Hamacher, K.; Weckesser, M.; Floeth, F.; Stoffels, G.; Bauer, D.; Coenen, H.H.; Pauleit, D. O-(2-^{18}F]fluoroethyl)-L-tyrosine: Uptake mechanisms and clinical applications. *Nucl. Med. Biol.* **2006**, *33*, 287–294. [CrossRef] [PubMed]

25. Hutterer, M.; Nowosielski, M.; Putzer, D.; Jansen, N.L.; Seiz, M.; Schocke, M.; McCoy, M.; Göbel, G.; la Fougère, C.; Virgolini, I.J.; et al. [18]F]-Fluoro-ethyl-L-tyrosine PET: A valuable diagnostic tool in neuro-oncology, but not all that glitters is glioma. *Neuro-Oncology* **2013**, *15*, 341–351. [CrossRef] [PubMed]

26. Galldiks, N.; Stoffels, G.; Filss, C.; Rapp, M.; Blau, T.; Tscherpel, C.; Ceccon, G.; Dunkl, V.; Weinzierl, M.; Stoffel, M.; et al. The use of dynamic O-(2-^{18}F-fluoroethyl)-L-tyrosine PET in the diagnosis of patients with progressive and recurrent glioma. *Neuro. Oncol.* **2015**, *17*, 1293–1300. [CrossRef] [PubMed]

27. Tsukada, H.; Sato, K.; Fukumoto, D.; Kakiuchi, T. Evaluation of D-isomers of O-^{18}F-fluoromethyl, O-^{18}F-fluoroethyl and O-^{18}F-fluoropropyl tyrosine as tumour imaging agents in mice. *Eur. J. Nucl. Med. Mol. Imaging* **2006**, *33*, 1017–1024. [CrossRef] [PubMed]

28. Pauleit, D.; Floeth, F.; Herzog, H.; Hamacher, K.; Tellmann, L.; Müller, H.-W.; Coenen, H.; Langen, K.-J. Whole-body distribution and dosimetry of O-(2-^{18}F]fluoroethyl)-L-tyrosine. *Eur. J. Nucl. Med. Mol. Imaging* **2003**, *30*, 519–524. [CrossRef] [PubMed]

29. Barret, O.; Thomae, D.; Tavares, A.; Alagille, D.; Papin, C.; Waterhouse, R.; McCarthy, T.; Jennings, D.; Marek, K.; Russell, D.; et al. In vivo assessment and dosimetry of 2 novel PDE10A PET radiotracers in humans: ^{18}F-MNI-659 and ^{18}F-MNI-654. *J. Nucl. Med.* **2014**, *55*, 1297–1304. [CrossRef] [PubMed]

30. Russell, D.S.; Barret, O.; Jennings, D.L.; Friedman, J.H.; Tamagnan, G.D.; Thomae, D.; Alagille, D.; Morley, T.J.; Papin, C.; Papapetropoulos, S.; et al. The phosphodiesterase 10 positron emission tomography tracer, ^{18}F]MNI-659, as a novel biomarker for early huntington disease. *J. Am. Med. Assoc. Neurol.* **2014**, *71*, 1520–1528. [CrossRef] [PubMed]

31. Russell, D.S.; Jennings, D.L.; Barret, O.; Tamagnan, G.D.; Carroll, V.M.; Caille, F.; Alagille, D.; Morley, T.J.; Papin, C.; Seibyl, J.P.; et al. Change in PDE10 across early huntington disease assessed by ^{18}F]MNI-659 and PET imaging. *Neurology* **2016**, *86*, 748–754. [CrossRef] [PubMed]

32. Nakada, Y.; Aicher, T.D.; Huerou, Y.L.; Turner, T.; Pratt, S.A.; Gonzales, S.S.; Boyd, S.A.; Miki, H.; Yamamoto, T.; Yamaguchi, H.; et al. Novel acyl coenzyme A (CoA): Diacylglycerol acyltransferase-1 inhibitors: Synthesis and biological activities of diacylethylenediamine derivatives. *Bioorg. Med. Chem.* **2010**, *18*, 2785–2795. [CrossRef] [PubMed]

33. Benton, F.L.; Dillon, T.E. The cleavage of ethers with boron bromide. I. Some common ethers. *J. Am. Chem. Soc.* **1942**, *64*, 1128–1129. [CrossRef]

34. Wuts, P.G.M.; Greene, T.W. Protection for phenols and catechols. In *Greene's Protective Groups in Organic Synthesis*; John Wiley & Sons, Inc.: Hoboken, NJ, USA, 2006; pp. 367–430.

35. Vickery, E.H.; Pahler, L.F.; Eisenbraun, E.J. Selective O-demethylation of catechol ethers. Comparison of boron tribromide and iodotrimethylsilane. *J. Org. Chem.* **1979**, *44*, 4444–4446. [CrossRef]

36. Bhatt, M.V.; Kulkarni, S.U. Cleavage of ethers. *Synthesis* **1983**, *1983*, 249–282. [CrossRef]

37. Pasquini, C.; Coniglio, A.; Bassetti, M. Controlled dealkylation by BBr$_3$: Efficient synthesis of para-alkoxy-phenols. *Tetrahedron Lett.* **2012**, *53*, 6191–6194. [CrossRef]

38. Robinson, P.D.; Groziak, M.P. A boron-containing estrogen mimic. *Acta Crystallogr. C* **1999**, *55*, 1701–1704. [CrossRef] [PubMed]

39. Shiao, M.J.; Lai, L.L.; Ku, W.S.; Lin, P.Y.; Hwu, J.R. Chlorotrimethylsilane in combination with sodium sulfide as the equivalent of sodium trimethylsilanethiolate in organic reactions. *J. Org. Chem.* **1993**, *58*, 4742–4744. [CrossRef]

40. Testaferri, L.; Tiecco, M.; Tingoli, M.; Bartoli, D.; Massoli, A. The reactions of some halogenated pyridines with methoxide and methanethiolate ions in dimethylformamide. *Tetrahedron* **1985**, *41*, 1373–1384. [CrossRef]

41. Beak, P.; Fry, F.S.; Lee, J.; Steele, F. Equilibration studies. Protomeric equilibria of 2- and 4-hydroxypyridines, 2- and 4-hydroxypyrimidines, 2- and 4-mercaptopyridines, and structurally related compounds in the gas phase. *J. Am. Chem. Soc.* **1976**, *98*, 171–179. [CrossRef]

42. Albert, A.; Phillips, J.N. Ionization constants of heterocyclic substances. Part II. Hydroxy-derivatives of nitrogenous six-membered ring-compounds. *J. Chem. Soc.* **1956**, 1294–1304. [CrossRef]

43. Hatherley, L.D.; Brown, R.D.; Godfrey, P.D.; Pierlot, A.P.; Caminati, W.; Damiani, D.; Melandri, S.; Favero, L.B. Gas-phase tautomeric equilibrium of 2-pyridinone and 2-hydroxypyridine by microwave spectroscopy. *J. Phys. Chem.* **1993**, *97*, 46–51. [CrossRef]

44. Comins, D.L.; Jianhua, G. *N*- vs. *O*-Alkylation in the mitsunobu reaction of 2-pyridone. *Tetrahedron Lett.* **1994**, *35*, 2819–2822. [CrossRef]

45. Chung, N.M.; Tieckelmann, H. Alkylations of heterocyclic ambient anions. Iv. Alkylation of 5-carbethoxy- and 5-nitro-2-pyridone salts. *J. Org. Chem.* **1970**, *35*, 2517–2520. [CrossRef]

46. Hopkins, G.; Jonak, J.; Minnemeyer, H.; Tieckelmann, H. Alkylations of heterocyclic ambient anions ii. Alkylation of 2-pyridone salts. *J. Org. Chem.* **1967**, *32*, 4040–4044. [CrossRef]

47. Waterhouse, R.N. Determination of lipophilicity and its use as a predictor of blood–brain barrier penetration of molecular imaging agents. *Mol. Imaging Biol.* **2003**, *5*, 376–389. [CrossRef] [PubMed]

48. Nakao, R.; Schou, M.; Halldin, C. Direct plasma metabolite analysis of positron emission tomography radioligands by micellar liquid chromatography with radiometric detection. *Anal. Chem.* **2012**, *84*, 3222–3230. [CrossRef] [PubMed]

49. Liu, J.; Wenzel, B.; Dukic-Stefanovic, S.; Teodoro, R.; Ludwig, F.-A.; Deuther-Conrad, W.; Schröder, S.; Chezal, J.-M.; Moreau, E.; Brust, P.; et al. Development of a new radiofluorinated quinoline analog for PET imaging of phosphodiesterase 5 (PDE5) in brain. *Pharmaceuticals* **2016**, *9*, 22. [CrossRef] [PubMed]

50. Lear, J.L.; Ackermann, R.F. Evaluation of radiolabeled acetate and fluoroacetate as potential tracers of cerebral oxidative metabolism. *Metab. Brain Dis.* **1990**, *5*, 45–56. [CrossRef] [PubMed]

51. Mori, T.; Sun, L.-Q.; Kobayashi, M.; Kiyono, Y.; Okazawa, H.; Furukawa, T.; Kawashima, H.; Welch, M.J.; Fujibayashi, Y. Preparation and evaluation of ethyl [^{18}F]fluoroacetate as a proradiotracer of [^{18}F]fluoroacetate for the measurement of glial metabolism by PET. *Nucl. Med. Biol.* **2009**, *36*, 155–162. [CrossRef] [PubMed]

52. Muir, D.; Berl, S.; Clarke, D.D. Acetate and fluoroacetate as possible markers for glial metabolism in vivo. *Brain Res.* **1986**, *380*, 336–340. [CrossRef]

53. Ponde, D.E.; Dence, C.S.; Oyama, N.; Kim, J.; Tai, Y.-C.; Laforest, R.; Siegel, B.A.; Welch, M.J. ^{18}F-Fluoroacetate: A potential acetate analog for prostate tumor imaging—In vivo evaluation of ^{18}F-fluoroacetate versus ^{11}C-acetate. *J. Nucl. Med.* **2007**, *48*, 420–428. [PubMed]

54. Park, B.K.; Kitteringham, N.R.; O'Neill, P.M. Metabolism of fluorine-containing drugs. *Annu. Rev. Pharmacol.* **2001**, *41*, 443–470. [CrossRef] [PubMed]

55. Shu, Y.-Z.; Johnson, B.M.; Yang, T.J. Role of biotransformation studies in minimizing metabolism-related liabilities in drug discovery. *AAPS J.* **2008**, *10*, 178–192. [CrossRef] [PubMed]

56. Pike, V.W. Pet radiotracers: Crossing the blood–brain barrier and surviving metabolism. *Trends Pharmacol. Sci.* **2009**, *30*, 431–440. [CrossRef] [PubMed]

57. Burns, D.H.; Chan, H.K.; Miller, J.D.; Jayne, C.L.; Eichhorn, D.M. Synthesis, modification, and characterization of a family of homologues of exo-calix[4]arene: Exo-[*n.m.n.m*]metacyclophanes, *n,m* ≥ 3. *J. Org. Chem.* **2000**, *65*, 5185–5196. [CrossRef] [PubMed]

molecules

Article

[18F]FEPPA a TSPO Radioligand: Optimized Radiosynthesis and Evaluation as a PET Radiotracer for Brain Inflammation in a Peripheral LPS-Injected Mouse Model

Nicolas Vignal [1,2], Salvatore Cisternino [2,3], Nathalie Rizzo-Padoin [1,2], Carine San [1], Fortune Hontonnou [4], Thibaut Gelé [1], Xavier Declèves [2,5], Laure Sarda-Mantel [1,6,7] and Benoît Hosten [1,2,*]

[1] Assistance Publique—Hôpitaux de Paris, Hôpital Saint-Louis, Unité Claude Kellershohn, 75010 Paris, France; nicolas.vignal@aphp.fr (N.V.); nathalie.rizzo@aphp.fr (N.R.-P.); carine.san@aphp.fr (C.S.); thibaut.gele@aphp.fr (T.G.); laure.sarda-mantel@aphp.fr (L.S.-M.)
[2] Inserm UMR-S 1144, Faculté de Pharmacie de Paris, Université Paris Descartes, 75006 Paris, France; salvatore.cisternino@aphp.fr (S.C.); xavier.decleves@aphp.fr (X.D.)
[3] Assistance Publique—Hôpitaux de Paris, Hôpital Universitaire Necker—Enfants Malades, 75015 Paris, France
[4] Institut Universitaire d'Hématologie, Université Paris Diderot, 75013 Paris, France; fortune.hontonnou@univ-paris-diderot.fr
[5] Assistance Publique—Hôpitaux de Paris, Hôpital Cochin, 75014 Paris, France
[6] Assistance Publique—Hôpitaux de Paris, Hôpital Lariboisière, Médecine Nucléaire, 75010 Paris, France
[7] Inserm UMR-S 942, Université Paris Diderot, 75013 Paris, France
*   Correspondence: benoit.hosten@aphp.fr; Tel.: +33-142-385-105

Received: 24 April 2018; Accepted: 4 June 2018; Published: 7 June 2018

Abstract: [18F]FEPPA is a specific ligand for the translocator protein of 18 kDa (TSPO) used as a positron emission tomography (PET) biomarker for glial activation and neuroinflammation. [18F]FEPPA radiosynthesis was optimized to assess in a mouse model the cerebral inflammation induced by an intraperitoneal injection of *Salmonella enterica* serovar *Typhimurium* lipopolysaccharides (LPS; 5 mg/kg) 24 h before PET imaging. [18F]FEPPA was synthesized by nucleophilic substitution (90 °C, 10 min) with tosylated precursor, followed by improved semi-preparative HPLC purification (retention time 14 min). [18F]FEPPA radiosynthesis were carried out in 55 min (from EOB). The non-decay corrected radiochemical yield were $34 \pm 2\%$ ($n = 17$), and the radiochemical purity greater than 99%, with a molar activity of 198 ± 125 GBq/µmol at the end of synthesis. Western blot analysis demonstrated a 2.2-fold increase in TSPO brain expression in the LPS treated mice compared to controls. This was consistent with the significant increase of [18F]FEPPA brain total volume of distribution (V_T) estimated with pharmacokinetic modelling. In conclusion, [18F]FEPPA radiosynthesis was implemented with high yields. The new purification/formulation with only class 3 solvents is more suitable for in vivo studies.

Keywords: [18F]FEPPA; TSPO; brain inflammation; small-animal PET imaging; radiolabeling; radiotracer metabolism

1. Introduction

The translocator protein of 18 kDa (TSPO) is the most advance target for the non-invasive and translational study of inflammatory processes using positron emission tomography (PET) imaging [1]. TSPO is an outer mitochondrial membrane protein found in many cell types but there is growing literature documenting its restricted brain expression in microglia and astrocytes [2].

Molecules **2018**, *23*, 1375

Cerebral inflammation is supported mainly by microglial cells, which are the resident immune cells of the central nervous system (CNS) [3]. Microglia are activated in response to endogenous signals such as beta-amyloid plaques, Tau protein [4–6], α-synuclein [6–8], or exogenous "danger" signals such as traumatic CNS injury [9] or bacterial lipopolysaccharide (LPS) [10]. This cellular activation leads to numerous functional and biochemical changes that involve cell morphology, metabolism, cytokine production and some de novo protein synthesis, for instance an increase in the ionized calcium binding adaptor molecule 1 (Iba1) and TSPO, both used as biomarkers for microglial activation. Microglia can adopt different morphologies and multiple functions: pro-inflammatory or reparative [11,12]. Other glial cells such as astrocytes are activated and may be involved in CNS inflammation [13,14].

Clinical PET imaging for brain inflammation in addition to conventional biological examinations may be a useful tool to support or reject the hypothesis of localized inflammation in the brain, to follow this process during the disease, and to study possible links between systemic and CNS inflammation. Neuroinflammation is known as a major factor in the pathophysiology of many neurodegenerative diseases [15–18]. TSPO PET studies have been conducted and showed an increase in TSPO signal in patients with Alzheimer disease [19,20], Parkinson disease [21,22], and multiple sclerosis [23,24]. More recently, clinical studies using [18F]FEPPA PET have demonstrated a brain inflammation component in some psychiatric diseases such as schizophrenia [25] and major depressions [26]. This technology is currently being applied in humans in psychosis [27,28] or Alzheimer disease [29].

PET imaging of neuroinflammation using TSPO radioligands is rather well documented, particularly for [11C]PK11195 and [18F]DPA-714. However, 11C-labeled radiotracers are more difficult to manage clinically, and [11C]PK11195 is characterized by rather broad non-specific binding, which may limit the quantification of the PET signal [30]. Hence specific radiotracers need to be developed and optimized for clinical use. [18F]FEPPA is one of the newest TSPO PET radiotracer with greater affinity for its target [31] and one of the two TSPO radioligand being the most advanced in clinical development with [18F]DPA-714 for PET imaging of neuroinflammation. Both radiotracers encountered limits to their applications, since it was shown in human three patterns of binding affinity (based on genetic polymorphism) [32] as well as an absence of reference tissue for their quantification. The radiosynthesis of [18F]FEPPA was first described by Wilson et al. in 2008 [31]. In this report, [18F]FEPPA purification was assessed using semi-preparative HPLC with methanol and formic acid followed by drying and recovery with ethanol and water, which could introduce traces of toxic solvents into the finished product and add steps to the radiotracer formulation process. In 2013, the same team modified their [18F]FEPPA radiosynthesis protocol by replacing the last step of the formulation with a cartridge capture followed by rinsing and ethanol/water elution [33]. An other team has used acetonitrile/ammonium formiate for HPLC purification which needs formulation with cartridge [34]. For human use, pharmaceutical products must limit the use of class 2 solvents (International Council for Harmonisation of Technical Requirements for Pharmaceuticals for Human Use Q3C) such as methanol or acetonitrile. To achieve this, Huang et al. [35] describe a radiosynthesis with an HPLC purification with only ethanol and water but with a great decrease in radiochemical yield.

PET studies of neuroinflammation have mostly investigated pathologies with an initially central inflammatory process, while very few have investigated those induced by a peripheral inflammatory stimulus. To our knowledge, the use of [18F]FEPPA to monitor brain inflammation induced by intraperitoneal (ip) LPS injection, has never been reported. Peripheral LPS injection is often used to induce neuroinflammation in mice [10]. However, the neuroinflammatory response may differ depending on the mouse and/or LPS strains [36]. Although the effects of *Salmonella enterica serovar typhimurium* LPS have been suggested to be more severe than *E. coli* LPS [37], LPS-injection models have mainly used *E. coli* LPS [10,36].

The objective of this work was to develop and optimize the fully automated radiosynthesis of [18F]FEPPA on the Allinone automat (Trasis®) with solvents more suitable for human use (class 3, ICH Q3C). The aim of this study was also to quantify metabolism of [18F]FEPPA in mice which is currently unknown, and to evaluate this radiotracer in a mouse model of neuroinflammation induced by peripheral *Salmonella enterica serovar typhimurium* LPS ip injection [38].

2. Results

2.1. Automated Radiosynthesis and Characterization of [18F]FEPPA

[18F]FEPPA was synthetized according to previously published methods [31,33] with some modifications and by using the AllInOne® module (Figure 1). [18F]FEPPA was produced with a non-decay corrected radiochemical yield of 34% ± 2% within 55 min (*n* = 17) from end of bombardment.

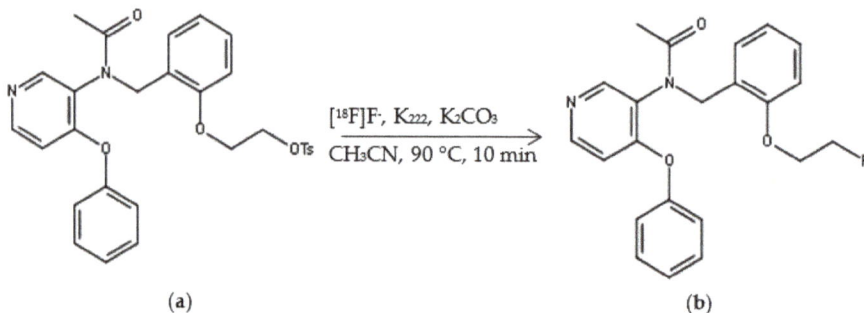

(a)   (b)

Figure 1. (a) FEPPA Precursor (2-(2-((N-4-phenoxypyridin-3-yl)acetamido)methyl)phenoxy)ethyl 4-methylbenzenesulfonate); (b) [18F]FEPPA.

The semi-preparative HPLC mobile phase was optimized from methanol and formic acid to ethanol and phosphoric acid. This new HPLC conditions allowed us to bypass the reformulation step and decrease the retention time (Figure 2) from 26 min [33] to 14 min, with a satisfactory separation of the [18F]FEPPA from its impurities. The useful fraction of the semi-preparative HPLC was collected in a vial containing 1.5 mL of sodium bicarbonate 8.4%. This allowed the final product to reach a physiological pH (between 6.8 and 7.3), which is more suitable for injection and stability.

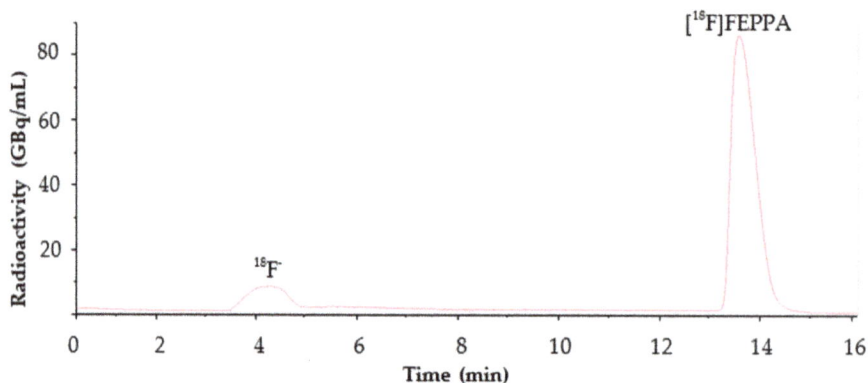

Figure 2. Semi-preparative HPLC radiochromatogram.

Quality control of the final product confirms that our semi-preparative HPLC allows a complete separation of [18F]FEPPA from synthesis impurities (Figure 3). Retenrion time for FEPPA and the precursor are respectively 3.5 and 4.6 min (resolution factor of 10.2). Our quality control method needs less amount of the final product (5 µL instead of 20 µL with a HPLC column similar to the method by Vasdev et al. [33]) and therefore reduces operator exposition during the procedure. Molar activity and activity concentration of the final product ranged respectively from 74 to 410 GBq/µmol and from 2.3 to 4.8 GBq/mL at the end of synthesis. For all produced batches, chemical and radiochemical purity were greater than 99% and the latter was maintained >98% over a period of 6 h in saline.

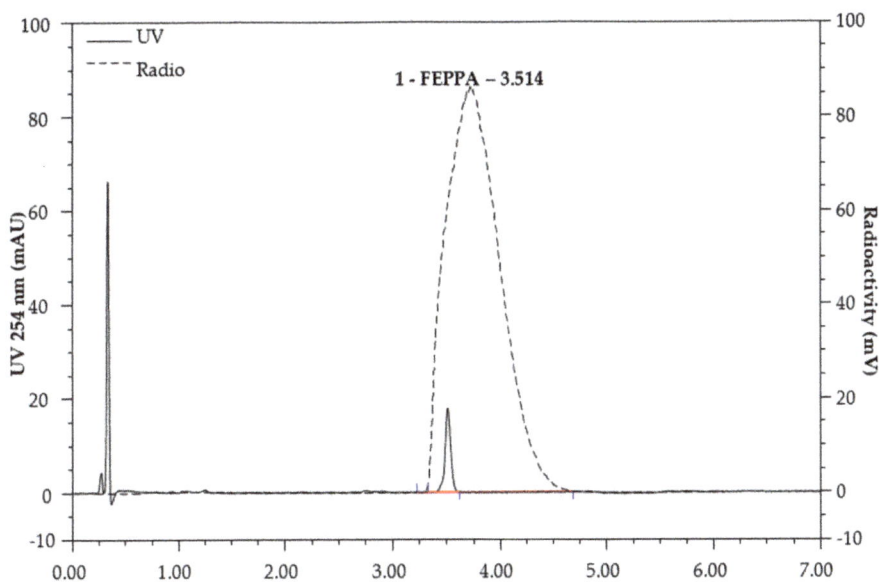

Figure 3. Analytical HPLC chromatogram (radio and UV).

2.2. Biodistribution

Figure 4 illustrates the whole body distribution of [18F]FEPPA over time from radiotracer injection up to 120 min, using microPET imaging coupled with computerized tomography (CT). Visual analysis demonstrated physiologic [18F]FEPPA rapid uptake in heart, lung, spleen, and kidney excretion. In contrast, brain uptake of [18F]FEPPA is low. At 15 min mean SUV were: 3.7 for lung, 5.1 for heart, 3.8 for kidney and 0.5 for brain. As expected, [18F]FEPPA distribution follow the basal TSPO distribution in mice which is found in most tissues and specially in heart and kidney [39] like [18F]DPA-714 [40]. The brain TSPO distribution explains the lack of reference tissue and the need for kinetic modelling for quantification.

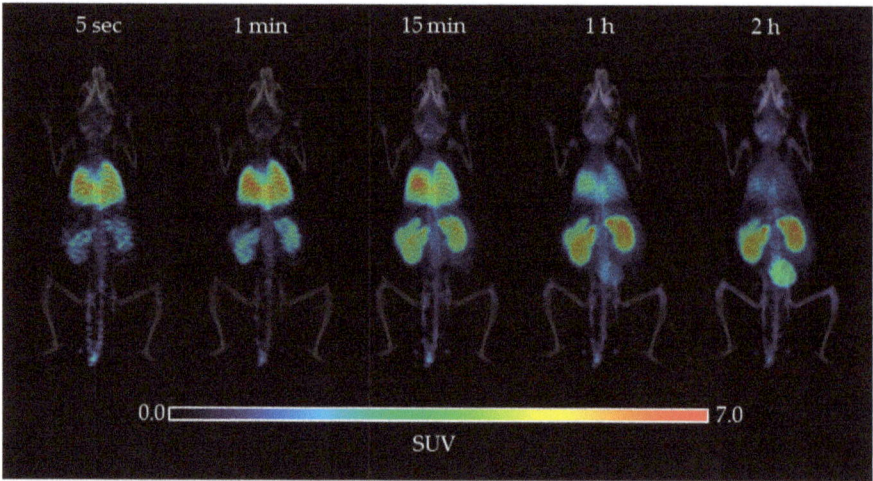

Figure 4. [^{18}F]FEPPA 3D PET/CT imaging from tracer injection to 120 min after in a C57BL/6 mouse. Image at 5 s, 1 min, 15 min, 1 h ,and 2 h post-injection.

2.3. TSPO Expression in Brain by Western Blot

TSPO expression is 2.16-fold greater in whole brain of LPS injected mice (mean TSPO/GAPDH = 0.93) as compared to control mice (mean TSPO/GAPDH = 0.43) (p = 0.004) (Figure 5). TSPO band has been compared to molecular weight scale and confirmed at 18 kDa. GAPDH has been used as loading reference.

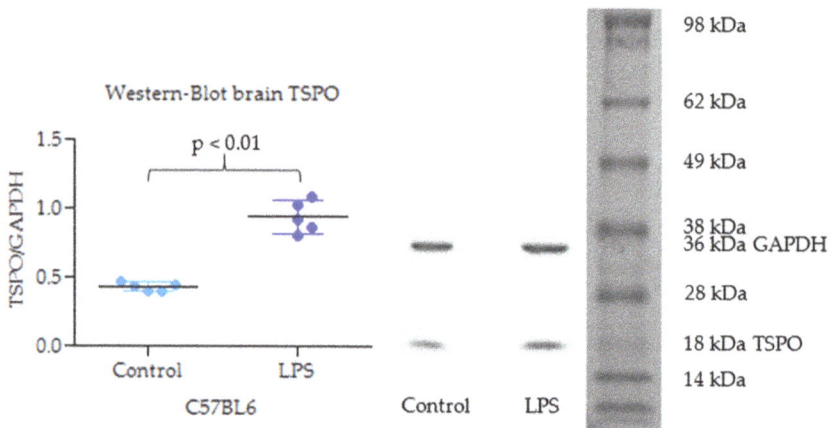

Figure 5. TSPO brain expression in C57Bl6 mice for control and lipopolysaccharides (LPS) conditions.

2.4. Metabolism Study

Metabolism study has shown a rapid decrease of the parent fraction in plasma (60% at 30 min) (Figure 6), with roughly 15% of the metabolite observed 15 min after injection, and 65% of the metabolite 120 min after radiotracer injection. In the brain, a lower metabolism was observed (96% and 77% of the parent fraction, respectively, at 15 and 120 min after [^{18}F]FEPPA injection). Only one metabolite was detected other than [^{18}F]FEPPA in both plasma and brain.

The percentage of parent fraction determined in brain was used to obtain the time activity curve of unmetabolized [18F]FEPPA in the brain. The percentage of plasma metabolite is part of the arterial input function for pharmacokinetics modelization.

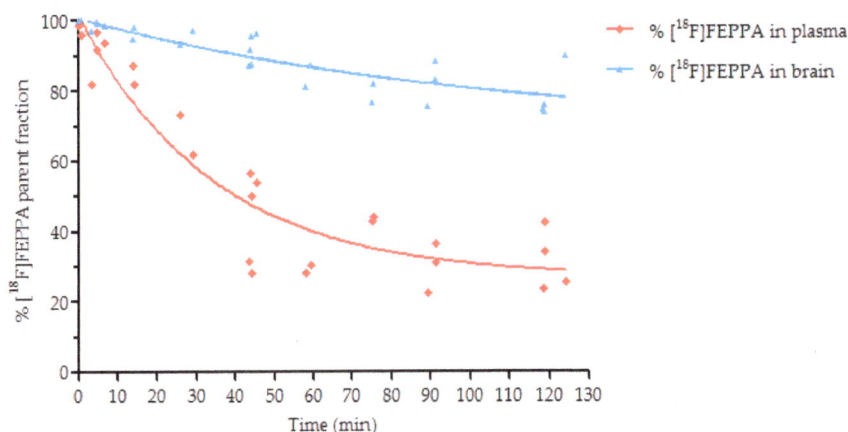

Figure 6. Percentage of parent fraction (unmetabolized) of [18F]FEPPA in plasma and brain for control condition.

2.5. [18F]FEPPA Brain Time Activity Curves

The regional [18F]FEPPA time activity curves corrected for radiometabolites for whole brain, cerebellum, cortex and hippocampus (Figure 7) showed a significantly higher cerebral distribution of [18F]FEPPA in LPS mice as compared to controls (area under curve (AUC) in different regions, LPS vs. control, t-test two-sided, $p = 0.0063$, $p = 0.0179$, $p = 0.0126$ and $p = 0.0266$, respectively). This significant increase of [18F]FEPPA AUC is consistent with the overexpression of its target in brain of LPS-injected mice despite the known and observed variability of the biological effect of LPS. The lack of reference tissue requires a pharmacokinetic modelling to compare [18F]FEPPA brain distribution.

In accordance with data of the literature [41–43], we used a two tissue compartment plus vascular trapping model in this study. In this model, K_1 and k_2 represent passage (through the blood-brain barrier) from the blood to a nonspecific compartment (free + non specific fixation), k_3 and k_4 are enter and exit from the non specific compartment to the specific one. K_b represents slow binding of the radiotracer to the endothelium of vasculature. The most relevant parameter to assert a modification in the distribution is the distribution volume V_T. In our study, the pharmacokinetics parameters in the whole mice brain (Table 1) showed a significant increase for K_1, V_T and AUC for LPS mice as compared to control mice (t test ± Welch's correction, one-sided, mice with a %SE > 50% for at least one parameter were excluded). An increase in k_3/k_4 was also observed, although not significant.

Table 1. Pharmacokinetics parameters for [18F]FEPPA in the whole mice brain.

Group	K_1 (mL·cm^{-3}·min^{-1})	k_2 (min^{-1})	k_3/k_4	K_b (min^{-1})	V_T (mL·cm^{-3})	AUC $_{0\ to\ 120\ min}$ (%ID/g s^{-1})
Control ($n = 6$)	0.58 ± 0.15	0.35 ± 0.06	0.34 ± 0.13	0.53 ± 0.05	2.25 ± 0.44	11,910 ± 934
LPS ($n = 5$)	0.86 ± 0.18	0.36 ± 0.13	0.61 ± 0.58	0.68 ± 0.23	3.77 ± 0.41	15,940 ± 1226
p	0.0112	0.3210	0.1818	0.1092	0.0001 ***	0.0032 **

** $p < 0.001$, *** $p < 0.0001$.

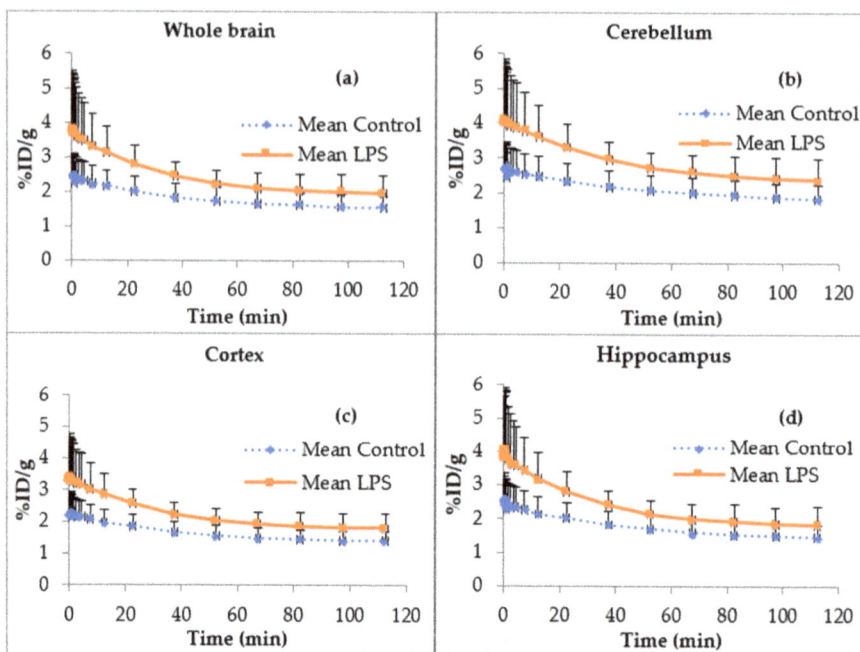

Figure 7. Parent fraction of [^{18}F]FEPPA Time Activity Curve in %ID/g ± SD control vs. LPS. (**a**) Mean in whole brain ($p = 0.0063$); (**b**) Mean in cerebellum ($p = 0.0179$); (**c**) Mean in cortex ($p = 0.0126$); (**d**) Mean in hippocampus ($p = 0.0266$) ($n = 8$ controls and 7 LPS).

3. Discussion

In this study, we developed and optimized an easy-to-perform automated [^{18}F]FEPPA radiosynthesis process suitable for clinical use based on previous work of Vasdev et al. [33]. Our study also documented plasma and brain metabolism of [^{18}F]FEPPA in mice, allowing for demonstration of significant changes in the brain distribution of unmetabolized [^{18}F]FEPPA in a murine model of peripheral inflammation using ip LPS injection.

In this radiosynthesis, a less toxic solvent was used in the mobile phase for semi-preparative HPLC than previous studies. Indeed, according to ICH (Q3C) and the FDA, ethanol is a class 3 solvent, whereas methanol, used previously [31,33], is class 2 and should be limited in pharmaceuticals due to its inherent toxicity. Use of methanol requires an additional reformulation step such as SPE cartridge capture [33] or drying and recovery with ethanol/water [31]. Besides, it still leaves residual traces of solvent in the final product and so requires residual solvent quantification, which is time-consuming and costly. Our new radiosynthesis conditions achieved a slightly better non-decay corrected radiochemical yield of 39% ± 3% in 34 min, as compared to 30% ± 6% in 36 min reached by Vasdev et al. [33], from end of azeotropic drying to the end of formulation. Berroterán-Infante et al. [34] decribed an optimized radiosynthesis process with a high nondecay corrected radiochemical yield of 38% ± 3% (EOB) but using acetonitrile for purification and followed by formulation with cartridge. Another study has reported a radiosynthesis process using also ethanol and water in the purification and formulation processes with a lower radiochemical yield of 13 ± 8% in 59 min (EOB) [35]. Our optimized radiosynthesis process use ethanol, water and phosphoric acid for purification allowing us to bypass the formulation step. This method achieved a nondecay corrected radiochemical yield of 34% ± 2% (EOB) in 55 min.

[^{18}F]FEPPA was produced with high radiochemical, chemical purity, and molar activity comparable to those reported in the literature [33,34] with class 3 solvent (ethanol/water) for purification.

To evaluate this radiotracer, a dynamic μPET/CT imaging study was carried out in vivo with a mouse model of inflammation induced by ip injection of 5 mg/kg of LPS as compared to saline-injected control mice. The intrastriatal injection of LPS or AMPA in mice or rats is a known model of brain inflammation. For example, a study using three differents PET TSPO radiotracers ([^{18}F]DPA-714, [^{11}C]PK11195, and [^{18}F]GE-180) in rats that have been stereotactically injected with 1 μg LPS in the right striatum have shown that [^{18}F]GE-180 was able to shows a higher core/contralateral ratio and BP_{ND} when compared to (R)-[^{11}C]PK11195, while [^{18}F]DPA-714 did not [44]. However, the act of introducing a needle into the brain itself may cause local inflammation that could interfere with the experiment. We therefore chose a less invasive and easier model consisting of a single ip injection of LPS (5 mg/kg). The *Salmonella enterica* serovar *typhimurium* LPS inflammation model is less documented than *E. coli* LPS model, and some studies suggest differences in the inflammation pathways induced. Indeed, *Salmonella enterica* serovar *typhimurium* LPS induces a greater increase in inflammatory cytokines through toll-like receptor TLR2 [45,46]. Western blot analysis showed a significant 2.2-fold increase in brain TSPO expression in the whole brain of LPS-injected mice as compared to control mice 24 h after LPS injection (Figure 6) which allowed to validate a positive control for the TSPO-PET brain imaging study.

Alongside dynamic PET/CT imaging with [^{18}F]FEPPA, the metabolism study showed the relatively important and rapid metabolism of the radiotracer in plasma, in accordance with previous studies [33]. Indeed, our metabolism study documented one main metabolite and a rapid decrease of the parent fraction in plasma (Figure 6). In contrast, our study showed that [^{18}F]FEPPA brain metabolism was less important than in the plasma, which is in accordance with one previous study in rats [31]. This metabolism in the plasma may be a disadvantage in terms of [^{18}F]FEPPA use since it makes its quantification more difficult. However, most of the TSPO radioligands available are also known to shown significant metabolism. For example, [^{18}F]DPA-714 parent fraction represents 11–15% of total radioactivity in the blood 120 min after injection in rats and baboons [47].

[^{18}F]FEPPA biodistribution in C57BL6 mice confirmed previously published results in rats and athymic mice: rapid uptake by the lungs, heart, and spleen and excretion by the kidneys [31,33]. The whole body distribution of [^{18}F]FEPPA (Figure 4) show the relative low uptake in brain as compared to peripheral tissue like heart or kidney as it was expected for a TSPO radiotracer [39]. The low [^{18}F]FEPPA brain uptake in our control mice (%ID/g in brain within 2–4%) is rather favorable for signal/background ratio i.e., the detection of cerebral processes (Figure 4). Furthermore, our study suggests that the time activity curve (TAC) of [^{18}F]FEPPA corrected from brain metabolism is significantly increased in the LPS-injected mouse group as compared to control mice (Figure 7). This result could be explained on one hand by the significant increase of the specific target TSPO protein expression in the brain of LPS-injected C57Bl6 mice as compared to control group, as confirmed by western blot analysis, and on the other hand by an opening of the blood-brain barrier (BBB) [38]. BBB disruption was shown in some brain regions sush as frontal cortex, thalamus, or cerebellum in male CD1 mice (6–10 weeks old) 24 h after one single ip injection of LPS 3 mg/kg [38].

This significant increase in [^{18}F]FEPPA distribution in the brain of LPS-injected mice as compared to control mice was observed consistently throughout the main brain structures/regions such as whole brain, cortex, cerebellum, and hippocampus (Figure 7).

Pharmacokinetic parameters for [^{18}F]FEPPA distribution in the whole brain of mice showed a significant increase in the total distribution volume V_T in LPS-injected mice as compared to control mice. There were no other statistically significant changes in constant rates between the plasma, brain and vascular compartments (Table 1), except for the K_1, which was significantly increased in the LPS group. K_1 is the uptake rate constant for compartment 1 which represents free and non-specifically bound tracer. This increase in K_1 suggests improved [^{18}F]FEPPA passage across the blood-brain barrier in LPS-injected mice which could be explained by a BBB disruption. K_b (the rate of binding of the radiotracer to the endothelium of the vasculature [41,42]) was not modified between LPS and control mice. This therefore suggests that the increased [^{18}F]FEPPA V_T observed in the brain is more likely due to an increase in the TSPO expression in the brain parenchyma of LPS-treated mice. The k_3/k_4 ratio, also named BP$_{ND}$, is the ratio at equilibrium of specifically-bound radioligand to non-displaceable radioligand in the tissue. The increased k_3/k_4 ratio in the LPS-treated mice suggests higher specific binding to TSPO. The fact that this increase in the k_3/k_4 ratio is not statistically significant is possibly due to its high variability. This high k_3/k_4 ratio variability we have observed is consistent with the relatively high standard deviation observed in [^{18}F]FEPPA TAC especially in the brain of LPS injected mice group (Figure 7). Indeed Rusjan et al. have reported that k_3/k_4 ratio is highly variable, and the most reliable parameter to document an increase in [^{18}F]FEPPA brain distribution is the increase in V_T [48]. Furthermore, the same team has reported, by Monte-Carlo simulation, that an increase or decrease of K_1 does not significantly modify V_T [48].

The hypothesis in a drug/compound brain distribution increase linked to a higher cerebral blood flow could be true only for molecules that are highly extracted by the brain. Indeed, according to the Renkin and Crone equation [49], one of the main pharmacokinetic principles of the distribution of molecules in tissues is that it is governed by blood flow and the ratio of blood concentration to tissue concentration. Depending on the value of this ratio, molecules behave in two ways: (i) for molecules that are poorly extracted by brain tissue (<30%), variations in blood flow have little or no influence on their brain distribution; these molecules are called blood-flow independent or permeability dependent; (ii) for molecules that are strongly extracted by brain tissue (>30%), such as glucose or oxygen, cerebral blood flow strongly influences their brain distribution. [^{18}F]FEPPA is a molecule with low brain extraction (2 to 4% ID/g), and the cerebral passage of this type of molecule does not depend on the cerebral blood flow. Similarly, the brain distribution of [^{11}C]verapamil, which also has low brain uptake between 1.5 and 3% of the injected dose, did not depend on cerebral blood flow [50]. Besides, in a clinical study in healthy subjects, Rusjan et al. have shown that V_T is the most robust parameter for quantifying [^{18}F]FEPPA uptake and that it is not modified by an increase or decrease in cerebral blood flow [48].

The increase in [^{18}F]FEPPA k_3/k_4 (although not significant) and V_T in the brains of LPS-treated mice is in accordance with an activation of pro-inflammatory microglial state with TSPO overexpression as confirmed by western-blot analysis. This phenotype overexpresses TSPO while reparative and mixed stages do not [51]. [^{18}F]FEPPA is one of the main second generation TSPO radioligands being the most advanced in clinical development to study the neuroinflammation process. It represents a useful tool to follow neuroinflammation in longitudinal study. The [^{18}F]FEPPA radio-synthesis method seems easier than that of [^{18}F]DPA-714, and the metabolism of [^{18}F]FEPPA is less important than that of [^{18}F]DPA-714 [47], which is more favorable for [^{18}F]FEPPA quantification in brain. However the dynamic acquisition and the three patterns of binding affinity in human are hindrance (both for [^{18}F]DPA-714 and [^{18}F]FEPPA) to their clinical use.

To summarize, in this study, we have developed an automated synthesis method for [^{18}F]FEPPA that includes HPLC purification/formulation easily performed on the Allinone® radiosynthetiser with a nondecay corrected radiochemical yield of 34% ± 2% (EOB) in 55 min and purification with only class 3 solvent (ethanol). The [^{18}F]FEPPA radiotracer obtained is more suitable for preclinical and clinical use than those obtained before. The PET imaging study using the TSPO ligand [^{18}F]FEPPA has allowed us to observe a significant increase in [^{18}F]FEPPA V_T in the brain of a murine neuroinflammation model 24 h after a single ip injection of *Salmonella enterica serovar typhimurium* LPS.

4. Materials and Methods

4.1. Radiochemical Synthesis of [^{18}F]FEPPA

4.1.1. General

All reagents and solvents were purchased from commercial suppliers (ABX, Radeberg, Germany or Sigma-Aldrich, St Quentin Fallavier, France) and were used without further purification. Sep-Pak QMA were purchased from ABX. [^{18}F]fluoride ion was produced via the [^{18}O(p,n)^{18}F] nuclear reaction (IBA Cyclone 18/9 cyclotron).

Radioactivity of the final product was measured with a dose calibrator (PET DOSE 5 Ci, COMECER®, Castel Bolognese, Italy).

4.1.2. Radiosynthesis

Radiosynthesis of [^{18}F]FEPPA was performed using an AllInOne® (Trasis, Ans, Belgium) synthesis module and a tosylated precursor for a one-step fluorine nucleophilic aliphatic substitution, based on the radiosynthesis previously described [31] (Figure 8).

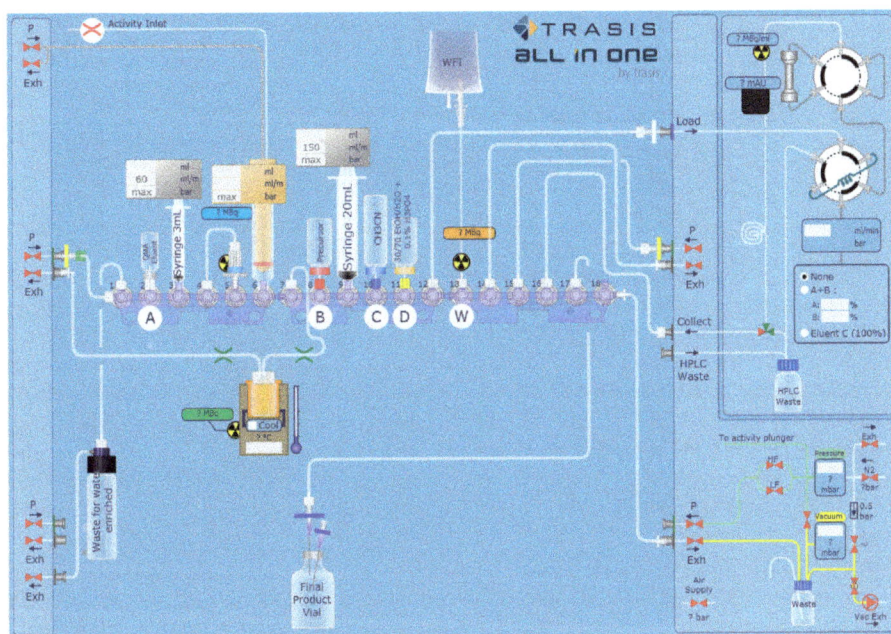

Figure 8. [^{18}F]FEPPA radiosynthesis layout on an AllInOne® module.

The list of reagents used in the automated procedure is presented in Table 2. [^{18}F]FEPPA was synthesized using an in-house reaction sequence described in Table 3.

Table 2. List of reagents.

Position	Reagents	Quantities
5	Pre-conditioned Sep-Pak® Light QMA	1
2 (vial A)	Eluent QMA (K_2CO_3/K_{222} in CH_3CN/H_2O, 80/20, v/v)	1 mL
8 (vial B)	Precursor	5 mg
10 (vial C)	CH_3CN anhydrous	15 mL
11 (vial D)	Mobile phase (30/70 $EtOH/H_2O$ + 0.1% phosphoric acid)	6 mL
13 (bag W)	WFI	250 mL

Table 3. Reaction sequence for [^{18}F]FEPPA radiosynthesis.

Fluorination of Precursor

1. [^{18}F]fluoride trapping on a pre-activated QMA cartridge
2. [^{18}F]fluoride desorption by eluent
3. Azeotropic evaporation at 110 °C for 10 min
4. Addition of precursor to the reactor vial
5. [^{18}F]fluorination at 90 °C for 10 min
6. Cooling the reactor vial
7. Addition of HPLC mobile phase to the reactor vial

Purification of [^{18}F]FEPPA

1. Injection on HPLC semi-preparative
2. Collection of [^{18}F]FEPPA in 1.5 mL 8.4% sodium bicarbonate

Formulation of [^{18}F]FEPPA

1. Dilution of the collected fraction with NaCl 0.9%
2. Sterile filtration

The aqueous [^{18}F]fluoride target solution was loaded on a QMA (Pre-conditioned Sep-Pak® Light QMA cartridge, ABX). The concentrated [^{18}F]fluoride was eluted into the reactor using a K_2CO_3 (3 mg) and Kryptofix (K_{222}, 15 mg) mixed solution (1 mL, CH_3CN/H_2O, 80/20, v/v). The solvents were evaporated under reduced pressure at 110 °C for 10 min. To the dry residue containing the K222/potassium [^{18}F]fluoride complex was added the tosylated precursor (2-(2-((*N*-4-phenoxypyridin-3-yl)acetamido)methyl)phenoxy)ethyl 4-methylbenzenesulfonate) (5 mg) in acetonitrile (1 mL), and the mixture was heated and maintained at 90 °C for 10 min. After cooling, the radiolabeling reaction is then stopped by the addition of 3 mL of mobile phase 30/70 v/v ethanol/water + 0.1% of phosphoric acid in the reactor. After passing through a 0.22 μm vent filter, this solution is transferred into the injection loop and followed by rinsing with 3 mL of mobile phase. The purification by semi-preparative HPLC (column Phenomenex Kinetex® C18 10 μm 250 × 10 mm; Le Pecq, France) is carried out in isocratic with a solution (30/70 v/v) of ethanol/water + 0.1% of phosphoric acid at 4 mL/min. The fraction containing [^{18}F]FEPPA, associated to a well-defined radioactive peak, was collected at 13–14 min. The finished product is then received in a flask containing 1.5 mL of sodium bicarbonate 8.4%. The radiotracer solution was finally passed through a 0.22 μm Millipore filter into a sterile vial for in vivo experiments. Production was diluted in saline to reach a volume activity of 1.8–2 GBq/mL.

4.1.3. Quality Control

The radiochemical and chemical purity, stability, and molar activity measurements were performed by analytical HPLC. The molar activity of the radiotracer was assessed by measurement of the injected radioactivity, and the FEPPA concentration in the sample was derived from the UV detection.

The identity of the labeled compound [^{18}F]FEPPA was confirmed by co-injection with a non-radioactive standard of FEPPA. The FEPPA concentration in the radioactive sample was obtained using the UV-peak area ratio between the radioactive product and the standard solution.

Analytical HPLCs were performed on a Dionex Ultimate 3000 (Thermofisher Scientific, Waltham, MA, USA) with a multi-wavelength UV Diode Array Detector (DAD) in series with a Bioscan gamma detector (Canberra, St Quentin Yvelines, France) using a Kinetex® C18 column (50 × 2.1 mm, 2.6 μm) H_2O/CH_3CN 75/25 at 0.6 mL/min. A delay time of 20 s was observed between the two detectors.

4.2. Animal Models

All animal experiments were performed in accordance with European guidelines for care of laboratory animals (2010/63/EU) and were approved by the Animal Ethics Committee of Paris Nord.

In vivo quality control of [^{18}F]FEPPA (biodistribution studies) was performed in 12-week-old C57Bl/6JRj mice (Janvier, France) (*n* = 16). Eight mice were injected with 500 μL 0.9% NaCl by ip, and 8 with LPS from *salmonella enterica* serotype typhimurium (Sigma-Aldrich, Saint Louis, MO, USA) 5 mg/kg in 500 μL 0.9% NaCl. Micro PET/CT imaging was held 24 h after the LPS injection. Brains were removed immediately after imaging and preserved for western blot. Metabolism study was performed in 12-week-old C57Bl/6JRj mice (Janvier, France) (*n* = 16).

4.3. Western Blot Analysis of TSPO

Adult mice brains were dissected and reduced in powder at −80 °C, immediately dissolved in PBS with 2% SDS, and 1× EDTA-free Complete Protease Inhibitor (Roche, Basel, Switzerland). Lysates were sonicated twice at 10 Hz (Vibra cell VCX130) and centrifuged for 30 min at 16,000 rcf at 4 °C. Supernatants were boiled in 5× Laemmli loading buffer. Protein content was measured using the BCA protein assay reagent (Thermo scientific, Waltham, MA, USA). Equal amounts of proteins (20 μg) were separated by denaturing electrophoresis in NuPAGE 4–12% Bis-Tris acetate gradient gel (Invitrogen, Carlsbad, CA, USA) and electrotransfered to nitrocellulose membranes. Membranes were analyzed using the following primary antibodies: rabbit anti-TSPO (Abcam, Cambridge, UK) (1:10,000); horseradish-peroxidase-conjugated (HRP) anti-GAPDH (1:10,000); Secondary antibody used was HRP-conjugated anti-rabbit antibodies (Amersham) (dilution 1:2000). HRP activity was visualized by enhanced chemiluminescence (ECL) using Western Lightning plus enhanced chemoluminescence system (Perkin Elmer, Waltham, MA, USA). Chemoluminescence imaging was performed on a LAS4000 (Fujifilm, Tokyo, Japan). GAPDH expression was used as a loading reference.

4.4. Metabolism Study

Adult mice were injected with 68.6 MBq ± 4.4 MBq of [^{18}F]FEPPA in the tail vein. Arterial blood sampling (intra cardiac) was done at different times (5, 15, 30, 45, 60, 75, 90, and 120 min) then followed by exsanguination and brain removal under anesthesia (ketamine/xylazine lethal dose). Samples were homogenized with CH_3CN and then centrifuged. Pellet and supernatant were counted separately on a gamma counter (WIZARD2®, Perkin-Elmer, Villebon-sur-Yvette, France).

Radiometabolites in supernatant were performed in HPLC on a Dionex Ultimate 3000 (Thermofisher Scientific, Courtaboeuf, France) with a multi-wavelength UV Diode Array Detector (DAD) in series with a Bioscan gamma detector (Canberra, St Quentin Yvelines, France) with a Kinetex® column (C18 250 × 4.6 mm, 5 μm), gradient H_2O/CH_3CN 60/40 to 50/50 in 5 min at 1 mL/min.

4.5. In Vivo PET/CT Imaging

PET/CT imaging was performed using Inveon micro PET/CT scanner (Siemens Medical Solutions, Erlangen, Germany) designed for small laboratory animals. Mice were anesthetized (isoflurane/oxygen, 2.5% for induction at 0.8–1.5 L/min, and 1–1.5% at 0.4–0.8 L/min thereafter) during injection of [^{18}F]FEPPA (9.9 ± 1.5 MBq) in a volume of 0.15 mL (0.22 ± 0.19 nmol of FEPPA) via the tail vein, and during PET/CT acquisitions.

For [^{18}F]FEPPA biodistribution studies, dynamic mod-list PET acquisitions were performed from time of radiotracer injection until 120 min after injection (*n* = 16). The dynamic list-mode contains 25 frames (5 s × 12; 60 s × 4; 5 min × 2; 15 min × 7).

The spatial resolution of Inveon PET device was 1.4 mm full-width at half-maximum at the center of the field of view. Images were reconstructed using a 3D ordered subset expectation maximization method including corrections for scanner dead time, scatter radiations, and randoms.

4.6. Data Analysis and Modeling

PET/CT images were visually assessed. Then quantitative analysis of PET/CT images was performed by PMOD version 3.806 image analysis software (PMOD Technologies, Zurich, Switzerland). For comparisons, all values of radioactivity concentrations were normalized by the injected dose and expressed as percentage of the injected dose per g of tissue (% ID/g).

PET images were automatically rigid matched with corresponding CT and then cropped to keep only the brain images. CT were automatically rigid matched with a T2 MRI template (M. Mirrione, included in PMOD), and then the transformation applied to the PET image was cropped. This method allowed us to use the atlas of the brain corresponding to the T2 MRI template.

The arterial input function was computed from plasma sampling and corrected for metabolism of the parent ligand. A two compartment + vascular trapping 4 rate-constant kinetic model was used to characterize [^{18}F]FEPPA pharmacokinetics as previously described [41,42]. Model parameters were estimated for influx constant K_1 (mL·cm^{-3}·min^{-1}), efflux (k_2) (min^{-1}) rate of radioligand diffusion between plasma and brain compartment. Exchange between compartments k_3 and k_4 (min^{-1}) were also estimated. Finally, the K_b (min^{-1}) parameter that describes the rate of binding to the TSPO in the endothelium and the macro-parameters V_T (mL·cm^{-3}) was also used to estimate the total distribution volume for the whole brain with goodness of fit evaluated by inspection [43].

4.7. Statistical Analysis

Data are presented as mean ± SD. Statistical analysis was performed using Prism 5, version 5.0.1 (La Jolla, CA, USA). A significance value of $p < 0.05$ was used.

Author Contributions: N.R.-P., L.S.-M., and B.H., conceived and designed the experiments; N.V., C.S., F.H., T.G., and B.H. performed the experiments; N.V., C.S., F.H., and B.H. analyzed the data; N.V., S.C., N.R.-P., X.D., and B.H. wrote the paper.

Funding: This research received no external funding.

Acknowledgments: We thanks INSERM U1144, AP-HP, Paris Descartes, and Paris Diderot universities for their funding supports in these experiments and INSERM U1144 and Paris Descartes for covering the costs to publish in open access.

Conflicts of Interest: The authors declare no conflict of interest.

References

1. Owen, D.R.; Narayan, N.; Wells, L.; Healy, L.; Smyth, E.; Rabiner, E.A.; Galloway, D.; Williams, J.B.; Lehr, J.; Mandhair, H.; et al. Pro-inflammatory activation of primary microglia and macrophages increases 18 kDa translocator protein expression in rodents but not humans. *J. Cereb. Blood Flow Metab. Off. J. Int. Soc. Cereb. Blood Flow Metab.* **2017**, *37*, 2679–2690. [CrossRef] [PubMed]
2. Li, F.; Liu, J.; Liu, N.; Kuhn, L.A.; Garavito, R.M.; Ferguson-Miller, S.M. Translocator protein 18 kDa (TSPO): An old protein with new functions? *Biochemistry* **2016**, *55*, 2821–2831. [CrossRef] [PubMed]
3. Ransohoff, R.M.; Perry, V.H. Microglial physiology: Unique stimuli, specialized responses. *Annu. Rev. Immunol.* **2009**, *27*, 119–145. [CrossRef] [PubMed]
4. Maezawa, I.; Zimin, P.I.; Wulff, H.; Jin, L.-W. Amyloid-beta protein oligomer at low nanomolar concentrations activates microglia and induces microglial neurotoxicity. *J. Biol. Chem.* **2011**, *286*, 3693–3706. [CrossRef] [PubMed]

5. Morales, I.; Jiménez, J.M.; Mancilla, M.; Maccioni, R.B. Tau oligomers and fibrils induce activation of microglial cells. *J. Alzheimers Dis.* **2013**, *37*, 849–856. [CrossRef] [PubMed]

6. Prinz, M.; Priller, J.; Sisodia, S.S.; Ransohoff, R.M. Heterogeneity of CNS myeloid cells and their roles in neurodegeneration. *Nat. Neurosci.* **2011**, *14*, 1227–1235. [CrossRef] [PubMed]

7. Molteni, M.; Rossetti, C. Neurodegenerative diseases: The immunological perspective. *J. Neuroimmunol.* **2017**, *313*, 109–115. [CrossRef] [PubMed]

8. Harms, A.S.; Delic, V.; Thome, A.D.; Bryant, N.; Liu, Z.; Chandra, S.; Jurkuvenaite, A.; West, A.B. α-Synuclein fibrils recruit peripheral immune cells in the rat brain prior to neurodegeneration. *Acta Neuropathol. Commun.* **2017**, *5*, 85. [CrossRef] [PubMed]

9. Donat, C.K.; Scott, G.; Gentleman, S.M.; Sastre, M. Microglial Activation in Traumatic Brain Injury. *Front. Aging Neurosci.* **2017**, *9*, 208. [CrossRef] [PubMed]

10. Hoogland, I.C.M.; Houbolt, C.; van Westerloo, D.J.; van Gool, W.A.; van de Beek, D. Systemic inflammation and microglial activation: Systematic review of animal experiments. *J. Neuroinflamm.* **2015**, *12*, 114. [CrossRef] [PubMed]

11. Benarroch, E.E. Microglia: Multiple roles in surveillance, circuit shaping, and response to injury. *Neurology* **2013**, *81*, 1079–1088. [CrossRef] [PubMed]

12. Durafourt, B.A.; Moore, C.S.; Zammit, D.A.; Johnson, T.A.; Zaguia, F.; Guiot, M.-C.; Bar-Or, A.; Antel, J.P. Comparison of polarization properties of human adult microglia and blood-derived macrophages. *Glia* **2012**, *60*, 717–727. [CrossRef] [PubMed]

13. Boulay, A.-C.; Mazeraud, A.; Cisternino, S.; Saubaméa, B.; Mailly, P.; Jourdren, L.; Blugeon, C.; Mignon, V.; Smirnova, M.; Cavallo, A.; et al. Immune quiescence of the brain is set by astroglial connexin 43. *J. Neurosci. Off. J. Soc. Neurosci.* **2015**, *35*, 4427–4439. [CrossRef] [PubMed]

14. Boulay, A.-C.; Cisternino, S.; Cohen-Salmon, M. Immunoregulation at the gliovascular unit in the healthy brain: A focus on Connexin 43. *Brain. Behav. Immun.* **2015**, *56*, 1–9. [CrossRef] [PubMed]

15. Heppner, F.L.; Ransohoff, R.M.; Becher, B. Immune attack: The role of inflammation in Alzheimer disease. *Nat. Rev. Neurosci.* **2015**, *16*, 358–372. [CrossRef] [PubMed]

16. Webster, S.J.; Van Eldik, L.J.; Watterson, D.M.; Bachstetter, A.D. Closed head injury in an age-related Alzheimer mouse model leads to an altered neuroinflammatory response and persistent cognitive impairment. *J. Neurosci. Off. J. Soc. Neurosci.* **2015**, *35*, 6554–6569. [CrossRef] [PubMed]

17. Macchi, B.; Di Paola, R.; Marino-Merlo, F.; Felice, M.R.; Cuzzocrea, S.; Mastino, A. Inflammatory and cell death pathways in brain and peripheral blood in Parkinson's disease. *CNS Neurol. Disord. Drug Targets* **2015**, *14*, 313–324. [CrossRef] [PubMed]

18. Van der Perren, A.; Macchi, F.; Toelen, J.; Carlon, M.S.; Maris, M.; de Loor, H.; Kuypers, D.R.J.; Gijsbers, R.; Van den Haute, C.; Debyser, Z.; et al. FK506 reduces neuroinflammation and dopaminergic neurodegeneration in an α-synuclein-based rat model for Parkinson's disease. *Neurobiol. Aging* **2015**, *36*, 1559–1568. [CrossRef] [PubMed]

19. Venneti, S.; Lopresti, B.J.; Wang, G.; Hamilton, R.L.; Mathis, C.A.; Klunk, W.E.; Apte, U.M.; Wiley, C.A. PK11195 labels activated microglia in Alzheimer's disease and in vivo in a mouse model using PET. *Neurobiol. Aging* **2009**, *30*, 1217–1226. [CrossRef] [PubMed]

20. Kreisl, W.C.; Lyoo, C.H.; McGwier, M.; Snow, J.; Jenko, K.J.; Kimura, N.; Corona, W.; Morse, C.L.; Zoghbi, S.S.; Pike, V.W.; et al. Biomarkers Consortium PET Radioligand Project Team In vivo radioligand binding to translocator protein correlates with severity of Alzheimer's disease. *Brain J. Neurol.* **2013**, *136*, 2228–2238. [CrossRef] [PubMed]

21. Ghadery, C.; Koshimori, Y.; Coakeley, S.; Harris, M.; Rusjan, P.; Kim, J.; Houle, S.; Strafella, A.P. Microglial activation in Parkinson's disease using [^{18}F]-FEPPA. *J. Neuroinflamm.* **2017**, *14*, 8. [CrossRef] [PubMed]

22. Gerhard, A.; Pavese, N.; Hotton, G.; Turkheimer, F.; Es, M.; Hammers, A.; Eggert, K.; Oertel, W.; Banati, R.B.; Brooks, D.J. In vivo imaging of microglial activation with [^{11}C](R)-PK11195 PET in idiopathic Parkinson's disease. *Neurobiol. Dis.* **2006**, *21*, 404–412. [CrossRef] [PubMed]

23. Debruyne, J.C.; Versijpt, J.; Van Laere, K.J.; De Vos, F.; Keppens, J.; Strijckmans, K.; Achten, E.; Slegers, G.; Dierckx, R.A.; Korf, J.; et al. PET visualization of microglia in multiple sclerosis patients using [^{11}C]PK11195. *Eur. J. Neurol.* **2003**, *10*, 257–264. [CrossRef] [PubMed]

24. Versijpt, J.; Debruyne, J.C.; Van Laere, K.J.; De Vos, F.; Keppens, J.; Strijckmans, K.; Achten, E.; Slegers, G.; Dierckx, R.A.; Korf, J.; et al. Microglial imaging with positron emission tomography and atrophy measurements with magnetic resonance imaging in multiple sclerosis: A correlative study. *Mult. Scler. Houndmills Basingstoke Engl.* **2005**, *11*, 127–134. [CrossRef] [PubMed]

25. Kenk, M.; Selvanathan, T.; Rao, N.; Suridjan, I.; Rusjan, P.; Remington, G.; Meyer, J.H.; Wilson, A.A.; Houle, S.; Mizrahi, R. Imaging Neuroinflammation in Gray and White Matter in Schizophrenia: An In-Vivo PET Study With [^{18}F]-FEPPA. *Schizophr. Bull.* **2015**, *41*, 85–93. [CrossRef] [PubMed]

26. Setiawan, E.; Wilson, A.A.; Mizrahi, R.; Rusjan, P.M.; Miler, L.; Rajkowska, G.; Suridjan, I.; Kennedy, J.L.; Rekkas, P.V.; Houle, S.; et al. Role of Translocator Protein Density, a Marker of Neuroinflammation, in the Brain During Major Depressive Episodes. *JAMA Psychiatry* **2015**, *72*, 268–275. [CrossRef] [PubMed]

27. Hafizi, S.; Da Silva, T.; Gerritsen, C.; Kiang, M.; Bagby, R.M.; Prce, I.; Wilson, A.A.; Houle, S.; Rusjan, P.M.; Mizrahi, R. Imaging Microglial Activation in Individuals at Clinical High Risk for Psychosis: An In Vivo PET Study with [^{18}F]FEPPA. *Neuropsychopharmacol. Off. Publ. Am. Coll. Neuropsychopharmacol.* **2017**, *42*, 2474–2481. [CrossRef] [PubMed]

28. Hafizi, S.; Tseng, H.-H.; Rao, N.; Selvanathan, T.; Kenk, M.; Bazinet, R.P.; Suridjan, I.; Wilson, A.A.; Meyer, J.H.; Remington, G.; et al. Imaging Microglial Activation in Untreated First-Episode Psychosis: A PET Study with [^{18}F]FEPPA. *Am. J. Psychiatry* **2017**, *174*, 118–124. [CrossRef] [PubMed]

29. Suridjan, I.; Pollock, B.G.; Verhoeff, N.P.L.G.; Voineskos, A.N.; Chow, T.; Rusjan, P.M.; Lobaugh, N.J.; Houle, S.; Mulsant, B.H.; Mizrahi, R. In-vivo imaging of grey and white matter neuroinflammation in Alzheimer's disease: A positron emission tomography study with a novel radioligand, [^{18}F]-FEPPA. *Mol. Psychiatry* **2015**, *20*, 1579–1587. [CrossRef] [PubMed]

30. Zanotti-Fregonara, P.; Zhang, Y.; Jenko, K.J.; Gladding, R.L.; Zoghbi, S.S.; Fujita, M.; Sbardella, G.; Castellano, S.; Taliani, S.; Martini, C.; et al. Synthesis and evaluation of translocator 18 kDa protein (TSPO) positron emission tomography (PET) radioligands with low binding sensitivity to human single nucleotide polymorphism rs6971. *ACS Chem. Neurosci.* **2014**, *5*, 963–971. [CrossRef] [PubMed]

31. Wilson, A.A.; Garcia, A.; Parkes, J.; McCormick, P.; Stephenson, K.A.; Houle, S.; Vasdev, N. Radiosynthesis and initial evaluation of [^{18}F]-FEPPA for PET imaging of peripheral benzodiazepine receptors. *Nucl. Med. Biol.* **2008**, *35*, 305–314. [CrossRef] [PubMed]

32. Mizrahi, R.; Rusjan, P.M.; Kennedy, J.; Pollock, B.; Mulsant, B.; Suridjan, I.; De Luca, V.; Wilson, A.A.; Houle, S. Translocator protein (18 kDa) polymorphism (rs6971) explains in-vivo brain binding affinity of the PET radioligand [^{18}F]-FEPPA. *J. Cereb. Blood Flow Metab. Off. J. Int. Soc. Cereb. Blood Flow Metab.* **2012**, *32*, 968–972. [CrossRef] [PubMed]

33. Vasdev, N.; Green, D.E.; Vines, D.C.; McLarty, K.; McCormick, P.N.; Moran, M.D.; Houle, S.; Wilson, A.A.; Reilly, R.M. Positron-Emission Tomography Imaging of the TSPO with [^{18}F]FEPPA in a Preclinical Breast Cancer Model †. *Cancer Biother. Radiopharm.* **2013**, *28*, 254–259. [CrossRef] [PubMed]

34. Berroterán-Infante, N.; Balber, T.; Fürlinger, P.; Bergmann, M.; Lanzenberger, R.; Hacker, M.; Mitterhauser, M.; Wadsak, W. [^{18}F]FEPPA: Improved Automated Radiosynthesis, Binding Affinity, and Preliminary in Vitro Evaluation in Colorectal Cancer. *ACS Med. Chem. Lett.* **2018**, *9*, 177–181. [CrossRef] [PubMed]

35. Huang, Y.-Y.; Huang, W.-S.; Wu, H.-M.; Kuo, Y.-Y.; Chang, Y.-N.; Lin, P.-Y.; Wu, C.-H.; Yen, R.-F.; Shiue, C.-Y. Automated Production of [^{18}F]FEPPA as a Neuroinflammation Imaging Agent. *J. Nucl. Med.* **2016**, *57*, 1033.

36. Catorce, M.N.; Gevorkian, G. LPS-induced Murine Neuroinflammation Model: Main Features and Suitability for Pre-clinical Assessment of Nutraceuticals. *Curr. Neuropharmacol.* **2016**, *14*, 155–164. [CrossRef] [PubMed]

37. Netea, M.G.; Kullberg, B.J.; Joosten, L.A.; Sprong, T.; Verschueren, I.; Boerman, O.C.; Amiot, F.; van den Berg, W.B.; Van der Meer, J.W. Lethal Escherichia coli and Salmonella typhimurium endotoxemia is mediated through different pathways. *Eur. J. Immunol.* **2001**, *31*, 2529–2538. [CrossRef]

38. Banks, W.A.; Gray, A.M.; Erickson, M.A.; Salameh, T.S.; Damodarasamy, M.; Sheibani, N.; Meabon, J.S.; Wing, E.E.; Morofuji, Y.; Cook, D.G.; et al. Lipopolysaccharide-induced blood-brain barrier disruption: Roles of cyclooxygenase, oxidative stress, neuroinflammation, and elements of the neurovascular unit. *J. Neuroinflamm.* **2015**, *12*, 223. [CrossRef] [PubMed]

39. Batarseh, A.; Papadopoulos, V. Regulation of translocator protein 18 kDa (TSPO) expression in health and disease states. *Mol. Cell. Endocrinol.* **2010**, *327*, 1–12. [CrossRef] [PubMed]

40. Arlicot, N.; Vercouillie, J.; Ribeiro, M.-J.; Tauber, C.; Venel, Y.; Baulieu, J.-L.; Maia, S.; Corcia, P.; Stabin, M.G.; Reynolds, A.; et al. Initial evaluation in healthy humans of [^{18}F]DPA-714, a potential PET biomarker for neuroinflammation. *Nucl. Med. Biol.* **2012**, 39, 570–578. [CrossRef] [PubMed]

41. Veronese, M.; Reis Marques, T.; Bloomfield, P.S.; Rizzo, G.; Singh, N.; Jones, D.; Agushi, E.; Mosses, D.; Bertoldo, A.; Howes, O.; et al. Kinetic modelling of [^{11}C]PBR28 for 18 kDa translocator protein PET data: A validation study of vascular modelling in the brain using XBD173 and tissue analysis. *J. Cereb. Blood Flow Metab. Off. J. Int. Soc. Cereb. Blood Flow Metab.* **2017**. [CrossRef] [PubMed]

42. Rizzo, G.; Veronese, M.; Tonietto, M.; Zanotti-Fregonara, P.; Turkheimer, F.E.; Bertoldo, A. Kinetic modeling without accounting for the vascular component impairs the quantification of [^{11}C]PBR28 brain PET data. *J. Cereb. Blood Flow Metab.* **2014**, 34, 1060–1069. [CrossRef] [PubMed]

43. Wimberley, C.; Lavisse, S.; Brulon, V.; Peyronneau, M.-A.; Leroy, C.; Bodini, B.; Remy, P.; Stankoff, B.; Buvat, I.; Bottlaender, M. Impact of endothelial TSPO on the quantification of ^{18}F-DPA-714. *J. Nucl. Med. Off. Publ. Soc. Nucl. Med.* **2017**. [CrossRef]

44. Sridharan, S.; Lepelletier, F.-X.; Trigg, W.; Banister, S.; Reekie, T.; Kassiou, M.; Gerhard, A.; Hinz, R.; Boutin, H. Comparative Evaluation of Three TSPO PET Radiotracers in a LPS-Induced Model of Mild Neuroinflammation in Rats. *Mol. Imaging Biol.* **2017**, 19, 77–89. [CrossRef] [PubMed]

45. Netea, M.G.; Fantuzzi, G.; Kullberg, B.J.; Stuyt, R.J.; Pulido, E.J.; McIntyre, R.C.; Joosten, L.A.; Van der Meer, J.W.; Dinarello, C.A. Neutralization of IL-18 reduces neutrophil tissue accumulation and protects mice against lethal Escherichia coli and Salmonella typhimurium endotoxemia. *J. Immunol. Baltim. Md 1950* **2000**, 164, 2644–2649. [CrossRef]

46. Yang, R.B.; Mark, M.R.; Gray, A.; Huang, A.; Xie, M.H.; Zhang, M.; Goddard, A.; Wood, W.I.; Gurney, A.L.; Godowski, P.J. Toll-like receptor-2 mediates lipopolysaccharide-induced cellular signalling. *Nature* **1998**, 395, 284–288. [CrossRef] [PubMed]

47. Peyronneau, M.-A.; Saba, W.; Goutal, S.; Damont, A.; Dolle, F.; Kassiou, M.; Bottlaender, M.; Valette, H. Metabolism and Quantification of [^{18}F]DPA-714, a New TSPO Positron Emission Tomography Radioligand. *Drug Metab. Dispos.* **2013**, 41, 122–131. [CrossRef] [PubMed]

48. Rusjan, P.M.; Wilson, A.A.; Bloomfield, P.M.; Vitcu, I.; Meyer, J.H.; Houle, S.; Mizrahi, R. Quantitation of translocator protein binding in human brain with the novel radioligand [^{18}F]-FEPPA and positron emission tomography. *J. Cereb. Blood Flow Metab.* **2011**, 31, 1807–1816. [CrossRef] [PubMed]

49. Smith, Q.R.; Takasato, Y.; Rapoport, S.I. Kinetic analysis of L-leucine transport across the blood-brain barrier. *Brain Res.* **1984**, 311, 167–170. [CrossRef]

50. Wanek, T.; Römermann, K.; Mairinger, S.; Stanek, J.; Sauberer, M.; Filip, T.; Traxl, A.; Kuntner, C.; Pahnke, J.; Bauer, F.; et al. Factors Governing P-Glycoprotein-Mediated Drug-Drug Interactions at the Blood-Brain Barrier Measured with Positron Emission Tomography. *Mol. Pharm.* **2015**, 12, 3214–3225. [CrossRef] [PubMed]

51. Beckers, L.; Ory, D.; Geric, I.; Declercq, L.; Koole, M.; Kassiou, M.; Bormans, G.; Baes, M. Increased Expression of Translocator Protein (TSPO) Marks Pro-inflammatory Microglia but Does Not Predict Neurodegeneration. *Mol. Imaging Biol. MIB Off. Publ. Acad. Mol. Imaging* **2018**, 20, 94–102. [CrossRef] [PubMed]

Sample Availability: All reagents and materials are commercially available.

molecules

MDPI

Review

PET Imaging of Microglial Activation—Beyond Targeting TSPO

Bieneke Janssen [1], Danielle J. Vugts [2], Albert D. Windhorst [2] and Robert H. Mach [1,*]

[1] Department of Radiology, Perelman School of Medicine, University of Pennsylvania, Philadelphia, PA 19104, USA; bieneke.janssen@pennmedicine.upenn.edu

[2] Department of Radiology & Nuclear Medicine, VU University Medical Center, 1081 HV Amsterdam, The Netherlands; d.vugts@vumc.nl (D.J.V.); ad.windhorst@vumc.nl (A.D.W.)

* Correspondence: rmach@pennmedicine.upenn.edu; Tel.: +1-215-746-8233

Received: 14 February 2018; Accepted: 6 March 2018; Published: 8 March 2018

Abstract: Neuroinflammation, which involves microglial activation, is thought to play a key role in the development and progression of neurodegenerative diseases and other brain pathologies. Positron emission tomography is an ideal imaging technique for studying biochemical processes in vivo, and particularly for studying the living brain. Neuroinflammation has been traditionally studied using radiotracers targeting the translocator protein 18 kDa, but this comes with certain limitations. The current review describes alternative biological targets that have gained interest for the imaging of microglial activation over recent years, such as the cannabinoid receptor type 2, cyclooxygenase-2, the $P2X_7$ receptor and reactive oxygen species, and some promising radiotracers for these targets. Although many advances have been made in the field of neuroinflammation imaging, current radiotracers all target the pro-inflammatory (M1) phenotype of activated microglia, since the number of known biological targets specific for the anti-inflammatory (M2) phenotype that are also suited as a target for radiotracer development is still limited. Next to proceeding the currently available tracers for M1 microglia into the clinic, the development of a suitable radiotracer for M2 microglia would mean a great advance in the field, as this would allow for imaging of the dynamics of microglial activation in different diseases.

Keywords: positron emission tomography; microglia; neuroinflammation

1. Microglial Activation—Focus on Imaging

Microglia are the resident immune cells of the brain and are involved in brain development, maintenance of homeostasis, neuroinflammation and neurodegeneration [1,2]. Microglia are highly dynamic cells, with processes surveilling the brain in homeostasis, while changing morphology to a more amoeboid shape upon activation, for instance when encountering a pathogen or injury to the brain. While altering morphology, cell surface receptor expression and secretion of chemokines and cytokines are altered as well. Although there is a spectrum of activation phenotypes, and transcriptome studies have shown microglial activation to be highly context dependent [2], microglial activation is roughly characterized as being either classic, pro-inflammatory activation (M1) or alternative, anti-inflammatory activation (M2). M1 microglia are recognized to produce pro-inflammatory cytokines such as interleukin (IL)-1β and tumor necrosis factor (TNF)-α and express nicotinamide adenine dinucleotide phosphate (NADPH) oxidase, which generates superoxide and reactive oxygen species (ROS) [2]. On the other hand, M2 microglia promote the healing process, as well as releasing anti-inflammatory factors like IL-10 and tumor growth factor (TGF)-β as well as other growth factors and neurotrophic factors [2]. In vitro, the different phenotypes of activated microglia can be characterized by looking at protein expression using, for example, immunohistochemical staining of post mortem brain samples [3–5]. Although a great deal has been learned from these in vitro studies in different neurodegenerative diseases, this has not been translated to an in vivo situation, mainly because of the difficulty of obtaining biopsy specimens from brain. For this purpose, positron emission tomography (PET) would be the ideal medical imaging technique, as it makes use of radioactively labeled molecules that can be specifically directed against a biological target. Traditionally, the target for PET imaging of microglial activation has been the translocator protein 18 kDa (TSPO) [6]. However, tracers targeting TSPO come with certain limitations, mainly due to their high non-specific binding, low brain uptake, and the fact that a TSPO polymorphism causes large differences in binding affinity between subjects [7–9]. In addition, TSPO has some drawbacks as a target itself, such as expression in multiple cell types, and upregulation of TSPO has been associated with the M1 activation status of microglia, and, to a lesser extent, with the M2 activation status [10]. Furthermore, it was recently reported that TSPO expression increases by about nine-fold under neuroinflammatory conditions in rodent macrophages and microglia, but in humans, no significant increase was found [11]. Moreover, under pro-inflammatory conditions, a decrease in TSPO protein expression was observed in human adult microglia and monocyte-derived macrophages [11]. Although more investigation, especially in vivo, is needed, these results suggest that increased signals obtained in humans using PET tracers targeting the TSPO do not necessarily reflect increased microglial activation, but merely provide a measure of microglial/macrophage density. For these reasons, more and more effort is being put into the search for new targets and tracers for PET imaging of microglial activation. Specific interest lies in the imaging of the different microglial phenotypes. Apart from alterations in the expression of TSPO, the expression of several other receptors and enzymes is altered during microglial activation (Figure 1). The current review focuses on the recent progress made in the development of PET tracers for neuroinflammation imaging, specifically for the cannabinoid receptor type 2 (CB2), cyclooxygenase-2 (COX-2), purinergic receptor P2X$_7$ and ROS.

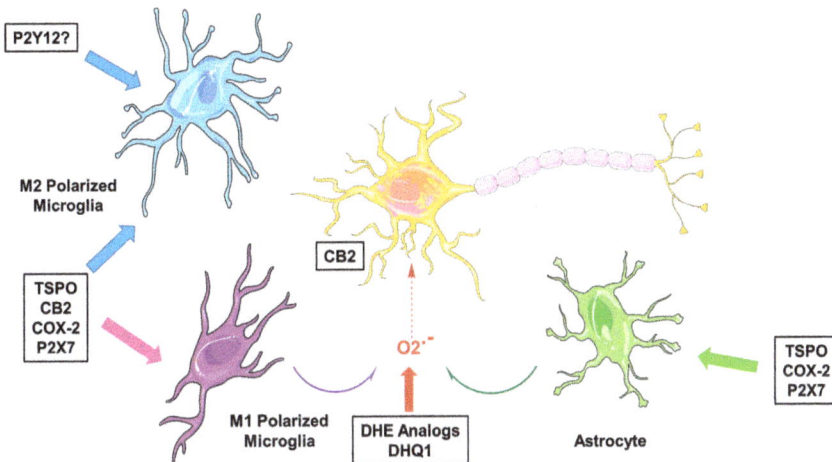

Figure 1. Molecular targets for imaging neuroinflammation in neurodegeneration. TSPO = translocator protein 18 kDa; CB2 = cannabinoid receptor type 2; COX-2 = cyclooxygenase-2; DHE = dihydroethidium; DHQ1 = dihydroquinoline analog.

2. PET Tracers Targeting the Cannabinoid Receptor Type 2

Cannabinoid receptors are G-protein coupled receptors, of which two subtypes are known to date. Whereas cannabinoid receptor type 1 (CB1) is constitutively expressed in the central nervous system (CNS), type 2 (CB2) is predominantly expressed in peripheral organs [12]. However, CB2 is also expressed on microglia and neurons [13], and its expression in particularly microglia increases significantly under neuroinflammatory conditions [12,13]. As a consequence, a number of tracers have been recently investigated in animal models of microglial activation. [¹¹C]A-836339 (Figure 2, K_i = 0.7 nM, 425-fold selective over CB1 [14,15]) showed an increase in specific brain uptake in mice, 5 days after systemic injection of lipopolysaccharide (LPS), compared with control mice [14]; however, in a later study with the same tracer, Pottier et al. could not reproduce these results in LPS-treated rats [16]. Two other rat models were included in the latter study (α-amino-3-hydroxy-5-methyl-4-isoxazolepropionic acid (AMPA) and experimental stroke) and neither model demonstrated increased tracer uptake [16]. Likewise, uptake of [¹¹C]NE40 (Figure 2, K_i = 9.6 nM, 100-fold selective over CB1 [17]) was not significantly increased in brains of rats after experimental stroke, compared with sham-operated animals [18]. In contrast to this, more recently, an increased uptake of [¹¹C]NE40 was observed one day after experimental stroke in rats, but uptake decreased to baseline levels at later time-points [19]. Using an ¹⁸F-labeled analog of A-836339, [¹⁸F]29 (Figure 2, K_i = 0.4 nM, 1000-fold selective over CB1 [20]), increased uptake was observed in brains of mice treated with LPS, but with only 7% of intact tracer left in plasma at 30 min post injection (p.i.) [20], its rapid metabolic degradation will hamper further application of this tracer.

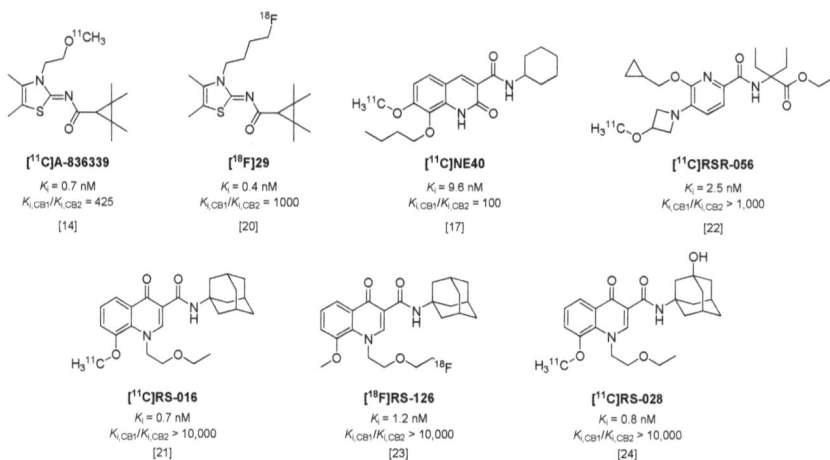

Figure 2. Radiotracers targeting the cannabinoid receptor type 2 (CB2).

Slavik et al. were able to show increased uptake in mouse brains after systemic administration of LPS with both [¹¹C]RS-016 (Figure 2, K_i = 0.7 nM, >10,000-fold selectivity over CB1 [21]) and [¹¹C]RSR-056 (K_i = 2.5 nM, >1000-fold selective over CB1 [22]), although uptake of [¹¹C]RSR-056 could only be blocked marginally, suggesting a high level of non-specific binding [22]. The same results could not be obtained with analog [¹⁸F]RS-126 (Figure 2, K_i = 1.2 nM, >10,000-fold selective over CB1) [23], and in a later in vitro study in post mortem tissue of amyotrophic lateral sclerosis (ALS) patients, a high amount of non-specific binding was observed [24]. Therefore, another ¹¹C-labeled analog was synthesized, which did show increased binding to ALS patient tissue compared with control tissue, but in vivo in mice [¹¹C]RS-028 (Figure 2, K_i = 0.8 nM, >10,000 selective over CB1) showed rapid washout of spleen (high abundance of CB2) and low brain uptake, leading authors to conclude this tracer is not optimal for in vivo application [24].

To date, one tracer targeting CB2 has been investigated in humans. [¹¹C]NE40 showed favorable fast brain uptake and washout in healthy human subjects [25]. Unfortunately, in a follow-up study in healthy controls and patients with Alzheimer's disease (AD), decreased brain uptake of [¹¹C]NE40 in AD patients compared with healthy control subjects was shown [26], even though upregulated expression of CB2 has been previously demonstrated in post mortem AD brain tissue [27].

The authors concluded that the decrease in uptake in the brain of AD patients is possibly a result of neuronal loss in late stage AD [26], leading to an overall decrease in CB2 expression, even though CB2 may be upregulated in microglia. This is supported by a study in transgenic AD mice using [¹¹C]A-836339, in which increased tracer binding was observed in brains of AD mice compared with control animals [13]. However, the transgenic model is not accompanied by neuronal loss, and extensive immunohistochemical staining studies showed that, in control mice, CB2 expression is mainly localized on neurons, whereas in transgenic AD mice, CB2 is mainly expressed in microglia [13]. Importantly, the intensity of the staining and thus, the expression of CB2 in neurons did not differ between control mice and transgenic AD mice, and therefore, increased tracer uptake could be attributed to upregulated CB2 expression in glial cells [13].

In summary, the upregulation of CB2 in neuroinflammation still needs to be further investigated. In particular with respect to neurodegenerative diseases, as in more advanced stages of these diseases, neuronal loss is known to occur. Therefore, CB2 may only be useful in the earliest stages of neurodegenerative diseases. In addition, as the above described studies have shown, high selectivity for CB2 over CB1 of radiotracers is necessary to eliminate a non-specific PET signal due to the high abundance of CB1 in CNS.

3. PET Tracers Targeting Cyclooxygenase-2

Cyclooxygenases (COX) are involved in the arachidonic acid cascade and activation of inflammatory pathways, such as Nuclear Factor (NF)-κB, leading to the release of prostaglandins, chemokines, cytokines and ROS [28]. Both isoforms (COX-1 and COX-2) are expressed in the brain, but regulatory functions and cell localization differ between the two. COX-2 has been of main interest as a target in neuroinflammation, as its expression is low in healthy brains but is rapidly overexpressed under inflammatory conditions [28,29]. However, COX-1 has also been implicated as having a role in neuroinflammation [28] and conflicting evidence exists for the expression and upregulation of COX-2 under inflammatory conditions [29]. To investigate the potential of COX-1 or COX-2 as targets for imaging of neuroinflammation, radiotracers should have high selectivity for one isoform over the other. Only a limited amount of isoform-selective tracers have been published, and COX-2 targeting tracers show high blood pool retention and limited uptake in target organs. These tracers also show high amounts of non-specific binding, relatively low affinities (>50 nM) and rapid metabolism or, in the case of ¹⁸F-labeled compounds, substantial defluorination [29,30]. Some promising COX-2 selective radiotracers are depicted in Figure 3. One of these tracers, [¹¹C]MC1 (IC$_{50}$ COX-2 = 3 nM; COX-1 > 1000 nM) [31], was recently used in a LPS-induced neuroinflammation model in rhesus monkeys together with COX-1 selective tracer [¹¹C]PS13 (IC$_{50}$ COX-1 = 1 nM; COX-2 > 1000 nM) [32]. Interestingly, this study showed an upregulation of COX-2, but not COX-1, after LPS-induced neuroinflammation in rhesus monkey brains [33].

An ¹⁸F-labeled analog of celecoxib ([¹⁸F]5; Figure 3) [34] was evaluated in colorectal cancer cells in vitro, but uptake of the tracer could not be blocked with either celecoxib or rofecoxib. Although [¹⁸F]5 showed increased selectivity for COX-2 over COX-1 (IC$_{50}$ > 100 μM) compared with celecoxib, its IC$_{50}$ value for COX-2 also decreased to 0.36 μM, which might explain the non-selectivity in blocking experiments. In addition, although [¹⁸F]5 was metabolically stable, the tracer was quickly cleared in vivo in COX-2 expressing tumor-bearing mice, which resulted in a lack of uptake in the tumor compared with the uptake observed in vitro, and is therefore not a suitable candidate for COX-2 imaging studies [34]. Another metabolically stable ¹⁸F-labeled analog of celecoxib, [¹⁸F]1 (IC$_{50}$ COX-2 = 1.7 nM; IC$_{50}$ COX-1 = 0.38 μM [35]), was obtained via the non-standard route of electrochemical radiofluorination [36] to attach the fluorine-18 atom to the pyrazole ring (Figure 3). Although this led to low yields (2% radiochemical yield) and low molar activity (~110 MBq/μmol), [¹⁸F]1 could be a valuable COX-2 radiotracer in vivo, once, as indicated by the authors, an efficient radiosynthesis method is identified, given its good in vitro affinity, and high metabolic stability in mice (>95% intact tracer at 60 min p.i. in blood), and its high brain uptake (2% ID/g in mouse brain at 60 min p.i.).

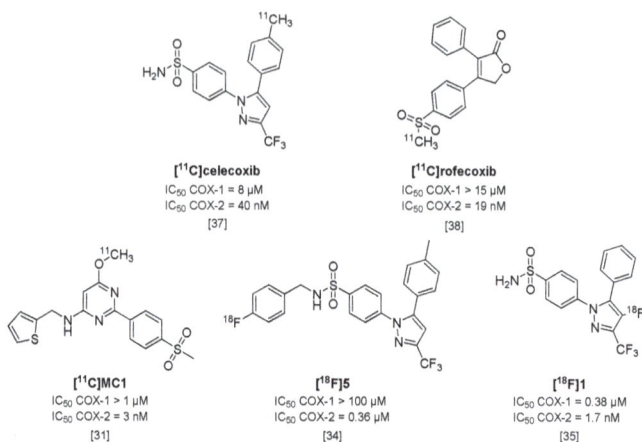

Figure 3. Radiolabeled inhibitors of cyclooxygenase-2 (COX-2).

To summarize, failure to show COX-2 upregulation in animal models of neuroinflammation and human disease may have been caused by the suboptimal properties of the tracers evaluated to date (i.e., [^{11}C]celecoxib [37] and [^{11}C]rofecoxib [38]), which includes either poor selectivity for COX-2 over COX-1, or a high level of non-specific binding. Therefore, the newly developed and selective radiotracers mentioned here (e.g., [^{11}C]MC1) should be evaluated in human disease, to assess their potential in a clinical setting.

4. PET Tracers Targeting the P2X$_7$ Receptor

The P2X$_7$ receptor (P2X$_7$R) is expressed in multiple cell types of the myeloid cell lineage, and although conflicting evidence exists for P2X$_7$R expression in astrocytes and neurons, in CNS the receptor is mainly expressed in microglia. The natural agonist of P2X$_7$R is adenosine triphosphate (ATP), but as the affinity of ATP for P2X$_7$R is low, the receptor is only activated at high (mM) concentrations of ATP. Therefore, the receptor is regarded as silent in normal physiology, but functionally upregulated in case of an imbalance in ATP concentration in pathological conditions [39]. Activation of P2X$_7$R is involved in a diverse series of signaling pathways that are linked to neuroinflammation [40] and is the key step in the activation of the inflammasome, leading to the release of pro-inflammatory cytokines like IL-1β [39]. In addition, the generation of ROS following P2X$_7$ receptor activation by ATP or BzATP has been described for multiple cell types, including microglia [40,41]. P2X$_7$R is therefore associated with the pro-inflammatory phenotype of microglia, and its functional expression is usually upregulated in CNS disease [39], which makes it an interesting target for both drugs and PET tracers.

Although P2X$_7$R antagonist [^{11}C]A-740003 (Figure 4) has already been shown to not enter the brain [42], recently, the tritiated analog of this potent P2X$_7$R antagonist ([^{3}H]A-740003; IC$_{50}$ hP2X$_7$R = 40 nM; IC$_{50}$ rP2X$_7$R = 18 nM [43]) was used in an in vitro study in post mortem brain sections of multiple sclerosis (MS) patients and rat brain sections of a rat model of MS (experimental autoimmune encephalomyelitis; EAE) [44]. P2X$_7$R was shown to be associated with the pro-inflammatory phenotype of microglia, and was highly expressed in active MS lesions in human brain compared with normal appearing white matter and chronic active lesions. In addition, in brain sections of the EAE rat model, [^{3}H]A-740003 binding increased during the peak of the disease (14 days after immunization). In both rat and human brain sections, increased tracer binding was confirmed with immunohistochemical staining for P2X$_7$R.

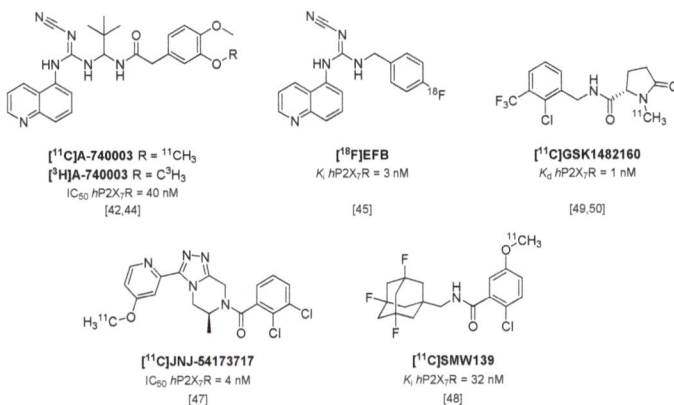

Figure 4. Radiolabeled P2X$_7$ receptor antagonists.

An analog of another cyanoguanidine containing compound (A-804598), was recently labeled with fluorine-18 [45]. [^{18}F]EFB (Figure 4) showed good affinity towards both human and rat P2X$_7$R (K_i of 3 and 36 nM, respectively), but low brain uptake was observed in both healthy rats and rats treated with LPS prior to PET scanning [45], and thus the application of [^{18}F]EFB in imaging of microglial activation will be limited.

Two carbon-11 labeled P2X$_7$R antagonists of different compound classes (Figure 4), [^{11}C]JNJ-54173717 (IC$_{50}$ hP2X$_7$R = 4 nM) and [^{11}C]SMW139 (K_i hP2X$_7$R = 32 nM [46]), were evaluated in a humanized rat model, in which the human P2X$_7$ receptor was locally expressed in striatum via an adeno-associated viral vector [47,48]. Both tracers entered the rat brain and showed excellent uptake in the hP2X$_7$R overexpressing striatum (standardized uptake value (SUV) 0.8 and 2.1 at 10 min p.i., respectively) compared with the contralateral striatum (SUV 0.6 and 1.4 at 10 min p.i., respectively). In addition, [^{11}C]JNJ-54173717 also showed high initial brain uptake (SUV 3.3) in non-human primates [47], which likely enables translation to humans, and [^{11}C]JNJ-54173717 is expected to proceed to clinical evaluation. Although P2X$_7$R overexpression in post mortem brain material of AD patients could not be shown using [^{11}C]SMW139 [48], a clinical study with this tracer is currently ongoing in patients diagnosed with MS, based on the findings of P2X$_7$R upregulation in active MS lesions described by Beaino et al. [44].

[^{11}C]GSK1482160 (K_d hP2X$_7$R = 1 nM / K_i hP2X$_7$R = 3 nM) was recently evaluated in a mouse model of LPS-induced neuroinflammation [49] and the EAE rat model [50]. In contrast to other studies using the LPS model of neuroinflammation [45,51] in which expression levels of Iba1 and P2X$_7$R were found to peak as early as 12 h post injection of LPS, the study by Territo et al. showed highest expression of Iba1 only at 72 h p.i. Biodistribution studies in LPS-treated (5 mg/kg i.p.) and saline-treated control mice revealed an increased uptake of [^{11}C]GSK1482160 in LPS-treated mice compared with saline-treated mice in all organs studied (2.9–5.7-fold) [49]. Small animal PET imaging revealed a stable uptake in tissue within 10 minutes after tracer injection, and increased uptake in the brains of LPS-treated mice (3.6-fold) was confirmed. In another study, [^{11}C]GSK1482160 was shown to enter the brains of rhesus macaques with an SUV maximum (2.7) at around 70 min p.i. [50]. The same group showed increased tracer binding in the lumbar spinal cord at the peak of the disease in EAE rats (12–14 days post immunization) compared with healthy rats in an autoradiography study, but failed to show this increase in vivo due to a stated insufficient affinity of [^{11}C]GSK1482160 for the rat P2X$_7$R [50].

Over the last few years, P2X$_7$R has gained interest as a target for PET imaging of microglial activation and may be a promising alternative for targeting TSPO. In addition, P2X$_7$R may even be more useful as a target for imaging of microglial activation, as it has been shown to be associated predominantly with the pro-inflammatory microglial phenotype [44,45,49]. In in vitro situations, in both animal models of disease and human tissue sections, P2X$_7$R tracers have shown promise in the ability to show the presence of neuroinflammation. However, the promising preclinical findings still need to be confirmed in a human situation, as P2X$_7$R overactivation may differ between diseases and possibly even disease stages. In addition, no data are available on the actual expression levels of P2X$_7$R in humans, or differences therein in health and disease. To uncover the potential of targeting P2X$_7$R for PET imaging of microglial activation, several P2X$_7$R tracers are currently being evaluated in clinical trials.

5. PET Radiotracers for Imaging ROS

Oxidative stress results from the formation of pro-inflammatory microglia and astrocyte activation, which are sources of nitric oxide (NO) and superoxide due to their increased expression of iNOS and high levels of NADPH oxidase activity (Figure 1). Superoxide is normally removed from cells by the action of superoxide dismutase (SOD). However, under high levels of NADPH oxidase activity, superoxide can overwhelm these "protective mechanisms" and react with NO to form peroxynitrite (ONOO-). Peroxynitrite is a highly reactive oxidant which can damage macromolecules within the cytoplasm and nucleus, including DNA strand breaks, lipid peroxidation and the oxidation of sulfur groups in proteins. The formation of DNA strand breaks is thought to lead to the activation of cell death pathways via caspase-mediated or noncaspase-mediated mechanisms (i.e., necroptosis or parthanatos) [52]. The development of PET radiotracers capable of imaging superoxide levels in the CNS is expected to provide a sensitive means for imaging pro-inflammatory neuroinflammation in neurodegenerative disorders such as AD [52]. In addition, the recognition of the role of NADPH oxidase as a key mediator of oxidative stress via its production of superoxide has led to the development of inhibitors of this enzyme as therapeutic targets for neurodegenerative diseases. Therefore, the availability of a PET radiotracer for imaging superoxide levels in the CNS is expected to serve as a sensitive measure of the therapeutic efficacy of a putative NADPH oxidase inhibitor.

The first PET radiotracer for imaging superoxide is [^{18}F]FDMT (Figure 5), an ^{18}F-labeled analog of the fluorescent probe dihydroethidium (DHE), that was synthesized using "click" chemistry [53]. DHE has been previously shown to provide a sensitive measure of superoxide levels in cells and tissues, using microscopy and optical imaging techniques [54,55]. Although [^{18}F]FDMT showed promising results in an animal model of adriamycin-induced model of cardiotoxicity, this radiotracer did not cross the blood-brain barrier (BBB) and is not capable of imaging the increased levels of superoxide that occur during neuroinflammation.

A number of radiolabeled analogs of dihydromethidium (a.k.a. hydromethidine), which is the corresponding N-methyl analog of DHE, have been synthesized and evaluated in vitro and in vivo [56,57]. [^{3}H]Dihydromethidium is oxidized by both superoxide and hydroxyl radicals and has been evaluated in animal models of stroke and cisplatin-induced nephrotoxicity [58,59]. A recent paper has also reported the radiosynthesis and preliminary in vivo evaluation of [^{11}C]dihydromethidium [57]. This compound shows high brain uptake in microPET imaging studies, but a detailed analysis of the mechanisms of uptake and trapping of this radiotracer was not conducted.

In a follow-up study to their earlier work with [^{18}F]FDMT, Mach and coworkers reported the synthesis, in vitro characterization, and in vivo evaluation of [^{18}F]ROStrace in an LPS-induced mouse model of neuroinflammation [60]. In this study, the investigators reported that replacing the corresponding triazole "click" moiety with the traditional [^{18}F]2-fluoroethoxy group resulted in a compound that freely crossed the BBB. A key observation in this study was the high variability in uptake of [^{18}F]ROStrace in the LPS-treated animals. However, when the investigators compared the uptake of [^{18}F]ROStrace with the degree of "sickness" following the LPS-treatment by using the scoring criteria outlined by Carstens and Moberg [61] for recognizing pain and distress in laboratory animals, there was a high correlation between radiotracer uptake and degree of sickness induced by LPS (Figure 6) [60]. The authors also conducted a detailed metabolite analysis study, confirming that the radioactive species in the LPS-treated animals was primarily the oxidized form of [18F]ROStrace, [18F]*ox*-ROStrace. Since [18F]*ox*-ROStrace does not cross the BBB, these results confirm the mechanism of uptake and trapping outlined in Figure 7A. Since ROStrace is oxidized by superoxide and not hydrogen peroxide and the hydroxyl radical, this trapping occurs via the oxidation of [18F]ROStrace to [18F]*ox*-ROStrace by superoxide.

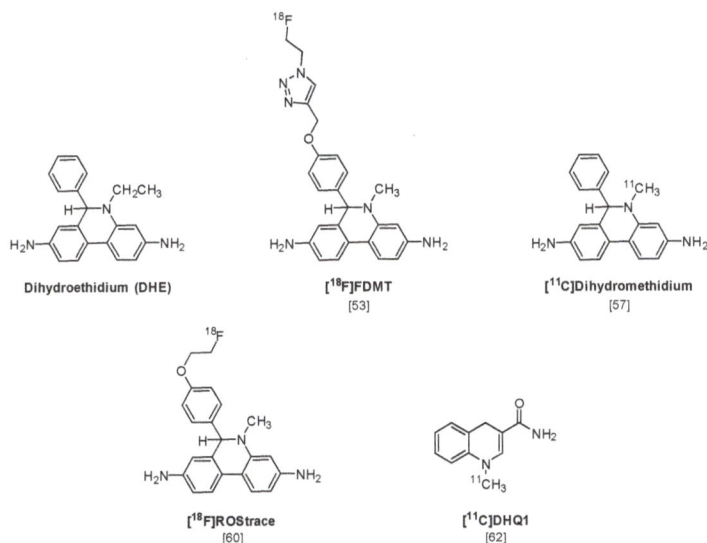

Figure 5. Structures of the positron emission tomography (PET) radiotracers for imaging increased levels of reactive oxygen species (ROS) in neuroinflammation.

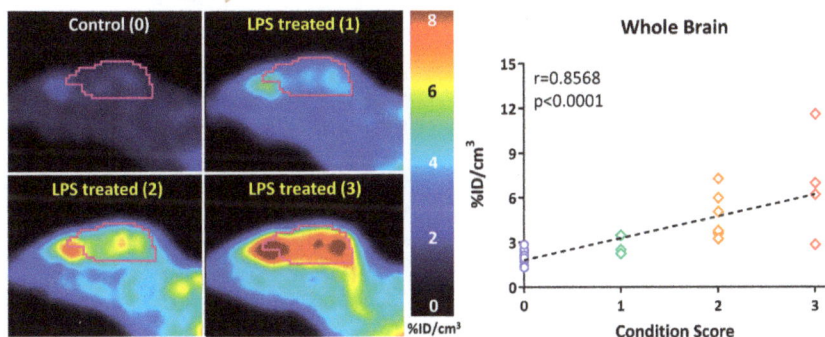

Figure 6. The uptake of [18F]ROStrace correlates with the condition score (degree of "sickness") following treatment with lipopolysaccharide (LPS).

A second compound that has shown promise in imaging oxidative stress is the dihydroquinoline analog, [11C]DHQ1 (Figure 5) [62]. [11C]DHQ1 is an analog of NADH/NADPH and has been used previously as a redox carrier in the delivery of drugs to the CNS. It is capable of crossing the BBB and, like the DHE analogs described above, is trapped in the brain by oxidation to a charged species

(Figure 7B). Since pretreatment with the NOX2 inhibitor, apocyanin, results in a reduction in uptake of [11C]DHQ1, part of its trapping mechanism can be attributed to the oxidation of the dihydro species to the corresponding *N*-methylquinolinium species. However, this compound is also thought to be a substrate for enzymes for the oxidation of NADH/NADPH in oxidative phosphorylation, so its trapping mechanism is not limited to the presence of elevated levels of superoxide. Furthermore, no information was provided on its relative reactivity to other oxidizing species, such as hydrogen peroxide (H_2O_2) or hypochlorous acid (HOCl). Therefore, this tracer is best described as providing a nonselective measure of oxidative stress.

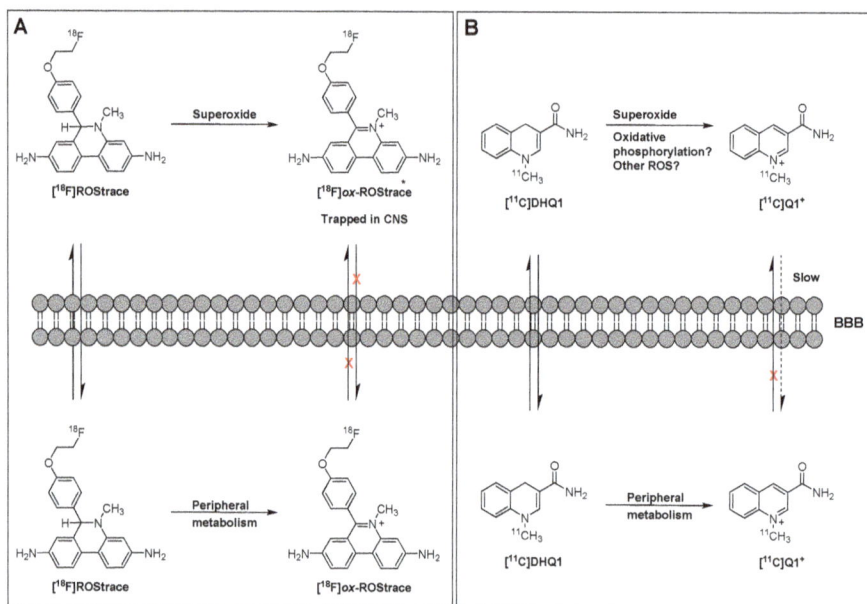

Figure 7. Mechanism of trapping of [18F]ROStrace (**A**) and [11C]DHQ1 (**B**).

6. Concluding Remarks and Future Directions

Whereas traditionally, the development of new PET tracers for imaging of neuroinflammation has been largely focused on improving the available ligands for TSPO, in recent years, attention has begun to shift towards the development of radiotracers for alternative targets. Since the activation of microglia is highly dynamic, and protein expression is dependent on both microglial phenotype and microglial environment, (over)expression of a certain target protein may differ per disease, or even be dependent on disease stage. Therefore, being able to visualize multiple targets in the living brain is of utmost importance to gain more insight in this dynamic process. This is especially true for the human situation, given the difficulty of obtaining biopsy specimens of brain tissue. As discussed in the current review, several promising new radiotracers have been developed, targeting CB2, COX-2, $P2X_7R$ and ROS. Although all of these targets are involved in the pro-inflammatory phenotype (M1) of microglia, it may very well be that not every tracer/target is equally suitable for imaging microglial activation in every disease, due to the aforementioned dynamics in protein expression in microglial phenotypes. In addition, while radiotracers are often evaluated in rodent models of excitotoxin-induced neuroinflammation or models of human disease, these are not necessarily representative of human disease, as was recently also reported for TSPO [11]. To evaluate the potential of the biological targets discussed in this review as valuable targets for PET imaging, it will be very interesting to see the outcome of the clinical studies that are currently ongoing for several tracers. Nevertheless, all of these

Molecules **2018**, *23*, 607

targets are specific for the pro-inflammatory phenotype (M1) of activated microglia, and, to get a complete view of the neuroinflammatory process, it is important to focus on the anti-inflammatory phenotype (M2) as well. However, tracer development is limited by the availability of targetable biomolecules specific for this phenotype. One target of high interest is the $P2Y_{12}$ receptor ($P2Y_{12}R$), a G-protein coupled receptor that is highly overexpressed in the anti-inflammatory phenotype compared with the pro-inflammatory and resting state phenotypes [2,63,64]. Moreover, the expression of $P2Y_{12}R$ in CNS is limited to microglia only [2] and could therefore exclude any PET signals from infiltrating monocytes and macrophages. Two recent autoradiography studies using a carbon-11 labeled $P2Y_{12}R$ antagonist on brain sections of the EAE rat model and MS patients [44] and rodent stroke models and a patient deceased from stroke [65] demonstrated the possibility of visualizing the anti-inflammatory subset of microglia cells (i.e., M2-polarized microglia). Unfortunately, despite the fact that many potent $P2Y_{12}$ antagonists have been developed due to their use as anti-coagulants, a (radiolabeled) compound that crosses the BBB is still lacking. The development of a brain-penetrant tracer that targets the anti-inflammatory phenotype of activated microglia would initiate a great advance in the field of neuroinflammation imaging, as a combined tracer study (pro- vs. anti-inflammatory) could provide new insights into the dynamics of microglial activation in health and disease.

Acknowledgments: A.D.W. and D.J.V. have received funding from European Union's Seventh Framework Programme (FP7/2007-2013) under grant agreement no. HEALTH-F2-2011-278850 (INMiND). R.H.M. has received funding from the Michael J. Fox Foundation.

Author Contributions: B.J., D.J.V., A.D.W. and R.H.M. wrote and proofread the manuscript.

Conflicts of Interest: The authors declare no conflict of interest.

References

1. Glass, C.K.; Saijo, K.; Winner, B.; Marchetto, M.C.; Gage, F.H. Mechanisms underlying inflammation in neurodegeneration. *Cell* **2010**, *140*, 918–934. [CrossRef] [PubMed]
2. Colonna, M.; Butovsky, O. Microglia function in the central nervous system during health and neurodegeneration. *Annu. Rev. Immunol.* **2017**, *35*, 441–468. [CrossRef] [PubMed]
3. Cherry, J.D.; Olschowka, J.A.; O'Banion, M.K. Neuroinflammation and M2 microglia: The good, the bad, and the inflamed. *J. Neuroinflamm.* **2014**, *11*, 98. [CrossRef] [PubMed]
4. Tang, Y.; Le, W. Differential roles of M1 and M2 microglia in neurodegenerative diseases. *Mol. Neurobiol.* **2016**, *53*, 1181–1194. [CrossRef] [PubMed]
5. Walker, D.G.; Lue, L.-F. Immune phenotypes of microglia in human neurodegenerative disease: Challenges to detecting microglial polarization in human brains. *Alzheimer's Res. Ther.* **2015**, *7*, 56. [CrossRef] [PubMed]
6. Alam, M.M.; Lee, J.; Lee, S.Y. Recent progress in the development of TSPO PET ligands for neuroinflammation imaging in neurological diseases. *Nucl. Med. Mol. Imaging* **2017**, *51*, 283–296. [CrossRef] [PubMed]
7. Guo, Q.; Colasanti, A.; Owen, D.R.; Onega, M.; Kamalakaran, A.; Bennacef, I.; Matthews, P.M.; Rabiner, E.A.; Turkheimer, F.E.; Gunn, R.N. Quantification of the specific translocator protein signal of ^{18}F-PBR111 in healthy humans: A genetic polymorphism effect on in vivo binding. *J. Nucl. Med.* **2013**, *54*, 1915–1923. [CrossRef] [PubMed]
8. Kreisl, W.C.; Jenko, K.J.; Hines, C.S.; Lyoo, C.H.; Corona, W.; Morse, C.L.; Zoghbi, S.S.; Hyde, T.; Kleinman, J.E.; Pike, V.W.; et al. A genetic polymorphism for translocator protein 18 kDa affects both in vitro and in vivo radioligand binding in human brain to this putative biomarker of neuroinflammation. *J. Cereb. Blood Flow Metab.* **2013**, *33*, 53–58. [CrossRef] [PubMed]
9. Owen, D.R.; Gunn, R.N.; Rabiner, E.A.; Bennacef, I.; Fujita, M.; Kreisl, W.C.; Innis, R.B.; Pike, V.W.; Reynolds, R.; Matthews, P.M.; et al. Mixed-affinity binding in humans with 18-kDa translocator protein ligands. *J. Nucl. Med.* **2011**, *52*, 24–32. [CrossRef] [PubMed]
10. Bonsack, F., IV; Alleyne, C.H., Jr.; Sukumari-Ramesh, S. Augmented expression of TSPO after intracerebral hemorrhage: A role in inflammation? *J. Neuroinflamm.* **2016**, *13*, 151. [CrossRef] [PubMed]

11. Owen, D.R.; Narayan, N.; Wells, L.; Healy, L.; Smyth, E.; Rabiner, E.A.; Galloway, D.; Williams, J.B.; Lehr, J.; Mandhair, H.; et al. Pro-inflammatory activation of primary microglia and macrophages increases 18 kDa translocator protein expression in rodents but not humans. *J. Cereb. Blood Flow Metab.* **2017**, *37*, 2679–2690. [CrossRef] [PubMed]

12. Navarro, G.; Morales, P.; Rodriguez-Cueto, C.; Fernandez-Ruiz, J.; Jagerovic, N.; Franco, R. Targeting cannabinoid CB_2 receptors in the central nervous system. Medicinal chemistry approaches with focus on neurodegenerative disorders. *Front. Neurosci.* **2016**, *10*, 11. [CrossRef] [PubMed]

13. Savonenko, A.V.; Melnikova, T.; Wang, Y.; Ravert, H.; Gao, Y.; Koppel, J.; Lee, D.; Pletnikova, O.; Cho, E.; Sayyida, N.; et al. Cannabinoid CB2 receptors in a mouse model of aβ amyloidosis: Immunohistochemical analysis and suitability as a PET biomarker of neuroinflammation. *PLoS ONE* **2015**, *10*, e0129618. [CrossRef] [PubMed]

14. Horti, A.G.; Gao, Y.; Ravert, H.T.; Finley, P.; Valentine, H.; Wong, D.F.; Endres, C.J.; Savonenko, A.V.; Dannals, R.F. Synthesis and biodistribution of [^{11}C]A-836339, a new potential radioligand for PET imaging of cannabinoid type 2 receptors (CB2). *Bioorg. Med. Chem.* **2010**, *18*, 5202–5207. [CrossRef] [PubMed]

15. Yao, B.B.; Hsieh, G.; Daza, A.V.; Fan, Y.; Grayson, G.K.; Garrison, T.R.; El Kouhen, O.; Hooker, B.A.; Pai, M.; Wensink, E.J.; et al. Characterization of a cannabinoid CB2 receptor-selective agonist, A-836339 [2,2,3,3-tetramethyl-cyclopropanecarboxylic acid [3-(2-methoxy-ethyl)-4,5-dimethyl -3H-thiazol-(2Z)-ylidene]-amide], using in vitro pharmacological assays, in vivo pain models, and pharmacological magnetic resonance imaging. *J. Pharmacol. Exp. Ther.* **2009**, *328*, 141–151. [PubMed]

16. Pottier, G.; Gomez-Vallejo, V.; Padro, D.; Boisgard, R.; Dolle, F.; Llop, J.; Winkeler, A.; Martin, A. PET imaging of cannabinoid type 2 receptors with [^{11}C]A-836339 did not evidence changes following neuroinflammation in rats. *J. Cereb. Blood Flow Metab.* **2017**, *37*, 1163–1178. [CrossRef] [PubMed]

17. Evens, N.; Vandeputte, C.; Coolen, C.; Janssen, P.; Sciot, R.; Baekelandt, V.; Verbruggen, A.M.; Debyser, Z.; Van Laere, K.; Bormans, G.M. Preclinical evaluation of [^{11}C]NE40, a type 2 cannabinoid receptor PET tracer. *Nucl. Med. Biol.* **2012**, *39*, 389–399. [CrossRef] [PubMed]

18. Vandeputte, C.; Casteels, C.; Struys, T.; Koole, M.; van Veghel, D.; Evens, N.; Gerits, A.; Dresselaers, T.; Lambrichts, I.; Himmelreich, U.; et al. Small-animal PET imaging of the type 1 and type 2 cannabinoid receptors in a photothrombotic stroke model. *Eur. J. Nucl. Med. Mol. Imaging* **2012**, *39*, 1796–1806. [CrossRef] [PubMed]

19. Hosoya, T.; Fukumoto, D.; Kakiuchi, T.; Nishiyama, S.; Yamamoto, S.; Ohba, H.; Tsukada, H.; Ueki, T.; Sato, K.; Ouchi, Y. In vivo TSPO and cannabinoid receptor type 2 availability early in post-stroke neuroinflammation in rats: A positron emission tomography study. *J. Neuroinflamm.* **2017**, *14*, 69. [CrossRef] [PubMed]

20. Moldovan, R.P.; Teodoro, R.; Gao, Y.J.; Deuther-Conrad, W.; Kranz, M.; Wang, Y.C.; Kuwabara, H.; Nakano, M.; Valentine, H.; Fischer, S.; et al. Development of a high-affinity PET radioligand for imaging cannabinoid subtype 2 receptor. *J. Med. Chem.* **2016**, *59*, 7840–7855. [CrossRef] [PubMed]

21. Slavik, R.; Herde, A.M.; Bieri, D.; Weber, M.; Schibli, R.; Kramer, S.D.; Ametamey, S.M.; Mu, L.J. Synthesis, radiolabeling and evaluation of novel 4-oxo-quinoline derivatives as PET tracers for imaging cannabinoid type 2 receptor. *Eur. J. Med. Chem.* **2015**, *92*, 554–564. [CrossRef] [PubMed]

22. Slavik, R.; Grether, U.; Herde, A.M.; Gobbi, L.; Fingerle, J.; Ullmer, C.; Kramer, S.D.; Schibli, R.; Mu, L.J.; Ametamey, S.M. Discovery of a high affinity and selective pyridine analog as a potential positron emission tomography imaging agent for cannabinoid type 2 receptor. *J. Med. Chem.* **2015**, *58*, 4266–4277. [CrossRef] [PubMed]

23. Slavik, R.; Muller Herde, A.; Haider, A.; Kramer, S.D.; Weber, M.; Schibli, R.; Ametamey, S.M.; Mu, L. Discovery of a fluorinated 4-oxo-quinoline derivative as a potential positron emission tomography radiotracer for imaging cannabinoid receptor type 2. *J. Neurochem.* **2016**, *138*, 874–886. [CrossRef] [PubMed]

24. Haider, A.; Spinelli, F.; Herde, A.M.; Mu, B.; Keller, C.; Margelisch, M.; Weber, M.; Schibli, R.; Mu, L.; Ametamey, S.M. Evaluation of 4-oxo-quinoline-based CB2 PET radioligands in R6/2 chorea huntington mouse model and human ALS spinal cord tissue. *Eur. J. Med. Chem.* **2018**, *145*, 746–759. [CrossRef] [PubMed]

25. Ahmad, R.; Koole, M.; Evens, N.; Serdons, K.; Verbruggen, A.; Bormans, G.; Van Laere, K. Whole-body biodistribution and radiation dosimetry of the cannabinoid type 2 receptor ligand [^{11}C]-NE40 in healthy subjects. *Mol. Imaging Biol.* **2013**, *15*, 384–390. [CrossRef] [PubMed]

26. Ahmad, R.; Postnov, A.; Bormans, G.; Versijpt, J.; Vandenbulcke, M.; Van Laere, K. Decreased in vivo availability of the cannabinoid type 2 receptor in Alzheimer's disease. *Eur. J. Nucl. Med. Mol. Imaging* **2016**, *43*, 2219–2227. [CrossRef] [PubMed]

27. Benito, C.; Nunez, E.; Tolon, R.M.; Carrier, E.J.; Rabano, A.; Hillard, C.J.; Romero, J. Cannabinoid CB2 receptors and fatty acid amide hydrolase are selectively overexpressed in neuritic plaque-associated glia in Alzheimer's disease brains. *J. Neurosci.* **2003**, *23*, 11136–11141. [PubMed]

28. Aid, S.; Bosetti, F. Targeting cyclooxygenases-1 and -2 in neuroinflammation: Therapeutic implications. *Biochimie* **2011**, *93*, 46–51. [CrossRef] [PubMed]

29. Tietz, O.; Marshall, A.; Wuest, M.; Wang, M.; Wuest, F. Radiotracers for molecular imaging of cyclooxygenase-2 (COX-2) enzyme. *Curr. Med. Chem.* **2013**, *20*, 4350–4369. [CrossRef] [PubMed]

30. Pacelli, A.; Greenman, J.; Cawthorne, C.; Smith, G. Imaging COX-2 expression in cancer using PET/SPECT radioligands: Current status and future directions. *J. Label. Compd. Radiopharm.* **2014**, *57*, 317–322. [CrossRef] [PubMed]

31. Cortes, M.; Singh, P.; Morse, C.; Kowalski, A.; Jenko, K.; Shrestha, S.; Zoghbi, S.; Fujita, M.; Innis, R.B.; Pike, V.W. Synthesis of a candidate brain-penetrant COX-2 PET radioligand as a potential probe for neuroinflammation. *J. Label. Compd. Radiopharm.* **2015**, *58*, S312.

32. Shrestha, S.; Singh, P.; Eldridge, M.; Cortes, M.; Gladding, R.; Morse, C.; Zoghbi, S.; Fujita, M.; Liow, J.-S.; Pike, V. A novel PET radioligand, [¹¹C]PS13, successfully images COX-1, a potential biomarker for neuroinflammation. *J. Nucl. Med.* **2016**, *57*, 115.

33. Kim, M.J.; Shrestha, S.; Eldridge, M.; Cortes, M.; Singh, P.; Liow, J.S.; Gladding, R.; Zoghbi, S.; Fujita, M.; Pike, V.; et al. Novel pet radioligands show that, in rhesus monkeys, cox-1 is constitutively expressed and cox-2 is induced by inflammation. *J. Nucl. Med.* **2017**, *58*, 2.

34. Kaur, J.; Tietz, O.; Bhardwaj, A.; Marshall, A.; Way, J.; Wuest, M.; Wuest, F. Design, synthesis, and evaluation of an ¹⁸F-labeled radiotracer based on celecoxib-NBD for positron emission tomography (PET) imaging of cyclooxygenase-2 (COX-2). *ChemMedChem* **2015**, *10*, 1635–1640. [CrossRef] [PubMed]

35. Penning, T.D.; Talley, J.J.; Bertenshaw, S.R.; Carter, J.S.; Collins, P.W.; Docter, S.; Graneto, M.J.; Lee, L.F.; Malecha, J.W.; Miyashiro, J.M.; et al. Synthesis and biological evaluation of the 1,5-diarylpyrazole class of cyclooxygenase-2 inhibitors: Identification of 4-[5-(4-methylphenyl)-3-(trifluoromethyl)-1h-pyrazol-1-yl]benze nesulfonamide (sc-58635, celecoxib). *J. Med. Chem.* **1997**, *40*, 1347–1365. [CrossRef] [PubMed]

36. Lebedev, A.; Jiao, J.; Lee, J.; Yang, F.; Allison, N.; Herschman, H.; Sadeghi, S. Radiochemistry on electrodes: Synthesis of an ¹⁸F-labelled and in vivo stable COX-2 inhibitor. *PLoS ONE* **2017**, *12*, e0176606. [CrossRef] [PubMed]

37. Prabhakaran, J.; Majo, V.J.; Simpson, N.R.; Van Heertum, R.L.; Mann, J.J.; Kumar, J.S.D. Synthesis of [11c]celecoxib: A potential PET probe for imaging COX-2 expression. *J. Label. Compd. Radiopharm.* **2005**, *48*, 887–895. [CrossRef]

38. Majo, V.J.; Prabhakaran, J.; Simpson, N.R.; Van Heertum, R.L.; Mann, J.J.; Dileep Kumar, J.S. A general method for the synthesis of aryl [¹¹C]methylsulfones: Potential PET probes for imaging cyclooxygenase-2 expression. *Bioorg. Med. Chem. Lett.* **2005**, *15*, 4268–4271. [CrossRef] [PubMed]

39. Bhattacharya, A.; Biber, K. The microglial ATP-gated ion channel P2X7 as a CNS drug target. *Glia* **2016**, *64*, 1772–1787. [CrossRef] [PubMed]

40. Bartlett, R.; Stokes, L.; Sluyter, R. The p2x7 receptor channel: Recent developments and the use of P2X7 antagonists in models of disease. *Pharm. Rev.* **2014**, *66*, 638–675. [CrossRef] [PubMed]

41. Bartlett, R.; Yerbury, J.J.; Sluyter, R. P2X7 receptor activation induces reactive oxygen species formation and cell death in murine EOC13 microglia. *Mediat. Inflamm.* **2013**, *2013*, 271813. [CrossRef] [PubMed]

42. Janssen, B.; Vugts, D.J.; Funke, U.; Spaans, A.; Schuit, R.C.; Kooijman, E.; Rongen, M.; Perk, L.R.; Lammertsma, A.A.; Windhorst, A.D. Synthesis and initial preclinical evaluation of the P2X7 receptor antagonist [¹¹C]A-740003 as a novel tracer of neuroinflammation. *J. Label. Compd. Radiopharm.* **2014**, *57*, 509–516. [CrossRef] [PubMed]

43. Honore, P.; Donnelly-Roberts, D.; Namovic, M.T.; Hsieh, G.; Zhu, C.Z.; Mikusa, J.P.; Hernandez, G.; Zhong, C.; Gauvin, D.M.; Chandran, P.; et al. A-740003 [n-(1-{[(cyanoimino)(5-quinolinylamino) methyl]amino}-2,2-dimethylpropyl)-2-(3,4-dimethoxyphenyl)acetamide], a novel and selective P2X7 receptor antagonist, dose-dependently reduces neuropathic pain in the rat. *J. Pharmacol. Exp. Ther.* **2006**, *319*, 1376–1385. [CrossRef] [PubMed]

44. Beaino, W.; Janssen, B.; Kooij, G.; van der Pol, S.M.A.; van Het Hof, B.; van Horssen, J.; Windhorst, A.D.; de Vries, H.E. Purinergic receptors P2Y12R and P2X7R: Potential targets for PET imaging of microglia phenotypes in multiple sclerosis. *J. Neuroinflamm.* **2017**, *14*, 259. [CrossRef] [PubMed]

45. Fantoni, E.R.; Dal Ben, D.; Falzoni, S.; Di Virgilio, F.; Lovestone, S.; Gee, A. Design, synthesis and evaluation in an LPS rodent model of neuroinflammation of a novel ^{18}F-labelled PET tracer targeting P2X7. *EJNMMI Res.* **2017**, *7*, 31. [CrossRef] [PubMed]

46. Wilkinson, S.M.; Barron, M.L.; O'Brien-Brown, J.; Janssen, B.; Stokes, L.; Werry, E.L.; Chishty, M.; Skarratt, K.K.; Ong, J.A.; Hibbs, D.E.; et al. Pharmacological evaluation of novel bioisosteres of an adamantanyl benzamide P2X7 receptor antagonist. *ACS Chem. Neurosci.* **2017**, *8*, 2374–2380. [CrossRef] [PubMed]

47. Ory, D.; Celen, S.; Gijsbers, R.; Van Den Haute, C.; Postnov, A.; Koole, M.; Vandeputte, C.; Andres, J.I.; Alcazar, J.; De Angelis, M.; et al. Preclinical evaluation of a P2X7 receptor-selective radiotracer: PET studies in a rat model with local overexpression of the human P2X7 receptor and in nonhuman primates. *J. Nucl. Med.* **2016**, *57*, 1436–1441. [CrossRef] [PubMed]

48. Janssen, B.; Vugts, D.J.; Wilkinson, S.M.; Ory, D.; Chalon, S.; Hoozemans, J.J.M.; Schuit, R.C.; Beaino, W.; Kooijman, E.J.M.; van den Hoek, J.; et al. Identification of the allosteric P2X7 receptor antagonist [11C]SMW139 as a PET tracer of microglial activation. *Sci. Rep.* **2018**, in revision.

49. Territo, P.R.; Meyer, J.A.; Peters, J.S.; Riley, A.A.; McCarthy, B.P.; Gao, M.; Wang, M.; Green, M.A.; Zheng, Q.H.; Hutchins, G.D. Characterization of ^{11}C-GSK1482160 for targeting the P2X7 receptor as a biomarker for neuroinflammation. *J. Nucl. Med.* **2017**, *58*, 458–465. [CrossRef] [PubMed]

50. Han, J.; Liu, H.; Liu, C.; Jin, H.; Perlmutter, J.S.; Egan, T.M.; Tu, Z. Pharmacologic characterizations of a P2X7 receptor-specific radioligand, [11C]GSK1482160 for neuroinflammatory response. *Nucl. Med. Commun.* **2017**, *38*, 372–382. [CrossRef] [PubMed]

51. Choi, H.B.; Ryu, J.K.; Kim, S.U.; McLarnon, J.G. Modulation of the purinergic P2X7 receptor attenuates lipopolysaccharide-mediated microglial activation and neuronal damage in inflamed brain. *J. Neurosci.* **2007**, *27*, 4957–4968. [CrossRef] [PubMed]

52. Makvandi, M.; Sellmyer, M.A.; Mach, R.H. Inflammation and DNA damage: Probing pathways to cancer and neurodegeneration. *Drug Discov. Today Technol.* **2017**, *25*, 37–43. [CrossRef] [PubMed]

53. Chu, W.; Chepetan, A.; Zhou, D.; Shoghi, K.I.; Xu, J.; Dugan, L.L.; Gropler, R.J.; Mintun, M.A.; Mach, R.H. Development of a PET radiotracer for non-invasive imaging of the reactive oxygen species, superoxide, in vivo. *Org. Biomol. Chem.* **2014**, *12*, 4421–4431. [CrossRef] [PubMed]

54. Dugan, L.L.; Ali, S.S.; Shekhtman, G.; Roberts, A.J.; Lucero, J.; Quick, K.L.; Behrens, M.M. IL-6 mediated degeneration of forebrain GABAergic interneurons and cognitive impairment in aged mice through activation of neuronal NADPH oxidase. *PLoS ONE* **2009**, *4*, e5518. [CrossRef] [PubMed]

55. Dugan, L.L.; Quick, K.L. Reactive oxygen species and aging: Evolving questions. *Sci. Aging Knowl. Environ.* **2005**, *2005*, pe20. [CrossRef] [PubMed]

56. Abe, K.; Takai, N.; Fukumoto, K.; Imamoto, N.; Tonomura, M.; Ito, M.; Kanegawa, N.; Sakai, K.; Morimoto, K.; Todoroki, K.; et al. In vivo imaging of reactive oxygen species in mouse brain by using [3H]hydromethidine as a potential radical trapping radiotracer. *J. Cereb. Blood Flow Metab.* **2014**, *34*, 1907–1913. [CrossRef] [PubMed]

57. Wilson, A.A.; Sadovski, O.; Nobrega, J.N.; Raymond, R.J.; Bambico, F.R.; Nashed, M.G.; Garcia, A.; Bloomfield, P.M.; Houle, S.; Mizrahi, R.; et al. Evaluation of a novel radiotracer for positron emission tomography imaging of reactive oxygen species in the central nervous system. *Nucl. Med. Biol.* **2017**, *53*, 14–20. [CrossRef] [PubMed]

58. Takai, N.; Abe, K.; Tonomura, M.; Imamoto, N.; Fukumoto, K.; Ito, M.; Momosaki, S.; Fujisawa, K.; Morimoto, K.; Takasu, N.; et al. Imaging of reactive oxygen species using [3H]hydromethidine in mice with cisplatin-induced nephrotoxicity. *EJNMMI Res.* **2015**, *5*, 116. [CrossRef] [PubMed]

59. Abe, K.; Tonomura, M.; Ito, M.; Takai, N.; Imamoto, N.; Rokugawa, T.; Momosaki, S.; Fukumoto, K.; Morimoto, K.; Inoue, O. Imaging of reactive oxygen species in focal ischemic mouse brain using a radical trapping tracer [3H]hydromethidine. *EJNMMI Res.* **2015**, *5*, 115. [CrossRef] [PubMed]

60. Hou, C.; Hsieh, C.J.; Li, S.; Lee, H.; Graham, T.J.; Xu, K.; Weng, C.C.; Doot, R.K.; Chu, W.; Chakraborty, S.K.; et al. Development of a positron emission tomography radiotracer for imaging elevated levels of superoxide in neuroinflammation. *ACS Chem. Neurosci.* **2017**. [CrossRef] [PubMed]

61. Carstens, E.; Moberg, G.P. Recognizing pain and distress in laboratory animals. *ILAR J.* **2000**, *41*, 62–71. [CrossRef] [PubMed]

62. Okamura, T.; Okada, M.; Kikuchi, T.; Wakizaka, H.; Zhang, M.R. A ^{11}C-labeled 1,4-dihydroquinoline derivative as a potential PET tracer for imaging of redox status in mouse brain. *J. Cereb. Blood Flow Metab.* **2015**, *35*, 1930–1936. [CrossRef] [PubMed]

63. Haynes, S.E.; Hollopeter, G.; Yang, G.; Kurpius, D.; Dailey, M.E.; Gan, W.B.; Julius, D. The P2Y12 receptor regulates microglial activation by extracellular nucleotides. *Nat. Neurosci.* **2006**, *9*, 1512–1519. [CrossRef] [PubMed]

64. Moore, C.S.; Ase, A.R.; Kinsara, A.; Rao, V.T.; Michell-Robinson, M.; Leong, S.Y.; Butovsky, O.; Ludwin, S.K.; Seguela, P.; Bar-Or, A.; et al. P2Y12 expression and function in alternatively activated human microglia. *Neurol. Neuroimmunol. Neuroinflamm.* **2015**, *2*, e80. [CrossRef] [PubMed]

65. Villa, A.; Klein, B.; Janssen, B.; Pedragosa, J.; Pepe, G.; Zinnhardt, B.; Vugts, D.J.; Gelosa, P.; Sironi, L.; Beaino, W.; et al. Identification of new molecular targets for PET imaging of microglial anti-inflammatory phenotype. *Theranostics* **2018**, submitted.

molecules

MDPI

Article

Exploring the Metabolism of (+)-[18F]Flubatine In Vitro and In Vivo: LC-MS/MS Aided Identification of Radiometabolites in a Clinical PET Study †

Friedrich-Alexander Ludwig [1,*], **Steffen Fischer** [1], **René Smits** [2], **Winnie Deuther-Conrad** [1], **Alexander Hoepping** [2], **Solveig Tiepolt** [3], **Marianne Patt** [3], **Osama Sabri** [3,‡] and **Peter Brust** [1,‡]

[1] Helmholtz-Zentrum Dresden-Rossendorf, Research Site Leipzig, Institute of Radiopharmaceutical Cancer Research, Permoserstraße 15, 04318 Leipzig, Germany; s.fischer@hzdr.de (S.F.); w.deuther-conrad@hzdr.de (W.D.-C.); p.brust@hzdr.de (P.B.)
[2] ABX advanced biochemical compounds GmbH, Heinrich-Gläser-Straße 10-14, 01454 Radeberg, Germany; smits@abx.de (R.S.); hoepping@abx.de (A.H.)
[3] Department of Nuclear Medicine, University Hospital Leipzig, Liebigstraße 18, 04103 Leipzig, Germany; solveig.tiepolt@medizin.uni-leipzig.de (S.T.); marianne.patt@medizin.uni-leipzig.de (M.P.); osama.sabri@medizin.uni-leipzig.de (O.S.)
* Correspondence: f.ludwig@hzdr.de; Tel.: +49-341-234179-4617; Fax: +49-341-234179-4699
† This publication is dedicated to Jörg Steinbach on the occasion of his 65th birthday.
‡ These authors contributed equally to this work.

Received: 29 January 2018; Accepted: 16 February 2018; Published: 20 February 2018

Abstract: Both (+)-[18F]flubatine and its enantiomer (−)-[18F]flubatine are radioligands for the neuroimaging of α4β2 nicotinic acetylcholine receptors (nAChRs) by positron emission tomography (PET). In a clinical study in patients with early Alzheimer's disease, (+)-[18F]flubatine ((+)-[18F]**1**) was examined regarding its metabolic fate, in particular by identification of degradation products detected in plasma and urine. The investigations included an in vivo study of (+)-flubatine ((+)-**1**) in pigs and structural elucidation of formed metabolites by LC-MS/MS. Incubations of (+)-**1** and (+)-[18F]**1** with human liver microsomes were performed to generate in vitro metabolites, as well as radiometabolites, which enabled an assignment of their structures by comparison of LC-MS/MS and radio-HPLC data. Plasma and urine samples taken after administration of (+)-[18F]**1** in humans were examined by radio-HPLC and, on the basis of results obtained in vitro and in vivo, formed radiometabolites were identified. In pigs, (+)-**1** was monohydroxylated at different sites of the azabicyclic ring system of the molecule. Additionally, one intermediate metabolite underwent glucuronidation, as also demonstrated in vitro. In humans, a fraction of 95.9 ± 1.9% (*n* = 10) of unchanged tracer remained in plasma, 30 min after injection. However, despite the low metabolic degradation, both radiometabolites formed in humans could be characterized as (i) a product of *C*-hydroxylation at the azabicyclic ring system, and (ii) a glucuronide conjugate of the precedingly-formed *N*8-hydroxylated (+)-[18F]**1**.

Keywords: [18F]flubatine; NCFHEB; [18F]FLBT; radiometabolites; glucuronides; liquid chromatography–tandem mass spectrometry (LC-MS/MS); liver microsomes; positron emission tomography (PET); nicotinic acetylcholine receptors (nAChRs)

1. Introduction

Both enantiomers (+)-[18F]flubatine ((+)-(1*S*,5*R*,6*R*)-6-(6-[18F]fluoro-pyridine-3-yl)-8-azabicyclo-[3.2.1]octane, (+)-[18F]**1**, Figure 1) and (−)-[18F]flubatine ((−)-[18F]**1**) are radioligands for imaging of α4β2 nicotinic acetylcholine receptors (nAChRs) in brain by positron emission tomography (PET). Their development starting from the alkaloid (−)-epibatidine [1], originally isolated from the poison dart frog *Epipedobates anthonyi*, has been widely reported [2–4]. In the framework of preclinical and clinical studies, the metabolism of (−)-[^{18}F]**1** in pigs [5], rhesus monkeys [6–8], and in humans [3,9–11] was examined mainly with the purpose of determination of the fraction of unchanged tracer over time to enable a metabolite correction of the PET data obtained. In human, (−)-[^{18}F]**1** showed a high metabolic stability [9].

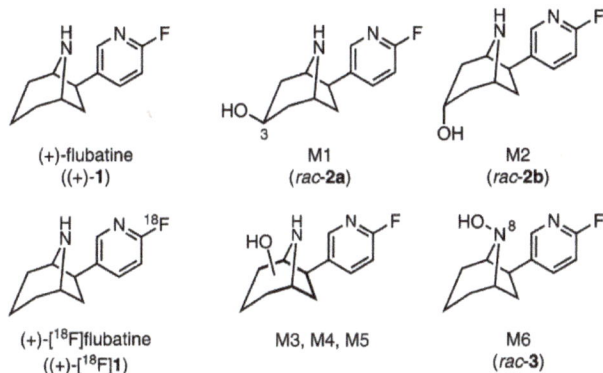

Figure 1. Chemical structures of (+)-flubatine ((+)-**1**), (+)-[^{18}F]flubatine ((+)-[^{18}F]**1**) and metabolites M1–M6, known from literature [4]. For M1, M2, and M6 the numbers in brackets refer to synthesized references (*rac*-**2a**, *rac*-**2b**, *rac*-**3**) used in this publication. References *rac*-**2a** and *rac*-**2b** have been described previously in [4].

For the determination of the fraction of unchanged tracer, samples of arterial blood are collected, prepared, and analyzed by high performance liquid chromatography followed by online radioactive detection (radio-HPLC). Alternatively, eluate fractions from HPLC separation can be collected and inspected by a gamma counter. By radio-HPLC, in addition to the unchanged tracer, metabolites, that still bear the radioactive nuclide, named radiometabolites, can also be detected. However, due to the low concentrations, characterization of fluorine-18 bearing radiometabolites regarding their chemical structures is usually not possible in a direct manner. For structural elucidation additional investigations have to be conducted, e.g., by using the non-labelled parent compound of the tracer and liquid chromatrography-mass spectrometry (LC-MS) or, more selective, liquid chromatrography-tandem mass spectrometry (LC-MS/MS), instead of radio-HPLC [12]. For studying the metabolism in vivo, the non-labelled parent compound can be administered to small animals, e.g., rodents, in appropriate dosage to obtain samples of tissue or body liquids for examination by LC-MS and LC-MS/MS [13]. In addition to the fact that species differences have to be considered [14,15], in particular when making conclusions on the metabolism in humans, this approach has the advantage of covering all possible metabolic transformations, still summarized as phase I and phase II metabolism, but more exactly termed as 'functionalization' and 'conjugation' [16]. Drawbacks are, for instance, a possible ion suppression in MS detection caused by residual biological matrix [17]. Furthermore, animal studies have to be approved by the responsible authorities. To circumvent some of these issues, different in vitro models for metabolism studies have been described [18,19]. For basic investigations, especially liver microsomes, which contain cytochrome P450 enzymes and different types of transferases [19], are a suitable means that is easy to use, and of minimal cost. Since microsomes are also available from humans, the problem of species differences can be avoided to some extent.

Such in vitro investigations have already been performed for (+)-**1** and (−)-flubatine ((−)-**1**), using liver microsomes from humans (HLM) and mice (MLM), as published by our group [4]. Incubations in the presence of β-nicotinamide adenine dinucleotide 2′-phosphate reduced tetrasodium salt (NADPH) resulted in the formation of hydroxylation products, provided in Figure 1. Published results from in vivo investigations of (+)-[^{18}F]**1** showed the formation of one metabolite, that was not characterized further, but assumed to possibly pass the blood-brain barrier (BBB) and, therefore, confound brain PET images [4,5].

Considering this presumption and the importance of knowledge about the metabolic fate of a new tracer [20], we aimed to support the evaluation of (+)-[^{18}F]**1** and investigated its metabolic pathways during a first clinical PET study, which compared the status of α4β2 nAChRs in patients with Alzheimer's disease (AD) and healthy controls [21]. As shown in the present publication, we studied the metabolism of (+)-**1** in pigs and elucidated the structures of metabolites. For the first time, we investigated both phase I and II metabolism and performed incubations of (+)-**1** and (+)-[^{18}F]**1** with HLM under conditions for oxidation and glucuronide conjugation. Resulting LC-MS/MS and radio-HPLC data provided the opportunity to conclude about the identity of formed in vitro radiometabolites. On that basis, we finally characterized radiometabolites formed in human after administration of (+)-[^{18}F]**1**.

2. Results

2.1. Radiosynthesis of (+)-[^{18}F]1

(+)-[^{18}F]**1** was synthesized by nucleophilic substitution [22] under GMP conditions for human application. The product was obtained with a radiochemical purity >98.0% and molar activity >1.0 × 10^6 GBq/mmol as described for (−)-[^{18}F]**1** [23,24].

2.2. Synthesis of rac-8-Hydroxy-flubatine (rac-3)

Rac-8-Hydroxy-flubatine was synthesized starting from *rac*-**1** [2] to enable identification of the corresponding metabolite, as well as the radiometabolite by HPLC co-injection (Figure 2). After treatment with peracetic acid the desired hydroxy compound was isolated in low yield due to low conversion of the starting material. Attempts to purify *rac*-**3**, e.g. by preparative HPLC, did not succeed. This can be explained by lack of stability of the N8-hydroxy function of the molecule, as it has been proposed in literature for metabolites that contain a secondary hydroxylamine structure [25].

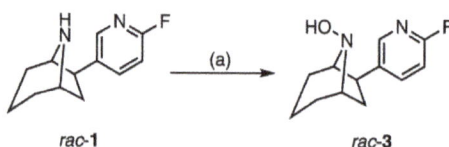

Figure 2. Synthesis of *rac*-8-hydroxy-flubatine (*rac*-**3**). Conditions: (**a**) HOOAc, 1 M NaHCO$_3$, room temperature, six days, 9%.

2.3. Investigation of (+)-1 in Pig–Identification of Phase I and Phase II Metabolites by LC-MS/MS

Since the structures of radiometabolites could not be identified in a direct manner, non-labelled (+)-**1** instead of (+)-[^{18}F]**1** was investigated and administered to pigs. Plasma and urine samples were taken before, as well as 30 and 45 min, respectively, after injection. Samples were prepared by protein precipitation with cold acetonitrile, subsequent centrifugation, and solvent evaporation. In a first survey, the samples were screened for a set of possible phase I and phase II metabolites with aid of the metabolite identification software LightSight (Version 2.3.0.152038, AB SCIEX, Framingham, MA, USA). On this basis, detailed analyses were performed using multi reaction monitoring (MRM), enhanced product ion (EPI) and MS^3 scan modes for selective detection and structure elucidation of metabolic degradation products.

In plasma and urine, (+)-**1** (*m/z* 207.1 [M + H]$^+$) was still detectable with high signal intensities which indicates a high metabolic stability (Figure 3a,b). However, a series of metabolites was detected, generated by cytochrome P_{450}-dependent hydroxylation reactions. Appropriate MRM scans revealed that monohydroxylations took place exclusively at the azabicyclic ring system of the flubatine molecule. In brief, the chromatograms (Figure 3c,d) were recorded observing an MRM transition of *m/z* 223.1/110.0, which selectively detected a fluoro-azatropylium ion (*m/z* 110.0), that originated from an unmodified fluoropyridyl moiety after fragmentation of the parent ion at *m/z* 223.1 [M$-_{flubatine}$ + O + H]$^+$. The pattern of metabolites in pigs detected in this way was similar to those previously reported for in vitro experiments using MLM and HLM in the presence of NADPH [4]. Since, in the presented study, the same LC conditions were used as described in the mentioned publication [4], a direct comparison of data was possible. In summary, all of the hydroxylated products M1–M6 found in pigs were also formed in MLM, but only M4 and M6 in HLM. For the purpose of identification and assignment of the metabolites detected in pig enhanced product ion (EPI) spectra were recorded (Table 1). The references *rac*-**2a** and *rac*-**2b** [4], as well as the newly-synthesized *rac*-**3** (Figure 1) were analysed in an analogue manner.

For plasma and urine samples, some of the metabolites (M3, M4) showed differing retention times. Since a mixture of both samples showed no additional peaks, this can be attributed to influences of the respective matrices. The metabolites M1 and M2 were identified as the C-3 *exo* and the C-3 *endo* alcohol, respectively, as their retention times and fragmentation patterns matched that of the references *rac*-**2a** and *rac*-**2b** (Figure 1). M3 and M4 had the same retention properties as two metabolites reported for MLM incubation [4] and were concluded to be identical to them. M4 also matches one main metabolite formed by HLM [4], additionally confirmed by similar EPI spectra. In contrast, M3 showed an expected MS/MS fragmentation but the exact pattern did not correspond to the equally eluting metabolite reported for MLM [4]. For M5 and M6 no EPI spectra could be recorded due to low concentrations. However, for M3 and M4 the characteristic fragment ion at *m/z* 188.1 [M + H-NH$_3$-H$_2$O]$^+$ could be detected, which gives evidence for C-hydroxylations at the azabicyclic ring system, according to the proposed fragmentation pathway already published [4]. The same has to be assumed for M5, since its retention time is equal to a C-hydroxylated metabolite previously detected in considerable amounts only after MLM incubation of (−)-flubatine [4]. Metabolite M6 eluted at the same time as the synthesized reference *rac*-**3** and accordingly was identified as N8-hydroxylated flubatine, which was also observed after incubations with MLM or HLM in the cited study [4]. For the C-hydroxy isomers M3, M4, and M5, a more exact assignment of their structures was not possible, also due to limitations in syntheses of the respective references.

Furthermore, the phase II metabolite M7a was found. It was detected in both plasma and urine, when an MRM transition of *m/z* 399.2/223.1 was monitored (Figures 3e,f and 5). The resulting neutral loss of *m/z* 176.0 ($C_6H_8O_6$) is characteristic for glucuronides. This proved that M7a was formed by hydroxylation and glucuronidation of (+)-**1**. By contrast, a product of a sole glucuronidation of (+)-**1** could not be detected. In MS^3 experiments for M7a, the primary fragment ion at *m/z* 223.1 [M + H-$C_6H_8O_6$]$^+$ showed a fragmentation pattern very similar to that of the 8-hydroxy-flubatine reference *rac*-**3**. In particular the occurrence of a secondary fragment ion at *m/z* 190.1 [M + H-$C_6H_8O_6$-NH$_2$OH]$^+$ gave evidence for an N8-hydroxyl substitution present in the detected glucuronide M7a. Additionally, another glucuronide isomer (M8a) was detectable in very small amounts in plasma, but not investigated further.

A few additional MRM measurements showed low signal intensities and gave only indications for further metabolites. For the MRM transition of *m/z* 255.1/126.0 one peak was detected at low retention time, which could be interpreted as a product of a two-fold hydroxylation at the azabicyclic ring system together with a hydroxylation or oxidation at the pyridine ring. Some data indicated cysteine conjugation of hydroxylation products, but they also were of only minor importance for the metabolic pathway.

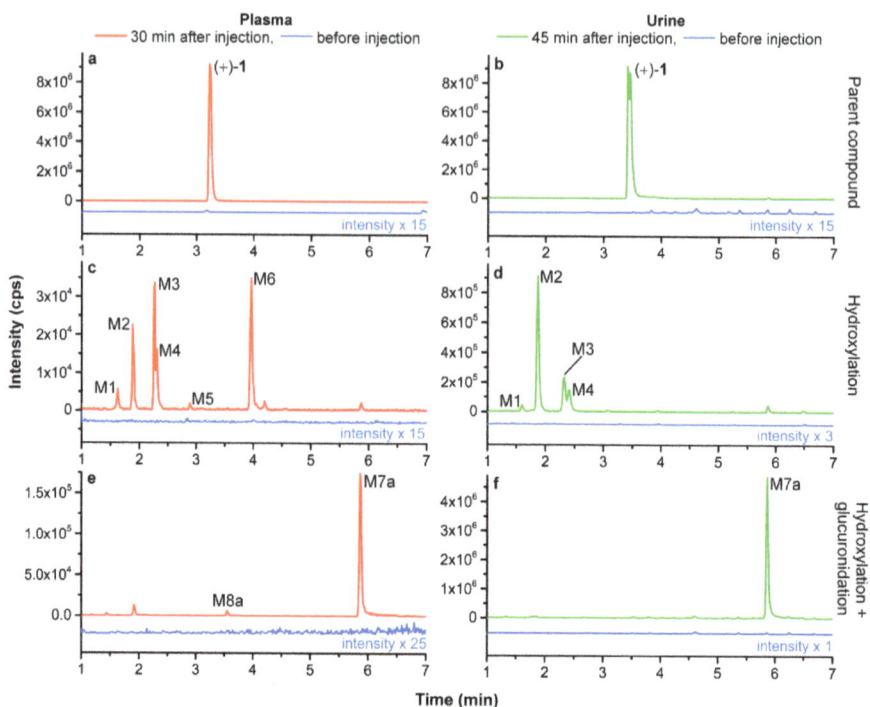

Figure 3. LC-MS/MS chromatograms of plasma and urine before and after injection of (+)-**1** into pigs; (**a**,**b**) detected (+)-**1** (MRM *m/z* 207.1/110.0), (**c**,**d**) detected monohydroxylated metabolites (MRM *m/z* 223.1/110.0), (**e**,**f**) detected monohydroxylated and glucuronidated metabolites (MRM *m/z* 399.2/223.1). Signals in (**c**,**d**) at 5.9 min and (**e**) at 1.9 min are supposed to result from ion-channel cross talks [26].

Table 1. Metabolites found in pig plasma and urine after injection of (+)-**1**.

Metabolite	t_R (min) [a] Plasma	t_R (min) [a] Urine	MRM Transition	EPI Fragmentation [b] (% Intensity in Brackets)	Identification
M1	1.62	1.60	223.1/110.0	81.9 (100), 205.0 (71), 223.0 (32), 163.1 (26), 162.1 (23), 110.0 (17), 136.0 (15), 131.0 (15), 103.9 (14), 188.1 (9)	
M2	1.89	1.87	223.1/110.0	223.1 (100), 136.0 (40), 110.0 (39), 162.1 (37), 180.1 (36), 188.0 (28), 179.1 (25), 114.0 (13), 124.0 (13), 206.0 (11), 160.0 (9), 138.0 (8), 142.0 (4), 205.1 (4)	
M3	2.26	2.32	223.1/110.0	223.1 (100), 150.0 (63), 164.1 (44), 110.0 (42), 188.1 (41), 82.0 (39), 135.1 (31), 206.1 (18), 176.1 (16), 120.1 (14), 162.1 (12), 124.0 (12), 205.1 (12), 134.0 (12), 136.1 (12), 163.1 (10)	Hydroxylation
M4	2.31	2.40	223.1/110.0	81.9 (100), 205.1 (60), 188.0 (39), 177.1 (31), 223.1 (23), 110.0 (22), 176.1 (9), 160.0 (7)	
M5	2.89	- [c]	223.1/110.0	no EPI spectrum due to low intensity	
M6	3.96	- [c]	223.1/110.0	no EPI spectrum due to low intensity	
M7a	5.87	5.86	399.2/223.1	205.2 (100), 223.2 (44), 136.1 (34), 203.2 (26), 110.0 (13), 162.1 (11), 177.2 (9), 83.0 (9), 207.2 (8), 189.1 (6), 109.1 (6), 190.2 (6)	Hydroxylation + glucuronidation

[a] Retention time (time/min); [b] parameters for data acquisition described in Section 4.7.; [c] not detected in urine.

2.4. *In Vitro Glucuronidation of (+)-[^{18}F]1, (+)-1, (−)-1, and rac-3 by HLM*

Since hydroxylation together with subsequent glucuronidation turned out as the significant metabolic pathway in pig, in vitro experiments were done to investigate both steps. Whilst hydroxylation of (+)-**1** and (−)-**1** has already been studied and reported [4], glucuronide conjugation was investigated for the first time. In preparation for analyses of samples from the clinical PET study radiolabeled (+)-[^{18}F]**1** was investigated similarly.

In brief, carrier added (+)-[^{18}F]**1** was incubated with HLM under conditions for oxidation and glucuronidation, that means in the presence of NADPH and uridine 5′-diphosphoglucuronic acid trisodium salt (UDPGA). Prepared samples were analysed by LC-MS/MS, as well as radio-HPLC. Furthermore, (−)-**1** was incubated in the same manner, whereas the N8-hydroxy reference *rac*-**3** was incubated without NADPH.

For both flubatine enantiomers (+)-**1** and (−)-**1** monohydroxylation products were observed by LC-MS/MS after HLM incubations as previously described by our group (Figure 4a,b) [4]. Originating from (+)-**1** and in accordance with results published, only two metabolites were detected, namely a C-hydroxylated metabolite (M4) and N8-hydroxylated flubatine (M6), which were also detected in pig in addition to other metabolites. Regarding the formation of glucuronides, studied for the first time, there was no glucuronidation of (+)-**1** or (−)-**1**, when each of them was incubated with UDPGA in absence of NADPH. Whereas in the presence of NADPH and UDPGA, under conditions for oxidation and glucuronidation, two products, M7a and M8a or M7b and M8b, were detected respectively by monitoring MRM transitions of *m/z* 399.2/223.1 or *m/z* 399.2/205.1 (Figure 4a,b, Table 2). As shown in Figure 4c, all of these four glucuronide isomers were formed when racemic 8-hydroxy-flubatine (*rac*-**3**) was incubated in the presence of UDPGA. This proves the N8-hydroxy metabolite M6, which corresponds to *rac*-**3**, to be the precursor metabolite for glucuronide conjugates. It should be noted that glucuronide M7a was formed in vitro as well as in vivo from (+)-**1**. Though having demonstrated the role of M6, further structural elucidation of the resulting glucuronide conjugates is challenging. Both the N8-hydroxy function and the pyridine-N of the molecule are possible sites for glucuronidation [27]. The finding that in the presence of UDPGA only, the N8-hydroxy derivative *rac*-**3** underwent glucuronidation, but in contrast not (+)-**1**, might give an indication for assignment. Due to the obvious necessity of the N8-hydroxy function for glucuronidations, the hypothesis can be proposed that the pyridine-N of the molecule remained unaffected. However, conjugation at the N8-hydroxy function gives the possibility of two structural isomers that can be formed: N-O-glucuronide and N$^+$(O$^-$)-glucuronide (Figure 5), which cannot be conclusively distinguished by mass spectrometry [28]. Therefore, the metabolites M7a, M7b, M8a, and M8b were tentatively identified as N-O-glucuronides and N$^+$(O$^-$)-glucuronides without an exact assignment. Remarkably, M7a and M7b eluted at later time points than the parent enantiomers, which is very untypical for glucuronides under RP-HPLC conditions, as it can be seen by comparison with retention properties of glucuronides reported in literature [29]. What kind of retention mechanism might be responsible for this phenomenon remains an open question.

Corresponding to the pattern of non-labelled metabolites detected by LC-MS/MS after HLM incubation of carrier added (+)-[^{18}F]**1** (Figure 4a), radio-HPLC data revealed the respective in vitro radiometabolites [^{18}F]M4, [^{18}F]M6, and [^{18}F]M7a (Figure 6). These data were used for the identification of radiometabolites formed in humans.

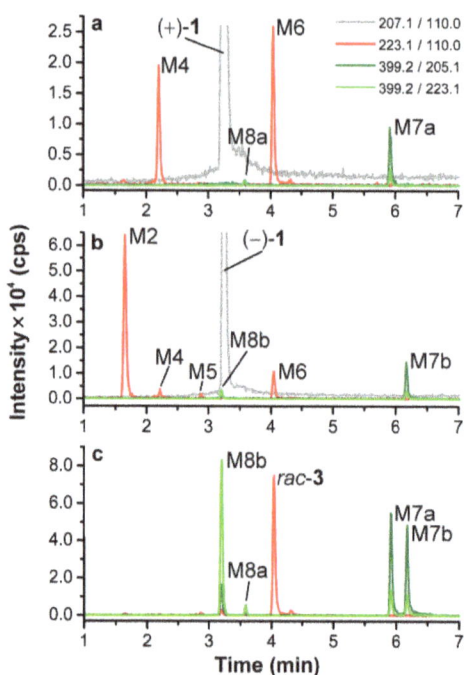

Figure 4. LC-MS/MS chromatograms after HLM incubation in presence of NADPH and UDPGA (37 °C, 120 min), (**a**) (+)-**1**, (**b**) (−)-**1**, (**c**) *rac*-**3** (without NADPH); Monitored MRM transitions: *m/z* 207.1/110.0 for (+)-**1** and (−)-**1**, *m/z* 223.1/110.0 for monohydroxylated metabolites (including *rac*-**3**), *m/z* 399.2/205.1, 399.1/223.1 for monohydroxylated and glucuronidated metabolites.

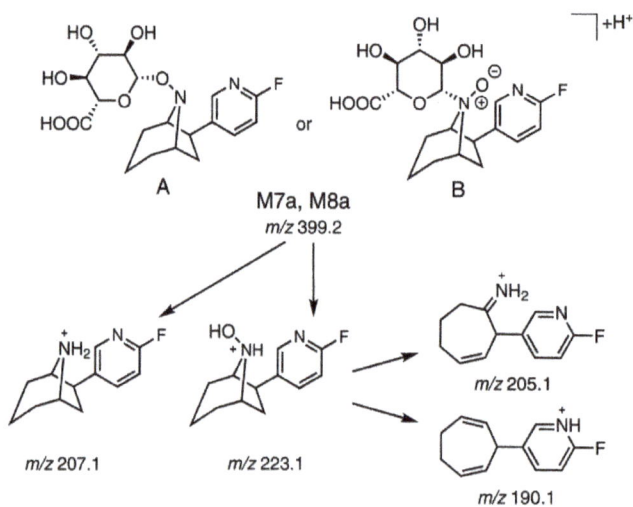

Figure 5. Proposed structures and fragmentation pathways shown for glucuronide conjugates M7a and M8a, drawn as N-O-glucuronide (**A**) and $N^+(O^-)$-glucuronide (**B**); similarly, for enantiomers M7b and M8b.

Table 2. Glucuronide conjugates formed by incubation with HLM.

Metabolite [a]	t_R (min) [b]	Substrate	MRM Transitions	EPI Fragmentation [c] (% Intensity in Brackets)	Identification
M7b	6.18	(−)-**1** or *rac*-**3**	399.2/223.1 399.2/205.1	205.1241 (100), 223.1 (60), 136.0 (37), 203.1 (20), 110.0 (18), 177.1 (16), 83.0 (11), 207.1 (11), 162.0 (11), 82.0 (10), 116.0 (8), 163.1 (8), 190.1 (7), 124.0 (7)	Hydroxylation + Glucuronidation [d]
M8a	3.60	(+)-**1** or *rac*-**3**	399.2/223.1 399.2/205.1	223.1 (100), 205.1 (21), 124.1 (6), 110.2 (6), 136.0 (6), 84.9 (4), 113.0 (4)	
M8b	3.22	(−)-**1** or *rac*-**3**	399.2/223.1 399.2/205.1	223.1 (100), 205.1 (27), 136.1 (16), 84.9 (9), 110.0 (8), 150.0 (6), 203.2 (6), 82.9 (6), 67.9 (4), 190.1 (4), 113.0 (3)	

[a] data for M7a, formed from (+)-**1** or *rac*-**3**, correspond to those stated in Table 1; [b] Retention time (time/min); [c] parameters for data acquisition described in Section 4.7.; [d] for *rac*-**3** glucuronidation only.

2.5. Metabolism Studies of (+)-[^{18}F]**1** in Humans

2.5.1. Metabolic Stability of (+)-[^{18}F]**1** and Detection of Radiometabolites

For identification of radiometabolites formed in human, arterial blood was received from two AD patients and eight healthy volunteers taken during PET measurements at 15 and 30 min after injection of of (+)-[^{18}F]**1**. Urine was collected from 14 subjects uniquely after 90–128 min. After centrifugation of blood, obtained plasma was extracted with acetonitrile and the recovery of total radioactivity was 95.4 ± 2.9% (mean ± SD, n = 17; samples: 15 and 30 min after injection). Prepared samples of plasma and untreated urine were analysed by radio-HPLC (Figure 6).

Fractions of non-metabolised (+)-[^{18}F]**1** remained very high in all samples investigated. In plasma, 15 min and 30 min after injection 97.4 ± 2.7% (mean ± SD, n = 10) and 95.9 ± 1.9% (n = 10) of (+)-[^{18}F]**1** remained unchanged, respectively, while 95.1 ± 4.5% (n = 14) were found for urine. Due to the low number of samples from patients with AD, data could not be compared statistically with that from human controls. However, analyses by radio-HPLC revealed no obvious differences in metabolic degradation between both groups.

One very fast eluting degradation product was detectable at very short retention times (3.5–4.0 min) only in seven of the plasma samples analysed in total and in five of the samples from urine. The signal represented, on average, 1.2% and 0.6% in plasma, 15 min and 30 min after injection, respectively, and 0.5% in urine. Due to its retention properties, it might correspond to metabolically-formed [^{18}F]fluoride, but was not investigated further. Generally, detection and quantification of degradation products were limited due to low peak intensities, as shown by representative radio-HPLC chromatograms in Figure 6. However, two formed radiometabolites could be detected. Radiometabolite h-M1, which had a shorter retention time than (+)-[^{18}F]**1**, represented 0.0−2.5%, 0.0−3.8%, and 0.0−4.9% in plasma, 15 and 30 min after injection, and urine, respectively. For radiometabolite h-M2, which eluted after a longer retention time, fractions of 0.0−2.6%, 0.0−4.0%, and 0.4–10.7% were determined in the same samples.

2.5.2. Identification of Radiometabolites of (+)-[^{18}F]**1** formed in Humans

Since samples obtained from human subjects were measured under the same chromatographic conditions as samples from HLM incubations of (+)-[^{18}F]**1**, radio-HPLC data were used for comparison and assignment of the radiometabolites found in vivo (Figure 6). Both h-M1 and h-M2 are in good accordance with the in vitro metabolites [^{18}F]M4 and [^{18}F]M7a, respectively, regarding their retention times. Therefore, both radiometabolites could be identified, namely h-M1 as a product of a C-hydroxylation at the azabicyclic ring system and h-M2 as a glucuronide conjugate of the N8-hydroxylated (+)-[^{18}F]**1** (Figure 6).

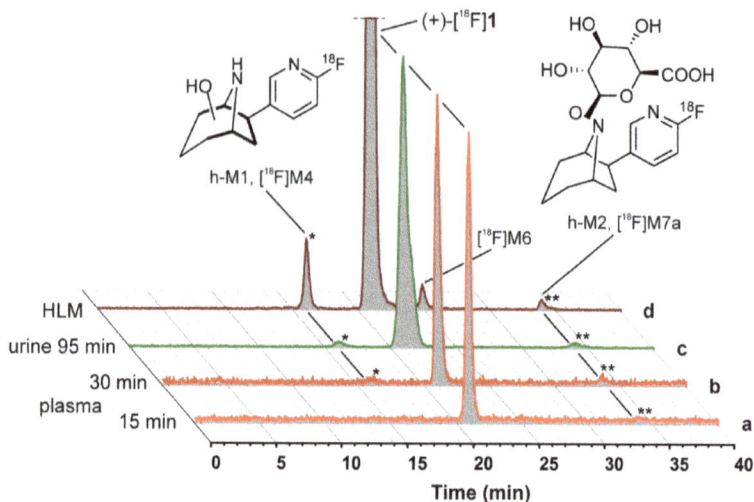

Figure 6. Comparison of metabolic profiles of (+)-[^{18}F]**1** (in vivo vs. in vitro) and identification of radiometabolites. Representative radio-HPLC chromatograms of samples obtained from human subjects: (**a**) plasma, 15 min, (**b**) plasma, 30 min, (**c**) urine, 95 min after injection, as well as (**d**) after HLM incubation (NADPH, UDPGA, TRIS, pH 8.4, 37 °C, 120 min). Scaling was adjusted for each chromatogram.

3. Discussion

The presented results should be seen in the context of previous metabolism studies on both enantiomers of flubatine and [^{18}F]flubatine [3–11]. In particular, the characterisation of phase I metabolites of (+)-**1** formed by HLM and MLM, as well as the assignment of corresponding incubated with radiometabolites in mouse, published by our group [4], are highly relevant for the discussion.

As shown, monohydroxylations at the azabicyclic ring system of (+)-**1** were the major phase I metabolism pathways in pig. This is in accordance with the previously reported NADPH-dependent degradation by HLM and MLM [4], while the metabolite pattern originating from pigs was more similar to the latter. In pigs, glucuronidation of the precedingly-formed N8-hydroxy-flubatine (M6) revealed as very dominant and could also be demonstrated, as reported here, by incubations with HLM in presence of NADPH and UDPGA. It can be concluded that the resulting glucuronide conjugate M7a plays a major role in the urinary excretion of (+)-**1**, which, however, exhibited considerable stability.

As described in the forementioned publication [4], after injection of (+)-[^{18}F]**1** in mouse, one major radiometabolite could not be characterized but was assumed to be a phase II metabolite. Having comparable chromatographic properties it is now plausible that this radiometabolite corresponds to the glucuronide [^{18}F]M7a (h-M2). The formation of [^{18}F]M7a can also be concluded for one radiometabolite, that was mentioned in a previous study, that investigated (+)-[^{18}F]**1** in pigs [5].

In human, radio-HPLC analyses from plasma and even urine samples revealed very high fractions of (+)-[^{18}F]**1**. In comparison, with results already published for (−)-[^{18}F]**1**, which showed a fraction of the unchanged tracer of ~91% [9] and ~93% [10], 30 min after injection, we found that (+)-[^{18}F]**1** exhibited on average a fraction of 96% of unchanged tracer at the same time point. In vitro and in vivo studies together with LC-MS/MS and radio-HPLC served successfully as means for characterization of radiometabolites formed. Thus, despite their very low fractions, two degradation products could be detected and characterized and statements about the metabolic pathway in human could be made: In addition to its high stability, (+)-[^{18}F]**1** underwent a C-hydroxylation at the azabicyclic ring system to form h-M1, which was equal to the in vitro formed [^{18}F]M4. Hydroxylation at the bridgehead nitrogen N8 and subsequent conjugation with glucuronic acid resulted in the phase II radiometabolite h-M2, which was equal to the in vitro formed [^{18}F]M7a. In summary, the radiometabolites detected in human were in good agreement with those formed by HLM. However, in pigs, a larger number of C-hydroxylation products was found.

The metabolism of the enantiomer (−)-[^{18}F]**1** has already been investigated in preclinical in vivo studies [5,7,8] and in human [3,9,10], but without structural elucidation of radiometabolites. It has been reported, that in human one single metabolite was detected in low amounts by radio-HPLC but not characterized [3]. On the basis of the results presented here, it can be assumed that this radiometabolite corresponds to M7b. That means it is a product of N-hydroxylation of (−)-[^{18}F]**1** and a subsequent glucuronidation.

Penetration of the blood-brain barrier by radiometabolites has a high impact on the quality of brain PET measurements. The structures of the radiometabolites of (+)-[^{18}F]**1**, namely h-M1 and h-M2, that were identified in the present paper, show no indications, e.g., high lipophilicity, to penetrate the human blood-brain barrier. This is in accordance with preclinical metabolism studies performed in mice, in which a low fraction of radiometabolites (< 7%) was detected in the brain, 30 min after injection [4].

Based on the present results and previous metabolism studies [4,5], (+)-[^{18}F]**1** could be conclusively assessed regarding its metabolism. Indications for high metabolic stability, as they were obtained previously from in vitro studies [4], could be validated in vivo. In particular, in human, a fraction of 95.9 ± 1.9% ($n = 10$) of the unchanged tracer was detected in plasma, 30 min after injection. Incubations with liver microsomes served as appropriate means to generate in vitro metabolites and radiometabolites, that revealed relevant in vivo. Subsequently, in human, the two radiometabolites detected could be characterized regarding their structures. However, due to the very low extent of metabolic degradation of (+)-[^{18}F]**1**, metabolism does not have to be considered in the description of the arterial input function for the kinetic modelling of PET data.

4. Materials and Methods

4.1. Chemicals and Reagents

Solvents for synthesis were purchased from Merck (Darmstadt, Germany) and Fisher Scientific (Schwerte, Germany). Chemicals were obtained from Merck, Fisher Scientific, Sigma-Aldrich (Steinheim, Germany), C. Roth (Karlsruhe, Germany) and Machery-Nagel (Düren, Germany). (+)-**1**, (−)-**1**, and *rac*-**1** were purchased from ABX (Radeberg, Germany). All chemical reagents were of highest commercially available quality and applied without further purification. The references *rac*-**2a** and *rac*-**2b** were synthesized as previously reported [4]. Acetonitrile (gradient grade) was purchased from VWR International (Darmstadt, Germany). Acetonitrile and water (both for LC-MS) were purchased from Fisher Scientific. Ammonium formate (for HPLC) was purchased from Acros Organics (Geel, Belgium). Formic acid and ammonium formate (both LC-MS), testosterone, NADPH, UDPGA, alamethicin and $MgCl_2$ were purchased from Sigma-Aldrich. GIBCO human liver microsomes (HLM, pooled donors, 20 mg/mL) were purchased from Life Technologies (Darmstadt, Germany). Dulbecco's phosphate-buffered saline (PBS) (without Ca^{2+}, Mg^{2+}) was purchased from Biochrom (Berlin, Germany).

4.2. Radiosynthesis of (+)-[^{18}F]**1**

(+)-[^{18}F]**1** was synthesized under GMP conditions for human application as described for (−)-[^{18}F]**1** [23].

4.3. Synthesis of Reference Compound rac-3

Reaction monitoring was performed by thin-layer chromatography (TLC) using TLC plastic sheets precoated with UV254 fluorescent indicator (Polygram SIL G/UV254, Machery-Nagel). Visualization of spots was effected by irradiation with an UV lamp (254 nm and 366 nm; Herolab, Wiesloch, Germany). ^{1}H Nuclear magnetic resonance (NMR) spectra were obtained with a Bruker AV500 spectrometer (Bruker Corporation, Billerica, MA, USA). Chemical shifts are reported as δ values. Coupling constants are reported in Hertz. Electrospray ionisation mass spectra were obtained using a Surveyor MSQ Plus mass detector (Thermo Fisher Scientific, Dreieich, Germany).

4.3.1. (+/−)-exo-6-(6-Fluoro-pyridin-3-yl)-8-aza-bicyclo[3.2.1]octan-8-ol ((*rac*)-8-hydroxy-flubatine, *rac*-3)

Racemic flubatine (*rac*-1) (75 mg, 0.36 mmol) was suspended in 10.7 mL 1M sodium bicarbonate. Peracetic acid (39%, 0.62 mL, 10 eq.) was added dropwise and the reaction mixture was stirred at room temperature. Further 10.7 mL 1 M sodium bicarbonate and 0.62 mL peracetic acid (39%) were added after 1, 2, and 3 days. The reaction mixture was diluted with dichloromethane after stirring for three further days. The phases were separated and the aqueous phase was extracted two times with dichloromethane. The combined organic phases were dried over sodium sulphate and the solvent was removed in vacuo. The crude product was purified by column chromatography (hexane: ethyl acetate = 1:1 + 1% triethylamine) to afford *rac*-3 (7.5 mg, 9%) and unreacted starting material *rac*-1 (36.2 mg, 48%). ^{1}H-NMR *rac*-3, purity ca. 80% (CDCl$_3$, 500 MHz): δ = 8.19 (d, *J* = 2.3 Hz, 1H), 8.11 (dt, *J* = 8.2 Hz, 2.6 Hz, 1H), 6.83 (dd, *J* = 8.6 Hz, 3.0 Hz, 1H), 3.70-3.77 (m, 1H), 3.62 (bd, *J* = 1.4 Hz, 1H), 3.27 (dd, *J* = 9.4 Hz, 6.9 Hz, 1H), 2.30-2.38 (m, 1H), 2.19 (dd, *J* = 13.0 Hz, 9.7 Hz, 1H), 1.43-1.87 (m, 7H). MS (ESI +): *m/z* 223.2 (M + H)$^+$.

4.4. In Vivo Metabolism Study of (+)-1 in Pigs

The animal experiment was conducted under procedures approved by the respective State Animal Care and Use Committee and in accordance with the German Law for the Protection of Animals and the EU directive 2010/63/EU.

A female piglet (German landrace, 15 kg, eight weeks of age, obtained from Medizinisch-Experimentelles Zentrum, Universität Leipzig, Leipzig, Germany) was used. It was deprived of food, but not of water, for 24 h before delivery to the laboratory. As premedication it received an i.m. injection of 15 mg midazolam and was initially anesthetized with 1.5% isoflurane in 70% nitrous oxide and 30% oxygen. Additionally, all incision sites were infiltrated with 1% lidocaine. The anaesthesia was maintained throughout the surgical procedure with 0.8% isoflurane. After blank samples (t = 0 min) of blood and urine were taken, (+)-1 was infused as bolus of 67 μg/kg in 50 mL saline into the left jugular vein over 6 min. After 30 and 45 min samples of blood and urine, respectively, were taken and stored on ice until further treatment. Plasma was obtained after centrifugation of blood at 4000 rpm (Megafuge 1.0R, HERAEUS) for 10 min and, using 14–20 aliquotes of 3 mL, respectively, extracted with acetonitrile (1:3 *v/v*, −20 °C). After shaking (5 min, Vortex-Genie 2, Bohemia, New York, NY, USA), cooling on ice and final shaking (3 min) the samples were centrifuged at 5000 rpm (Centrifuge 5424, Eppendorf, Hamburg, Germany) for 15 min. Supernatants of combined aliquots were concentrated at 75 °C under a flow of nitrogen using the DB-3D TECHNE Sample Concentrator (Biostep, Jahnsdorf, Germany) to obtain residual volumes of ~150 μL. After filtration (Multoclear 0.45 μm, PTFE, CS-Chromatographie Service, Langerwehe, Germany) samples were stored at 4 °C until investigated by LC-MS/MS. Urine samples were prepared in the same manner.

4.5. Microsomal Incubations

All incubations with (+)-1, (−)-1, (+)-[^{18}F]1, *rac*-2a, *rac*-2b, and *rac*-3 as substrate, had a final volume of 250 μL and were performed in TRIS (pH 8.4). In the following, final concentrations are provided in brackets. TRIS, HLM (1 mg/mL), and alamethicin (50 μg/mL, from methanolic solution) were mixed and kept on ice for 15 min. Substrate (10 μM) and MgCl$_2$ (2 mM) were added and the mixture was preincubated at 37 °C for 3 min. After addition of analogously-preincubated NADPH (2 mM) and UDPGA (5 mM), incubations were proceeded by gentle shaking at 37 °C for 120 min using the BioShake iQ (QUANTIFOIL Instruments, Jena, Germany). After termination by adding

1.0 mL of cold acetonitrile ($-20\ °C$) and vigorous mixing for 30 s, the mixtures were stored at $4\ °C$ for 5 min. Thereafter, centrifugation at 14,000 rpm (Eppendorf Centrifuge 5424) was performed for 10 min, followed by concentration of the supernatants at $50\ °C$ under a flow of nitrogen (DB-3D TECHNE Sample Concentrator) to provide residual volumes of 40–70 μL, which were reconditioned by adding water to obtain samples of 100 μL, which were stored at $4\ °C$ until analysis by LC-MS/MS. For HLM incubations of carrier added (+)-[^{18}F]**1**, 6.7 MBq of the synthesized tracer in 20 μL TRIS was used together with (+)-**1** (10 μM) as the substrate. Prepared samples were immediately analyzed by radio-HPLC, and by LC-MS/MS at a later time. Incubations without HLM, NADPH, UDPGA, and substrates, respectively, were performed as negative controls, as well as to provide conditions only for oxidation and not glucuronidation, and vice versa. As positive controls testosterone (for oxidation) and 4-nitrophenol (for glucuronidation) were incubated at appropriate concentrations, similarly to the protocol described above. Complete conversions of both were confirmed by RP-HPLC analyses with UV detection.

4.6. Investigation of (+)-[^{18}F]**1** in Humans

All investigations were conducted in the framework of an approved and registered clinical study [21].

After injection of 259–308 MBq (mean: 285 MBq, $n = 14$) of (+)-[^{18}F]**1** into two AD patients and 12 healthy controls, respectively, 16 mL of arterial blood were taken from 10 subjects at 15 min and 30 min. The samples were collected directly into S-Monovettes 9 mL K3E (SARSTEDT, Nümbrecht, Germany) and stored on ice. From 14 subjects, after 90–128 min, circa 8 mL of urine were collected uniquely and stored on ice. Plasma was obtained by centrifugation of blood samples at 7000 rpm (UNIVERSAL 320 R, Hettich, Germany) for 7 min and extracted with acetonitrile (1:4 v/v, 10–15 aliquotes of 400 μL each). After addition of the cold extraction solvent ($-35\ °C$) and shaking for 3 min, samples were cooled at $4\ °C$, shaken for a further 3 min, and centrifuged at 7000 rpm (Eppendorf Centrifuge 5424) for 5 min. The concentration of the supernatants at $50\ °C$ under a flow of nitrogen (Sample Concentrator DB-3D TECHNE) provided residual volumes of 40–70 μL, which were reconditioned by adding water to obtain samples of 100 μL, which were immediately analyzed by radio-HPLC. Urine samples were analyzed without any pretreatment.

4.7. LC-MS/MS Analyses

Analyses were performed as previously described [4] on an Agilent 1260 Infinity Quarternary LC system (Agilent Technologies, Böblingen, Germany) coupled with a QTRAP 5500 hybrid linear ion-trap triple quadrupole mass spectrometer (AB SCIEX, Concord, ON, Canada). Data were acquired and processed using Analyst software (Version 1.6.1, AB SCIEX) and for further data processing Origin Pro 8.5.0G (OriginLab, Northampton, MA, USA) was used. For chromatographic separations a Poroshell 120 EC-C18-column, 50 mm × 4.6 mm, 2.7 μm (Agilent Technologies) was used. The solvent system consisted of eluent A: aq. ammonium formate, 2.5 mM, pH 3 and eluent B: water/acetonitrile, 20:80 (v/v), containing ammonium formate, 2.5 mM, pH 3. Gradient elution (% acetonitrile) at $25\ °C$ and a flow rate of 1.0 mL/min: 0–9.0 min, 5–37%; 9.0–9.1 min, 37–80%; 9.1–11.0 min 80%; 11.0–11.1 min, 80–5%; 11.1–14.0 min, 5%. The mass spectrometer was operated in positive electrospray ionization mode and the following parameters were: curtain gas (CUR) 40, collision gas (CAD) high, ion spray voltage (IS) 5500, temperature (TEM) 650, ion source gas 1 (GS1) 60, and ion source gas 2 (GS2) 60. For the multiple reaction monitoring (MRM) scan type: different MRM transitions, scan time. 40 ms; declustering potential (DP), 126; entrance potential (EP), 10; collision energy (CE), 37; collision cell exit potential (CXP), 14. For the enhanced product ion (EPI) scan type: product of m/z, 223.1; scan rate, 10,000 Da/s; dynamic fill time; DP, 110; EP, 10; CE, 33; collision energy spread (CES), 0. For the EPI scan type: product of m/z, 399.2; scan rate, 10,000 Da/s; dynamic fill time; DP, 110; EP, 10; CE, 50; CES, 0. For the MS^3 scan type the excitation energy (AF2) was optimized prior to data acquisition.

Molecules **2018**, 23, 464

4.8. Radio-HPLC

Analyses were performed on a JASCO LC-2000 system (JASCO Labor- und Datentechnik, Gross-Umstadt, Germany) including a UV-2070 UV–VIS detector (monitoring at 254 nm) online with a GABI Star radioactivity flow detector (raytest Isotopenmessgeräte, Straubenhardt, Germany) with a NaI detector (2 × 2″ pinhole, 16 mm × 30 mm). Chromatographic separations were achieved using a Multosphere 120 RP 18 AQ-5μ-column, 250 mm × 4.6 mm, 5 μm, including precolumn, 10 mm × 4 mm (CS-Chromatographie Service). The solvent system consisted of eluent A: aq. ammonium formate, 25 mM, pH 3/acetonitrile, 95:5 (*v/v*) and eluent B: aq. ammonium formate, 25 mM, pH 3/acetonitrile, 20:80 (*v/v*). Gradient elution (% acetonitrile) at a flow rate of 1.0 mL/min: 0–5 min, 5%; 5–40 min, 5–30%; 40–41 min, 30–80%; 41–51 min, 80%; 51–52 min, 80–5%; 52–62 min, 5%.

5. Conclusions

By metabolism studies in pigs and incubations with liver microsomes, supported by LC-MS/MS and radio-HPLC, radiometabolites of (+)-[^{18}F]**1** formed in humans could be identified: one C-hydroxylation product (h-M1) and one product of N-hydroxylation, and subsequent glucuronide conjugation (h-M2). Due to the very low occurrence of these metabolites, (+)-[^{18}F]**1** exhibits a tremendously high metabolic stability in human. (+)-[^{18}F]**1** is considered as a very appropriate tracer for the molecular imaging of $\alpha_4\beta_2$ nAChRs by PET.

Acknowledgments: The work was financially supported by the Helmholtz-Validierungsfonds (HVF). We further thank Tina Spalholz for the conduction of microsomal incubations.

Author Contributions: F.-A.L., S.F., R.S., W.D.-C., A.H., M.P., O.S., and P.B. conceived and designed the experiments; F.-A.L., S.F., R.S., W.D.-C., S.T., M.P., O.S., and P.B. performed the experiments; F.-A.L., S.F., and R.S. analyzed the data; and F.-A.L., S.F., R.S., and P.B. wrote the paper.

Conflicts of Interest: The authors declare no conflict of interest. The founding sponsor had no role in the design of the study; in the collection, analyses, or interpretation of data; in the writing of the manuscript, and in the decision to publish the results.

References

1. Badio, B.; Garraffo, H.M.; Spande, T.F.; Daly, J.W. Epibatidine: Discovery and definition as a potent analgesic and nicotinic agonist. *Med. Chem. Res.* **1994**, *4*, 440–448.
2. Smits, R.; Fischer, S.; Hiller, A.; Deuther-Conrad, W.; Wenzel, B.; Patt, M.; Cumming, P.; Steinbach, J.; Sabri, O.; Brust, P.; et al. Synthesis and biological evaluation of both enantiomers of [^{18}F]flubatine, promising radiotracers with fast kinetics for the imaging of α4β2-nicotinic acetylcholine receptors. *Bioorg. Med. Chem.* **2014**, *22*, 804–812. [CrossRef] [PubMed]
3. Sabri, O.; Becker, G.-A.; Meyer, P.M.; Hesse, S.; Wilke, S.; Graef, S.; Patt, M.; Luthardt, J.; Wagenknecht, G.; Hoepping, A.; et al. First-in-human PET quantification study of cerebral α4β2* nicotinic acetylcholine receptors using the novel specific radioligand (−)-[^{18}F]Flubatine. *NeuroImage* **2015**, *118*, 199–208. [CrossRef] [PubMed]
4. Ludwig, F.-A.; Smits, R.; Fischer, S.; Donat, C.K.; Hoepping, A.; Brust, P.; Steinbach, J. LC-MS Supported Studies on the in Vitro Metabolism of both Enantiomers of Flubatine and the in vivo Metabolism of (+)-[^{18}F]Flubatine—A Positron Emission Tomography Radioligand for Imaging α4β2 Nicotinic Acetylcholine Receptors. *Molecules* **2016**, *21*, 1200. [CrossRef] [PubMed]
5. Brust, P.; Patt, J.T.; Deuther-Conrad, W.; Becker, G.; Patt, M.; Schildan, A.; Sorger, D.; Kendziorra, K.; Meyer, P.; Steinbach, J.; et al. In vivo measurement of nicotinic acetylcholine receptors with [^{18}F]norchloro-fluoro-homoepibatidine. *Synapse* **2008**, *62*, 205–218. [CrossRef] [PubMed]
6. Hockley, B.G.; Stewart, M.N.; Sherman, P.; Quesada, C.; Kilbourn, M.R.; Albin, R.L.; Scott, P.J.H. (−)-[^{18}F]Flubatine: Evaluation in rhesus monkeys and a report of the first fully automated radiosynthesis validated for clinical use. *J. Labelled Comp. Radiopharm.* **2013**, *56*, 595–599. [CrossRef] [PubMed]
7. Gallezot, J.-D.; Esterlis, I.; Bois, F.; Zheng, M.-Q.; Lin, S.-F.; Kloczynski, T.; Krystal, J.H.; Huang, Y.; Sabri, O.; Carson, R.E.; et al. Evaluation of the sensitivity of the novel α4β2* nicotinic acetylcholine receptor PET radioligand ^{18}F-(−)-NCFHEB to increases in synaptic acetylcholine levels in rhesus monkeys. *Synapse* **2014**, *68*, 556–564. [CrossRef] [PubMed]

8. Bois, F.; Gallezot, J.-D.; Zheng, M.-Q.; Lin, S.-F.; Esterlis, I.; Cosgrove, K.P.; Carson, R.E.; Huang, Y. Evaluation of [^{18}F]-(–)-norchlorofluorohomoepibatidine ([^{18}F]-(–)-NCFHEB) as a PET radioligand to image the nicotinic acetylcholine receptors in non-human primates. *Nucl. Med. Biol.* **2015**, *42*, 570–577. [CrossRef] [PubMed]

9. Patt, M.; Becker, G.A.; Grossmann, U.; Habermann, B.; Schildan, A.; Wilke, S.; Deuther-Conrad, W.; Graef, S.; Fischer, S.; Smits, R.; et al. Evaluation of metabolism, plasma protein binding and other biological parameters after administration of (–)-[^{18}F]Flubatine in humans. *Nucl. Med. Biol.* **2014**, *41*, 489–494. [CrossRef] [PubMed]

10. Hillmer, A.T.; Esterlis, I.; Gallezot, J.D.; Bois, F.; Zheng, M.Q.; Nabulsi, N.; Lin, S.F.; Papke, R.L.; Huang, Y.; Sabri, O.; et al. Imaging of cerebral α4β2* nicotinic acetylcholine receptors with (–)-[^{18}F]Flubatine PET: Implementation of bolus plus constant infusion and sensitivity to acetylcholine in human brain. *NeuroImage* **2016**, *141*, 71–80. [CrossRef] [PubMed]

11. Bhatt, S.; Hillmer, A.T.; Nabulsi, N.; Matuskey, D.; Lim, K.; Lin, S.-F.; Esterlis, I.; Carson, R.E.; Huang, Y.; Cosgrove, K.P. Evaluation of (–)-[^{18}F]Flubatine-specific binding: Implications for reference region approaches. *Synapse* **2017**, *72*. [CrossRef]

12. Ma, Y.; Kiesewetter, D.O.; Lang, L.; Gu, D.; Chen, X. Applications of LC-MS in PET Radioligand Development and Metabolic Elucidation. *Curr. Drug Metab.* **2010**, *11*, 483–493. [CrossRef] [PubMed]

13. Lee, M.S.; Kerns, E.H. LC/MS applications in drug development. *Mass. Spectrom. Rev.* **1999**, *18*, 187–279. [CrossRef]

14. Lin, J.H.; Lu, A. Y. H. Role of Pharmacokinetics and Metabolism in Drug Discovery and Development. *Pharmacol. Rev.* **1997**, *49*, 403. [PubMed]

15. Martignoni, M.; Groothuis, G.M.M.; de Kanter, R. Species differences between mouse, rat, dog, monkey and human CYP-mediated drug metabolism, inhibition and induction. *Expert Opin. Drug Metab. Toxicol.* **2006**, *2*, 875–894. [CrossRef] [PubMed]

16. Testa, B.; Krämer, S.D. The Biochemistry of Drug Metabolism—An Introduction: Part 1. Principles and Overview. *Chem. Biodivers.* **2006**, *3*, 1053–1101. [CrossRef] [PubMed]

17. Trufelli, H.; Palma, P.; Famiglini, G.; Cappiello, A. An overview of matrix effects in liquid chromatography-mass spectrometry. *Mass. Spectrom. Rev.* **2011**, *30*, 491–509. [CrossRef] [PubMed]

18. Jia, L.; Liu, X. The Conduct of Drug Metabolism Studies Considered Good Practice (II): In vitro Experiments. *Curr. Drug Metab.* **2007**, *8*, 822–829. [CrossRef] [PubMed]

19. Asha, S.; Vidyavathi, M. Role of Human Liver Microsomes in In Vitro Metabolism of Drugs-A Review. *Appl. Biochem. Biotechnol.* **2010**, *160*, 1699–1722. [CrossRef] [PubMed]

20. Lever, S.Z.; Fan, K.-H.; Lever, J.R. Tactics for preclinical validation of receptor-binding radiotracers. *Nucl. Med. Biol.* **2016**, *44*, 4–30. [CrossRef] [PubMed]

21. German Clinical Trials Register/Deutsches Register Klinischer Studien (DRKS): DRKS00005819. Available online: https://www.drks.de/drks_web/navigate.do?navigationId=trial.HTML&TRIAL_ID=DRKS00005819 (accessed on 14 February 2018).

22. Fischer, S.; Hiller, A.; Smits, R.; Hoepping, A.; Funke, U.; Wenzel, B.; Cumming, P.; Sabri, O.; Steinbach, J.; Brust, P. Radiosynthesis of racemic and enantiomerically pure (−)-[^{18}F]flubatine—A promising PET radiotracer for neuroimaging of α₄β₂ nicotinic acetylcholine receptors. *Appl. Radiat. Isot.* **2013**, *74*, 128–136. [CrossRef] [PubMed]

23. Patt, M.; Schildan, A.; Habermann, B.; Fischer, S.; Hiller, A.; Deuther-Conrad, W.; Wilke, S.; Smits, R.; Hoepping, A.; Wagenknecht, G.; et al. Fully automated radiosynthesis of both enantiomers of [^{18}F]Flubatine under GMP conditions for human application. *Appl. Radiat. Isot.* **2013**, *80*, 7–11. [CrossRef] [PubMed]

24. Stewart, M.N.; Hockley, B.G.; Scott, P.J.H. Synthesis of (−)-[^{18}F]Flubatine ([^{18}F]FLBT). In *Radiochemical Syntheses: Further Radiopharmaceuticals for Positron Emission Tomography and New Strategies for Their Production*; Scott, P.J.H., Ed.; John Wiley & Sons, Inc: Hoboken, NJ, USA, 2015; pp. 1–11.

25. Miller, R.R.; Doss, G.A.; Stearns, R. Identification of a Hydroxylamine Glucuronide Metabolite of an Oral Hypoglycemic Agent. *Drug Metab. Disp.* **2004**, *32*, 178–185. [CrossRef] [PubMed]

26. Trontelj, T. Quantification of Glucuronide Metabolites in Biological Matrices by LC-MS/MS. In *Tandem Mass Spectrometry-Applications and Principles*; Prasain, J., Ed.; INTECHOPEN: London, UK. Available online: https://www.intechopen.com/books/tandem-mass-spectrometry-applications-and-principles/quantification-of-glucuronide-metabolites-in-biological-matrices-by-lc-ms-ms (accessed on 14 February 2018). [CrossRef]

27. Kaivosaari, S.; Finel, M.; Koskinen, M. N-glucuronidation of drugs and other xenobiotics by human and animal UDP-glucuronosyltransferases. *Xenobiotica* **2011**, *41*, 652–669. [CrossRef] [PubMed]
28. Bolleddula, J.; DeMent, K.; Driscoll, J.P.; Worboys, P.; Brassil, P.J.; Bourdet, D.L. Biotransformation and bioactivation reactions of alicyclic amines in drug molecules. *Drug Metab. Rev.* **2014**, *46*, 379–419. [CrossRef] [PubMed]
29. Holčapek, M.; Kolářová, L.; Nobilis, M. High-performance liquid chromatography–tandem mass spectrometry in the identification and determination of phase I and phase II drug metabolites. *Anal. Bioanal. Chem.* **2008**, *391*, 59–78. [CrossRef] [PubMed]

Sample Availability: Samples of the compounds are not available from the authors.

molecules

MDPI

Article

[18F]Fallypride-PET/CT Analysis of the Dopamine D$_2$/D$_3$ Receptor in the Hemiparkinsonian Rat Brain Following Intrastriatal Botulinum Neurotoxin A Injection

Teresa Mann [1,*], Jens Kurth [2], Alexander Hawlitschka [1], Jan Stenzel [3], Tobias Lindner [3], Stefan Polei [3], Alexander Hohn [2], Bernd J. Krause [2] and Andreas Wree [1]

[1] Institute of Anatomy, Rostock University Medical Center, Gertrudenstrasse 9, 18057 Rostock, Germany; alexander.hawlitschka@med.uni-rostock.de (A.H.); andreas.wree@med.uni-rostock.de (A.W.)

[2] Department of Nuclear Medicine, Rostock University Medical Centre, Gertrudenplatz 1, 18057 Rostock, Germany; jens.kurth@med.uni-rostock.de (J.K.); alexander.hohn@med.uni-rostock.de (A.H.); bernd.krause@med.uni-rostock.de (B.J.K.)

[3] Core Facility Multimodal Small Animal Imaging, Rostock University Medical Center, Schillingallee 69a, 18057 Rostock, Germany; Jan.Stenzel@med.uni-rostock.de (J.S.); tobias.lindner@med.uni-rostock.de (T.L.); stefan.polei@gmx.de (S.P.)

*   Correspondence: teresa.mann@med.uni-rostock.de; Tel.: +49-381-494-8433

Received: 30 January 2018; Accepted: 4 March 2018; Published: 6 March 2018

Abstract: Intrastriatal injection of botulinum neurotoxin A (BoNT-A) results in improved motor behavior of hemiparkinsonian (hemi-PD) rats, an animal model for Parkinson's disease. The caudate–putamen (CPu), as the main input nucleus of the basal ganglia loop, is fundamentally involved in motor function and directly interacts with the dopaminergic system. To determine receptor-mediated explanations for the BoNT-A effect, we analyzed the dopamine D$_2$/D$_3$ receptor (D$_2$/D$_3$R) in the CPu of 6-hydroxydopamine (6-OHDA)-induced hemi-PD rats by [18F]fallypride-PET/CT scans one, three, and six months post-BoNT-A or -sham-BoNT-A injection. Male Wistar rats were assigned to three different groups: controls, sham-injected hemi-PD rats, and BoNT-A-injected hemi-PD rats. Disease-specific motor impairment was verified by apomorphine and amphetamine rotation testing. Animal-specific magnetic resonance imaging was performed for co-registration and anatomical reference. PET quantification was achieved using PMOD software with the simplified reference tissue model 2. Hemi-PD rats exhibited a constant increase of 23% in D$_2$/D$_3$R availability in the CPu, which was almost normalized by intrastriatal application of BoNT-A. Importantly, the BoNT-A effect on striatal D$_2$/D$_3$R significantly correlated with behavioral results in the apomorphine rotation test. Our results suggest a therapeutic effect of BoNT-A on the impaired motor behavior of hemi-PD rats by reducing interhemispheric changes of striatal D$_2$/D$_3$R.

Keywords: D$_2$/D$_3$ receptors; hemiparkinsonian rat model; Botulinum neurotoxin A; basal ganglia; striatum; Parkinson's disease; small animal imaging; PET/CT; [18F]fallypride; MRI

---

1. Introduction

Positron emission tomography (PET) using the radioligand [18F]fallypride enables in vivo detection of disease-specific alterations of the dopaminergic system, more precisely of D$_2$/D$_3$ receptor (D$_2$/D$_3$R) availability. Small animal PET hybrid tomographs allow imaging and quantification of D$_2$/D$_3$R binding by the use of [18F]fallypride in rodent models of neurodegenerative disorders like Parkinson's disease (PD) [1–3]. [18F]fallypride is characterized by high sensitivity and selectivity for D$_2$R; for instance, administration of the D$_2$R antagonist haloperidol blocked specific [18F]fallypride binding in the mouse caudate–putamen (CPu) by 95% [4,5]. Besides specific binding to D_2R, [^{18}F]fallypride displays affinity to the D_2-like D_3R, about 20% of the radioligand bind to D_3R in vivo in small animals [6]. In PD a drastic loss of striatal dopamine (DA) caused by progressing degeneration of dopaminergic neurons in the substantia nigra pars compacta (SNpc) results in imbalanced neurotransmitter systems and underlies motor complications. Moreover, due to DA depletion and the subsequent missing inhibition of striatal cholinergic interneurons, hypercholinism is an additional feature of PD exacerbating motor impairment [7,8]. Interestingly, there are physiological bidirectional regulating effects of the cholinergic and dopaminergic system [9] via complex involvement of muscarinic and nicotinic receptors on DA release from dopaminergic terminals [10,11].

For causal analysis and development of novel therapeutics for PD in preclinical research the experimental hemiparkinsonian (hemi-PD) is an accepted animal model [12]. Unilateral injection of 6-hydroxdopamine (6-OHDA) into the medial forebrain bundle (MFB) of rats provokes rapid dopaminergic depletion by auto-oxidation and consequent oxidative stress [13]. The near-complete loss of dopaminergic neurons in the SNpc via retrograde axonal transport of the toxin 6-OHDA after stereotaxic injection into the MFB mimics a late stage of PD [14]. Current therapeutic strategies for PD focus primarily on compensation of DA in the striatum (caudate–putamen, CPu) by either DA precursors [15,16] or DA receptor agonists [17]. Though, clinical efficiency is limited and chronic administrations leads to severe side effects like motor fluctuations and dyskinesia [18,19]. Other therapeutic options target the cholinergic system mostly by blocking muscarinic receptors or by inhibition of cholinesterase [20,21]. Systemic administration of anticholinergic substances causes severe side effects like confusion, dry mouth, blurred vision, and cognitive impairment [22].

Recently, we demonstrated that local injection of the anticholinergic Botulinum neurotoxin A (BoNT-A) significantly improved D_2R agonist-induced asymmetric rotational behavior in hemi-PD rats [23–26]. BoNT-A acts mainly on cholinergic neurons and inhibits distribution of acetylcholine into the synaptic cleft via cleavage of the synaptosomal-associated protein of 25-kDa (SNAP25) [27–29]. The intracerebral injection of BoNT-A avoids severe side effects in both the central and peripheral nervous system [23]. Notably, intrastriatal application of BoNT-A does not cause cytotoxicity [30] or impaired cognition [24] in rats. As known from other medical implementation, BoNT-A demonstrates a transient therapeutic effect in hemi-PD rats that lasts up to six months post-injection [23,31]. To examine the longitudinal cellular mechanisms of the positive BoNT-A effect on receptor level, we performed [^{18}F]fallypride-PET/CT scans one, three, and six months post-BoNT-A or -sham-BoNT-A injection and quantified D_2/D_3R availability in controls, sham-injected hemi-PD rats, and BoNT-A-injected hemi-PD rats.

2. Results

D_2/D_3R availability was analyzed longitudinally in controls (sham-6-OHDA + sham-BoNT-A, $n = 9$), sham-injected hemi-PD rats (6-OHDA + sham-BoNT-A, $n = 7$) and BoNT-A-injected hemi-PD rats (6-OHDA + BoNT-A, $n = 10$) by dynamic [^{18}F]fallypride-PET/CT scans.

2.1. Immunohistochemistry and Behavioral Testing

To qualitatively verify successful 6-OHDA-induced dopaminergic deafferentation we performed tyrosine hydroxylase (TH) immunostaining (for dopaminergic neurons). TH-reaction in the left and right CPu and SN of control rats (sham-6-OHDA + sham-BoNT-A) showed no loss of TH-reaction (Figure 1a,b). In hemi-PD rats (6-OHDA + sham-BoNT-A) an ipsilateral loss of almost all TH-immunoreactivity was visible in the CPu and SN, indicating dopaminergic deafferentation in the CPu due to dopaminergic cell loss in the SN (Figure 1c,d), and BoNT-A injection in hemi-PD rats (6-OHDA + BoNT-A) did not demonstrate an additive effect on these reaction patterns (Figure 1e,f).

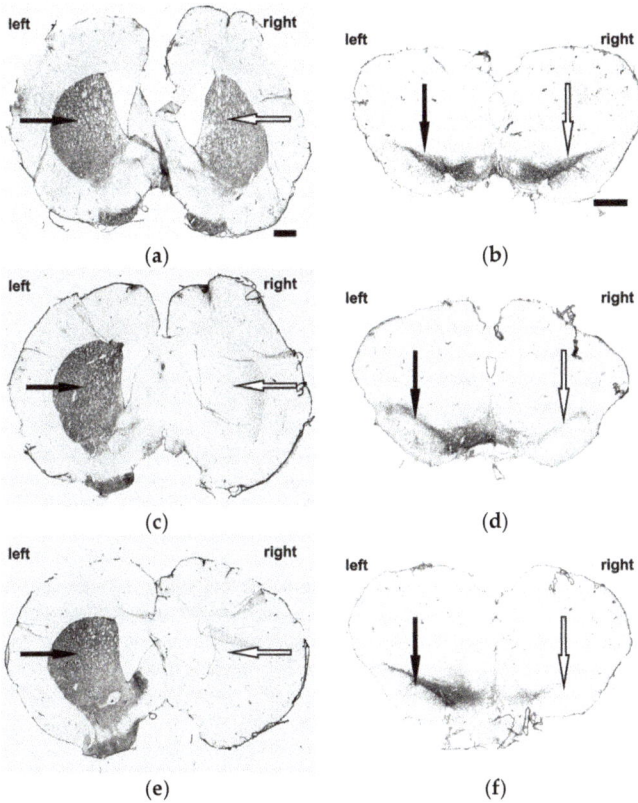

Figure 1. TH-immunoreactivity in the telencephalon (left column) and the mesencephalon (right column) of (**a,b**) controls (sham-6-OHDA + sham-BoNT-A) (**c,d**) sham-injected hemi-PD rats (6-OHDA + sham-BoNT-A) and (**e,f**) BoNT-A-injected hemi-PD rats (6-OHDA + BoNT-A). 6-OHDA or sham-6-OHDA was unilaterally injected into the MFB of the right hemisphere and BoNT-A was injected ipsilateral at two sites into the CPu. Controls showed symmetric TH pattern in the CPu (**a,c,e**; black (**left**) and white (**right**) arrow) and SN (**b,d,f**; black (**left**) and white (**right**) arrow), sham-injected hemi-PD rats demonstrated an almost complete loss of TH-positive cells in the CPu and SN and BoNT-A injection did not influence these findings in hemi-PD rats. The scale bar applies for **a–f** = 1 mm.

Asymmetric rotations of hemi-PD rats were tested using apomorphine- and amphetamine-induced rotations one month after 6-OHDA lesion. Also, the positive effect of intrastriatally injected BoNT-A on drug-induced rotations was analyzed two weeks after administration. All hemi-PD rats exhibited distinct apomorphine-induced rotations contralateral to the 6-OHDA lesion before BoNT-A or sham-BoNT-A injection of 8.2 ± 3.6 rpm (6-OHDA + sham-BoNT-A) and 9.2 ± 3.0 rpm (6-OHDA + BoNT-A), controls did not show rotational behavior (sham-6-OHDA + sham-BoNT-A) (Figure 2a). Sham injection in hemi-PD rats slightly decreased rotations to 4.6 ± 2.5 rpm (6-OHDA + sham-BoNT-A) and did not affect the behavior of controls (sham-6-OHDA + sham-BoNT-A). Following BoNT-A injection rotational behavior was reversed in hemi-PD rats to 2.2 ± 2.1 rpm (6-OHDA + BoNT-A). The positive motor effect of BoNT-A in hemi-PD rats was significant compared to sham injections ($p = 0.021$) (Figure 2a). Before BoNT-A or sham-BoNT-A injection amphetamine administration caused ipsilateral rotations in hemi-PD rats of -7.3 ± 3.4 rpm (6-OHDA + sham-BoNT-A) and -6.6 ± 2.9 rpm (6-OHDA + BoNT-A) but not in controls (sham-6-OHDA + sham-BoNT-A) (Figure 2b). After BoNT-A or sham-BoNT-A injection asymmetric rotational behavior was with -12.0 ± 6.3 rpm (6-OHDA + BoNT-A) and -10.3 ± 3.5 rpm (6-OHDA + sham-BoNT-A) further increased and BoNT-A effect was compared to sham injection not significantly abolished. Sham-BoNT-A injection in controls (sham-6-OHDA + sham-BoNT-A) did not result in ipsilateral rotations (Figure 2b).

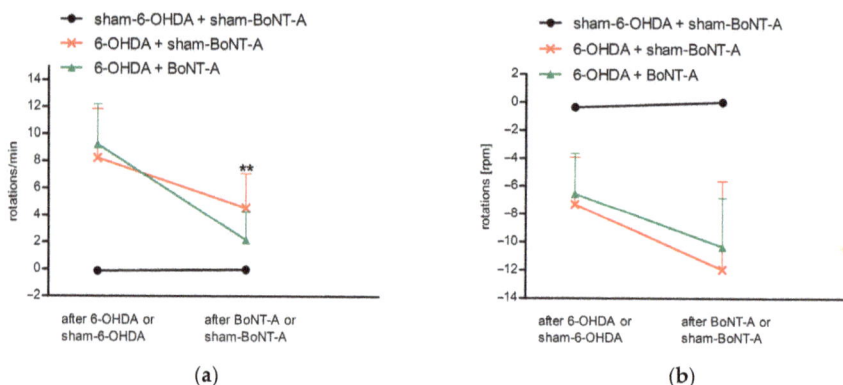

Figure 2. Results of the rotational behavior in (**a**) apomorphine- and (**b**) amphetamine-induced testing for controls (sham-6-OHDA + sham-BoNT-A), sham-injected hemi-PD rats (6-OHDA + sham-BoNT-A) and BoNT-A-injected hemi-PD rats (6-OHDA + BoNT-A) displayed after 6-OHDA or sham-6-OHDA and BoNT-A or sham-BoNT-A injection. Rotations contralateral to the injection side (clockwise) are displayed with negative algebraic signs, anti-clockwise rotations with positive algebraic signs. Controls did not demonstrate designated rotational behavior, hemi-PD rats exhibited strong asymmetric drug-induced rotations that were almost completely abolished after BoNT-A injection in apomorphine testing and slightly abolished in amphetamine testing. Significance is displayed as ** $p < 0.01$.

2.2. Striatal D_2/D_3R Availability

Qualitative analysis with parametric mapping of non-displaceable binding potential (BP_{nd}) revealed no obvious interhemispheric differences for controls (sham-6-OHDA + sham-BoNT-A). However, increased signals in the right CPu of hemi-PD rats were visible (6-OHDA + sham-BoNT-A) compared to the unaffected side. This visual right–left difference was diminished after BoNT-A injection into the right CPu of hemi-PD rats (6-OHDA + BoNT-A) (Figure 3a–c).

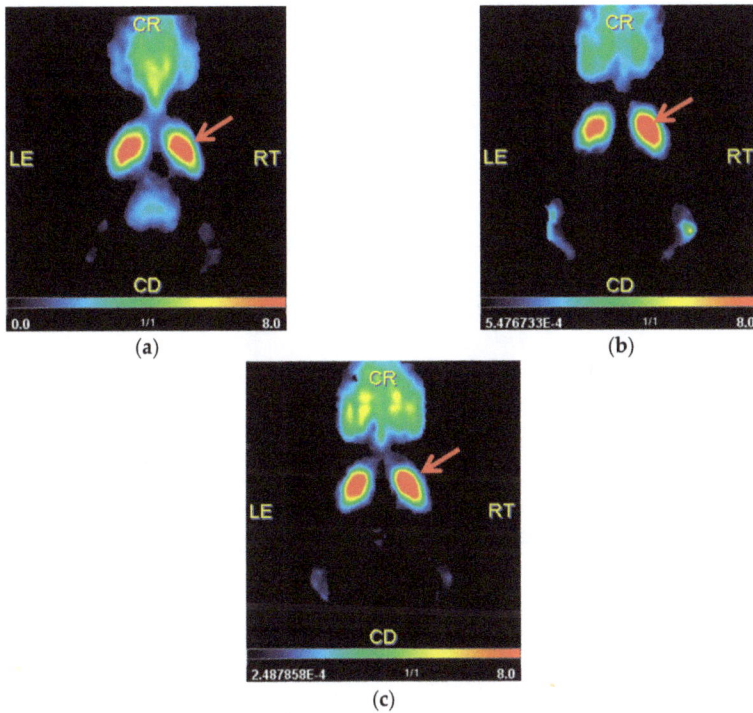

Figure 3. Qualitative analysis using pixel-wised parametric mapping for the parameter BP_{nd} showing the left and right CPu in transversal sections for (**a**) controls (sham-6-OHDA + sham-BoNT-A) (**b**) sham-injected hemi-PD rats (6-OHDA + sham-BoNT-A) and (**c**) BoNT-A-injected hemi-PD rats (6-OHDA + BoNT-A) one month after BoNT-A or sham-BoNT-A injection. A representative animal of each experimental group was used. Controls did not reveal visual side differences, an increased signal of BP_{nd} distribution in the right CPu of hemi-PD rats was clearly visible and BoNT-A injection normalized the increased signal in hemi-PD rats. The right CPu is marked with a red arrow.

For quantification the simplified reference tissue model 2 (SRTM2) was applied and BP_{nd} was estimated separately for the left and right striatum. Controls (sham-6-OHDA + sham-BoNT-A) revealed no relative interhemispheric right–left differences: mean BP_{nd} of $4.2 \pm 0.8/4.2 \pm 0.8$ (left/right CPu) one month post-sham-BoNT-A injection, $3.8 \pm 0.8/3.7 \pm 0.9$ (left/right CPu) three months post-sham-BoNT-A injection and $4.1 \pm 0.7/4.1 \pm 0.7$ (left/right CPu) six months post-sham-BoNT-A injection were found (Figure 4a,c; Table 1). Hemi-PD rats that received sham-BoNT-A injection (6-OHDA + sham-BoNT-A) exhibited strong interhemispheric right–left differences of about 23%: kinetic analysis resulted in mean BP_{nd} values of $4.4 \pm 0.6/5.5 \pm 0.9$ (left/right CPu) one month post-sham-BoNT-A injection, $4.4 \pm 1.1/5.5 \pm 1.3$ (left/right CPu) three months post-sham-BoNT-A injection, and $4.8 \pm 1.0/5.7 \pm 1.0$ (left/right CPu) six months post-sham-BoNT-A injection. The contralateral CPu was never affected and displayed very stable BP_{nd} throughout all experimental groups and scanning time points (Figure 4a–c; Table 1). The increase in BP_{nd} in the right CPu of hemi-PD rats (6-OHDA + sham-BoNT-A) was significant compared to the left CPu of the same experimental group one month post-sham-BoNT-A injection ($p = 0.047$) (Figure 4a) and compared to the right CPu of controls (sham-6-OHDA + sham-BoNT-A) at all 3 examined time points ($p = 0.03$, $p = 0.064$, $p = 0.039$) (Figure 4a–c). BoNT-A injection in hemi-PD rats (6-OHDA + BoNT-A) reduced relative interhemispheric right–left difference to about 13.4%: quantification revealed mean BP_{nd} of

4.7 ± 0.4/5.3 ± 0.5 (left/right CPu) one month post-BoNT-A injection, 4.4 ± 1.0/5.0 ± 1.2 (left/right CPu) three months post-BoNT-A injection and 4.4 ± 1.0/5.1 ± 1.1 (left/right CPu) six months post-BoNT-A injection (Figure 4a–c; Table 1). The BoNT-A effect was significant compared to the ipsilateral CPu of controls (sham-6-OHDA + sham-BoNT-A) one month post-BoNT-A injection ($p = 0.0087$) and showed a transient course throughout the timeline (Figure 4a–c). A list of all individual values for BP$_{nd}$ expressing D$_2$/D$_3$R availability separately for the left and right CPu and the relative interhemispheric right–left difference in each of the 26 analyzed rats is displayed in Table 1.

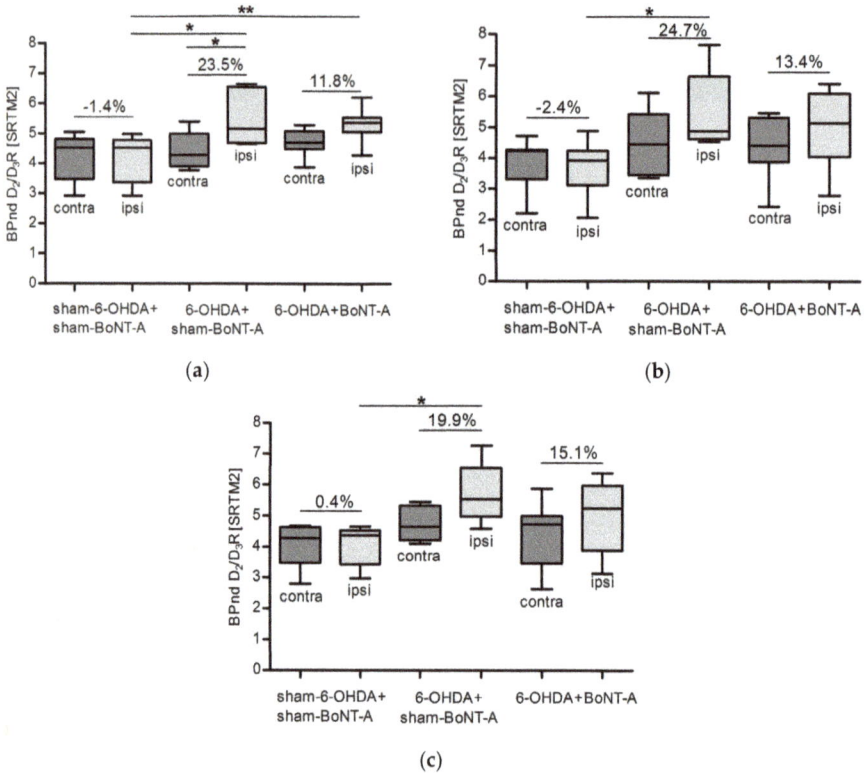

(a)

(b)

(c)

Figure 4. Box plots for BP$_{nd}$ values of D$_2$/D$_3$R depicting median and interquartile ranges separately for the contralateral (dark grey) and ipsilateral (light grey) CPu for controls (sham-6-OHDA + sham-BoNT-A), sham-injected hemi-PD rats (6-OHDA + sham-BoNT-A) and BoNT-A-injected hemi-PD rats (6-OHDA + BoNT-A) (**a**) one month, (**b**) three months, and (**c**) six months post-BoNT-A or -sham-BoNT-A injection. D$_2$/D$_3$R availability was consistently symmetric in controls, increased in sham-injected hemi-PD rats at all analyzed time points and was reduced to nearly normal values after BoNT-A injection in hemi-PD rats. Significance is displayed as * $p < 0.05$, ** $p < 0.01$.

Table 1. Summary of all single BP_{nd} values of D_2/D_3R for the left and right CPu and the interhemispheric difference relative to the left hemisphere in (%) analyzed in controls (sham-6-OHDA + sham-BoNT-A), sham-injected hemi-PD rats (6-OHDA + sham-BoNT-A) and BoNT-A-injected hemi-PD rats (6-OHDA + BoNT-A). Data are shown for all three PET/CT scans (PET/CT 1: one month post-BoNT-A or sham-BoNT-A, PET/CT 2: three months post-BoNT-A or -sham-BoNT-A, PET/CT 3: six months post-BoNT-A or -sham-BoNT-A). * indicate that no data were analyzed due to incorrect tracer injection or no data acquisition.

Group	PET/CT 1			PET/CT 2			PET/CT 3		
	BP_{nd} Left	BP_{nd} Right	(%)	BP_{nd} Left	BP_{nd} Right	(%)	BP_{nd} Left	BP_{nd} Right	(%)
sham-6-OHDA + sham-BoNT-A	5.06	4.98	−1.66	2.21	2.07	−6.35	*	*	*
sham-6-OHDA + sham-BoNT-A	2.92	2.92	−0.01	3.74	3.93	5.21	4.56	4.44	−2.66
sham-6-OHDA + sham-BoNT-A	4.53	4.53	−0.04	2.87	2.78	−3.37	3.29	3.23	−1.84
sham-6-OHDA + sham-BoNT-A	4.58	4.59	0.28	4.25	4.15	−2.28	4.65	4.53	−2.46
sham-6-OHDA + sham-BoNT-A	4.03	3.80	−5.71	4.24	3.93	−7.22	4.07	4.02	−1.36
sham-6-OHDA + sham-BoNT-A	*	*	*	3.81	3.46	−9.32	4.68	4.52	−3.47
sham-6-OHDA + sham-BoNT-A	*	*	*	4.73	4.89	3.34	2.80	2.97	6.23
sham-6-OHDA + sham-BoNT-A	*	*	*	4.27	4.32	1.04	4.50	4.66	3.60
sham-6-OHDA + sham-BoNT-A	*	*	*	4.23	4.13	−2.48	4.07	4.30	5.66
|Mean| ± SD	4.2 ± 0.8	4.2 ± 0.8	1.4 ± 2.5	3.8 ± 0.8	3.7 ± 0.9	2.4 ± 4.9	4.1 ± 0.7	4.1 ± 0.7	0.4 ± 4.0
6-OHDA + sham-BoNT-A	4.86	6.53	34.37	6.11	7.67	25.53	4.11	4.59	11.53
6-OHDA + sham-BoNT-A	5.41	6.65	22.90	*	*	*	4.66	5.54	19.07
6-OHDA + sham-BoNT-A	3.77	4.67	23.84	3.35	4.54	35.56	5.45	7.26	33.13
6-OHDA + sham-BoNT-A	4.08	4.71	15.53	4.44	4.88	10.04	5.21	5.83	11.98
6-OHDA + sham-BoNT-A	3.96	4.79	21.02	*	*	*	*	*	*
6-OHDA + sham-BoNT-A	4.49	5.54	23.50	4.74	5.64	18.80	4.33	5.37	23.95
6-OHDA + sham-BoNT-A	*	*	*	3.53	4.71	33.46	*	*	*
|Mean| ± SD	4.4 ± 0.6	5.5 ± 0.9	23.5 ± 6.1	4.4 ± 1.1	5.5 ± 1.3	24.7 ± 14.8	4.8 ± 1.0	5.7 ± 1.0	19.9 ± 11.4
6-OHDA + BoNT-A	4.58	5.33	16.19	*	*	*	3.30	3.92	18.62
6-OHDA + BoNT-A	4.54	4.90	7.94	5.47	6.10	11.59	*	*	*
6-OHDA + BoNT-A	5.07	5.54	9.23	4.41	5.15	16.68	2.63	3.13	19.11
6-OHDA + BoNT-A	5.09	5.42	6.39	2.43	2.79	14.93	4.74	5.25	10.77
6-OHDA + BoNT-A	4.85	5.43	11.80	*	*	*	5.88	6.38	8.56
6-OHDA + BoNT-A	4.37	5.15	17.86	*	*	*	4.86	6.03	24.05
6-OHDA + BoNT-A	4.59	5.12	11.55	3.88	4.06	4.77	3.63	3.87	6.45
6-OHDA + BoNT-A	3.88	4.28	10.26	4.33	5.13	18.59	5.00	5.93	18.53
6-OHDA + BoNT-A	5.30	6.22	17.22	4.94	5.26	6.49	4.58	5.23	14.22
6-OHDA + BoNT-A	5.10	5.59	9.74	5.32	6.41	20.38	5.01	5.78	15.38
|Mean| ± SD	4.7 ± 0.4	5.3 ± 0.5	11.8 ± 4.0	4.4 ± 1.0	5.0 ± 1.2	13.4 ± 6.0	4.4 ± 1.0	5.1 ± 1.1	15.1 ± 5.7

2.3. Correlation of D_2/D_3R Side Differences and Behavior

A possible correlation of the degree of interhemispheric differences in D_2/D_3R availability and apomorphine-induced rotations for controls (sham-6-OHDA + sham-BoNT-A), sham-injected hemi-PD rats (6-OHDA + sham-BoNT-A) and BoNT-A-injected hemi-PD rats (6-OHDA + BoNT-A) was examined one month post-BoNT-A or -sham-BoNT-A injection. Controls (sham-6-OHDA + sham-BoNT-A) did not demonstrate right–left differences or rotational behavior. With increasing right–left differences contralateral rotations of hemi-PD rats (6-OHDA + sham-BoNT-A) increased and also the normalizing effect on interhemispheric D_2/D_3R differences after BoNT-A injection (6-OHDA + BoNT-A) was connected with behavior. A highly significant relationship between increasing right–left differences, expressing a higher D_2/D_3R availability in the right CPu, and the apomorphine-induced rotational behavior was found ($p = 0.0007$) (Figure 5).

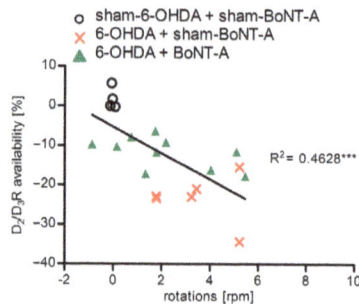

Figure 5. Linear correlation of right–left differences of D_2/D_3R availability in (%) and apomorphine-induced rotations one month after BoNT-A or sham-BoNT-A injection for controls (sham-6-OHDA + sham-BoNT-A), sham-injected hemi-PD rats (6-OHDA + sham-BoNT-A) and BoNT-A-injected hemi-PD rats (6-OHDA + BoNT-A). Controls did neither demonstrate interhemispheric differences nor rotational behavior. With increasing right–left differences also asymmetric rotations increased in hemi-PD rats and both was normalized after BoNT-A injection. Significance is displayed as *** $p < 0.001$.

3. Discussion

In this study we examined cellular mechanisms of the positive motor effect of intrastriatally injected BoNT-A by [18F]fallypride PET/CT scans in hemi-PD rats, as BoNT-A was previously demonstrated to abolish apomorphine-induced rotational behavior in 6-OHDA-lesioned [23–26]. The control group respected the entire surgical procedure as the minimal lesion caused by the insertion of the syringe could lead to changes in receptor binding sites [32,33]. We did not include an experimental group studying BoNT-A in sham-lesioned rats in our design as we assumed that BoNT-A would not alter per se the expression of D_2/D_3 receptors. Indeed, we have previously performed extensive in vitro analysis of D_2/D_3 receptors as well as apomorphine-induced rotational behavior in BoNT-A-injected rats earlier and did not find major effects [31].

Hemi-PD rats demonstrated a constant contralateral rotational behavior after apomorphine injection and rather inconsistent amphetamine-induced rotations four weeks after 6-OHDA lesion (Figure 2a,b). A period of four weeks before behavioral testing was left to ensure maximum dopaminergic deafferentation, as dopaminergic cell death [34] as well as consequent plasticity effects [31] last up to four weeks after injection of 6-OHDA. Notably, increasing right–left differences of D_2/D_3R availability significantly correlated with increasing asymmetry in apomorphine-induced rotations (Figure 5). Rotational tests using the D_2 agonist apomorphine or the DA releaser amphetamine are commonly used to detect the degree of dopaminergic deafferentation in hemi-PD rats. As apomorphine acts on increased striatal D_2/D_3R in the DA-depleted CPu of hemi-PD rats [35–37], it leads to a larger inhibition of the right CPu and as a consequence to an elevated motor urge to the contralateral side. Resulting rotations to the left of more than four rotations per minute confirm dopaminergic degeneration of more than 90% [25,38–40]. Amphetamine induces DA release form nerve terminals and strongly affects the non-lesioned hemisphere of hemi-PD rats, which begin to turn to the ipsilateral side of the 6-OHDA lesion [41]. In line with our findings, another study demonstrated that apomorphine but not amphetamine is a reliable indicator for maximal dopaminergic cell death in hemi-PD rats [42].

Dynamic [^{18}F]fallypride PET/CT scans over 90 min revealed an increase of 23% in D_2/D_3R availability being consistent up to six months post-6-OHDA lesion and a normalization of this pathological imbalance after BoNT-A injection into the CPu of hemi-PD rats (Figures 3 and 4; Table 1). Unlike [^{11}C]raclopride, [^{18}F]fallypride is not easily displaced by endogenous DA, as demonstrated in monkey [43], human [44] and rat brain [45]. To cover the transient effect of BoNT-A demonstrated previously [23], we performed longitudinal measurements using [^{18}F]fallypride in the same rodent. This seemed feasible as repeated measurements with [^{18}F]fallypride PET/CT exhibited only small variations in mice [5] and also in our study, controls (sham-6-OHDA + sham-BoNT-A) did not show variations in BP_{nd} comparing the three PET/CT scanning time points (Figures 3a and 4a–c ; Table 1).

Our finding of a constant increase of about 23% in D_2/D_3R availability in hemi-PD rats is in line with a number of similar studies both in vitro and in vivo. Unilateral injection of 6-OHDA into the MFB or SNpc of rats resulted in a consistent increase of D_2R density of 20% to 40% subject to the injected dosage and survival time analyzed using in vitro autoradiography [35,46–50]. Also, in vivo PET/CT analyses with [^{11}C]raclopride and [^{18}F]fallypride are in accordance with our results. [^{11}C]raclopride PET demonstrated an ipsilateral increase of 17% to 27.7% [2] and approximately 35% [51] in D_2R availability in hemi-PD rats after MFB injection and an increase of 23% [52] and 16.6% [53] after 6-OHDA lesion of the SNpc. [^{18}F]fallypride PET/CT scans revealed an 12% increase in D_2/D_3R availability in hemi-PD rats after injection of 6-OHDA into the CPu [2].

Intrastriatally injections with BoNT-A significantly reduced the pathologically increased D_2/D_3R availability in hemi-PD rats (Figures 3c and 4) and significantly abolished apomorphine-induced rotations (Figure 2a). Apomorphine-induced rotations were also moderately decreased after sham-BoNT-A injection in hemi-PD rats. This effect is likely to be caused by minimal mechanical damage caused by insertion of the cannula into the CPu and injection of sham solution. One might argue that the positive BoNT-A effect is caused by simple striatal cell death after BoNT-A injection, as D_2/D_3R are localized on medium spiny neurons (MSN), presynapses of cholinergic interneurons and boutons of dopaminergic afferents in the CPu [54–56]. Previously we demonstrated that BoNT-A injection did neither cause striatal neuronal loss or reduced volume [26] nor death of cholinergic interneurons in the CPu [30]. The positive BoNT-A effect diminished with increasing post-injection time. The BoNT-A effect on D_2/D_3R in hemi-PD rats has been analyzed before in a quantitative in vitro autoradiography study demonstrating a normalizing effect, which significantly correlates with apomorphine-induced rotations [31].

A preceding study performed [^{11}C]raclopride PET to analyze the effect of BoNT-A on pathological increased D_2/D_3R and found a positive effect on pathological increased D_2R availability in hemi-PD rats [3]. Here, we extended this experimental setup by introducing a control group (sham-6-OHDA + sham-BoNT-A) to investigate both ipsi- and contralateral effects of tissue damage by cannula injection and increased group size to substantiate possible significant effects. Moreover, we used the more specific radioligand [^{18}F]fallypride instead of [^{11}C]raclopride and conducted animal-specific MRI scans for co-registration with CT-corrected PET data to improve data analysis by making use of the high morphologic resolution of MRI.

Quantification was subsequently performed using SRTM2 [57] for kinetic modeling, having the advantage of no need for arterial blood sampling and being established especially for neuroimaging studies [58]. Application of kinetic models allows quantitative determination of transfer rates and provides in depth understanding of physiological parameters. The cerebellum was used as the reference region being devoid of D_2/D_3R [59] and being validated as a suitable reference region for D_2/D_3R before [60–62]. As spatial resolution of PET scans is a limiting factor especially in small animals [63], we used animal-specific MRI scans for spatial normalization, thus high precision of Voxels of interest (VOI) delineation and almost full recovery could be guaranteed in our study. Partial volume effects (PVE) occur mostly in structures being smaller than about 2 times of the width at half maximum of the scanner [64,65]. As our rats exhibited striatal volumes of about 35 to 50 mm^2 [26,66] and the utilized scanner is characterized by a spatial resolution of 1.5 mm, the CPu as the target region exceeded the critical size for the occurrence of PVE. Moreover, the CPu seems to be only minimally effected by extracranial gland and skull activity resulting from its deep localization within the brain [6]. Due to predominantly relative comparisons between hemispheres within treatment groups in our study setup, a potentially emerging PVE would be mathematically shortened. Altogether, in this study we decided to omit PVE correction for quantification of [^{18}F]fallypride uptake in the CPu of hemi-PD rats. Also in comparable studies, no correction of the PVE seemed necessary for quantification of radioligands demonstrating very specific and region-limited binding kinetics like [^{18}F]fallypride [2] or [^{11}C]raclopride [67]. Nevertheless, a methodical consideration of this aspect might be of high interest for further investigations. Another critical point of small animal imaging is the need for anesthesia, commonly realized by respiration of isoflurane. For analyzing the availability of D_2/D_3R effects of isoflurane seem not pivotal, as the BP of D_2R measured with [^{11}C]raclopride was only marginally changed in mice [5] and [^{18}F]fallypride uptake did not differ in awake rats that received late isoflurane anesthesia compared to rats under continuous anesthesia [61].

4. Materials and Methods

4.1. Animals

Twenty-six male Wistar rats (Charles River WIGA, Sulzfeld, Germany; RRID: RGD_737929) either assigned to controls (sham-6-OHDA + sham-BoNT-A, n = 9), sham-injected hemi-PD rats (6-OHDA + sham-BoNT-A, n = 7) or BoNT-A-injected hemi-PD rats (6-OHDA + BoNT-A, n = 10) were included in the experiments. Housing was conducted under standard conditions (22 ± 2 °C, 12 h day-and-night cycle) in a fully air-conditioned room with access to water and food ad libitum. The research protocol and all experimental procedures fulfilled legal obligations of the animal welfare act and were approved by the state Animal Research Committee of Mecklenburg–Western Pomerania (LALLF M-V/7221.3-1-005/16, approval: 03/08/2016). The timeline of the experimental setup is presented in Figure 6.

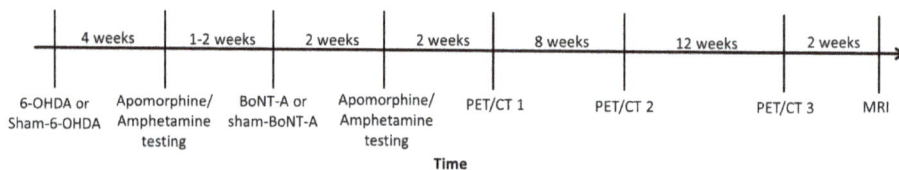

Figure 6. Timeline of the study design. Hemi-PD was unilaterally induced by 6-OHDA injection into the right MFB. Controls received sham-6-OHDA injection. The degree of dopaminergic cell loss was verified with apomorphine- and amphetamine-induced behavioral testing. Five to six weeks after the 6-OHDA or sham-6-OHDA injection, rats obtained BoNT-A or sham-BoNT-A injection into the ipsilateral CPu. The positive effect on the motor behavior of BoNT-A was then controlled in rotation tests. Subsequently, each rat was scanned by [^{18}F]fallypride-PET/CT analysis one, three and six months post-BoNT-A or -sham-BoNT-A injection. A final MRI scan was performed as anatomical reference for PET/CT imaging.

4.2. 6-OHDA and BoNT-A Injection

Animals received 6-OHDA or sham-6-OHDA injection at a weight of 285–305 g in a stereotaxic operation and BoNT-A or sham-BoNT-A injection five to six weeks later. Anesthesia was induced with a ketamine-xylazine mixture (ketamine: 50 mg/kg, xylazine: 4 mg/kg). Dopaminergic cell death was provoked by unilateral injection of 24 µg of 6-OHDA (Sigma-Aldrich, St. Louis, MO, USA) dissolved in 4 µL 0.1 M citrate buffer into the right MFB. Sham-6-OHDA animals received only citrate buffer. Exact coordinates of 6-OHDA or sham-6-OHDA injection were anterior-posterior = −2.3 mm, lateral = −1.5 mm and ventral = −9.0 mm [68]. BoNT-A (Lot#13029A1; List, Campbell, CA, USA; purchased via Quadratech, Epsom, UK) supplemented with phosphate-buffered saline (PBS) and 0.1% bovine serum albumin (BSA) was injected at two sites into the right CPu (total dose: 1 ng). Sham-BoNT-A animals received only PBS + BSA. Exact coordinates of BoNT-A or sham-BoNT-A injection were anterior-posterior = +1.3 mm/−0.4 mm, lateral = −2.6 mm/−3.6 mm and ventral = −5.5 mm/−5.5 mm [68]. For application of either 6-OHDA, BoNT-A or sham solution a 5 µL Hamilton syringe was used and the respective volume was continuously delivered over a time span of 4 min. Afterwards, the needle was left in place for another 5 min to avoid reflux.

4.3. Immunohistochemistry

Serial brain sections showing CPu and SN were immunohistochemically reacted for TH to verify successful 6-OHDA lesioning and to exclude an additive effect of BoNT-A. Brains were fixed with 3.7% paraformaldehyde overnight and stained with monoclonal mouse anti-TH antibody (clone TH2, Sigma-Aldrich) following biotinylated horse anti-mouse IgG (Vector Laboratories, Burlingame, CA, USA, 1:67). For details of the procedure see [69].

4.4. Behavioral Testing

The degree of dopaminergic cell loss and the positive motor effect of BoNT-A was evaluated using apomorphine- and amphetamine-induced rotational behavior. Testing was performed in a rotometer [41] four weeks after 6-OHDA or sham-6-OHDA injection and again two weeks after BoNT-A or sham-BoNT-A injection (Figure 6). Drugs were solved in 0.9% NaCl and injected i.p. (apomorphine: 0.25 mg/kg, amphetamine: 2.5 mg/kg). Following apomorphine injection and a waiting time of 5 min to ensure cerebral uptake, rotations were monitored for 40 min. Rotational behavior induced by amphetamine application was analyzed after a waiting time of 15 min throughout a period of 60 min.

4.5. Radioligand Preparation and PET/CT Imaging

Synthesis of [^{18}F]fallypride ([^{18}F](S)-N-((1-allylpyrrolidin-2-yl)methyl-5-(3-fluoropropyl)-2,3-dimethoxybenzamide) and semi-preparative HPLC for purification was conducted according to the protocol of [70], followed by an extensive quality control. D_2/D_3R availability was analyzed by dynamic [^{18}F]fallypride PET/CT imaging over 90 min, each animal was measured one, three, and six months post-BoNT-A or -sham-BoNT-A injection. Anesthesia was initially administered with 5% isoflurane (AbbVie, North Chicago, IL, USA) vaporized in oxygen gas and maintained during scanning time with 1.5–3%. Body temperature was held constant at 38 °C via a heating pad and respiration rate of the animals was monitored throughout the PET/CT measurement. Each rat was placed in head-prone position centered in the field of view of a commercially available preclinical PET/CT scanner (Inveon®, Siemens Healthcare, Knoxville, TN, USA); performance evaluation of the system was described by [71,72]. [^{18}F]fallypride was injected as a bolus over 1 min via a microcatheter into the lateral tail vein in a mean dose of 23.44 ± 1.75/24.06 ± 1.76/22.81 ± 2.03 MBq (sham-6-OHDA + sham-BoNT-A), PET/CT 1-3), 24.52 ± 2.48, 23.84 ± 2.28, 23.88 ± 2.46 MBq (6-OHDA + Sham-BoNT-A, PET/CT 1–3) and 22.99 ± 2.50, 22.55 ± 3.93, 21.61 ± 2.27 MBq (6-OHDA + BoNT-A, PET/CT 1–3). The acquisition of dynamic PET as list mode data set was started immediately with the injection. PET studies were reconstructed as series 3D PET images of multiple frames with various time durations (6×10 s, 8×30 s, 5×300 s, 5×1800 s) with a voxel size of 0.86 mm $\times$ 0.86 mm $\times$ 0.79 mm using a 2D-ordered subsets expectation maximization algorithm (four iterations, six subsets). Attenuation correction was performed on the basis of whole body CT scan and PET studies were also corrected for random coincidences, dead time, scatter and radioactive decay.

4.6. MRI Imaging

MRI was performed on anesthetized rats (1.5–3% isoflurane in oxygen) at least 10 days after the last PET/CT examination and about eight months post-6-OHDA lesion. MRI of the rats was conducted using a 7 Tesla small animal MRI scanner (BioSpec 70/30 AVANCE III, 7.0 T, 440 mT/m gradient strength, Paravision software v6.01., Bruker BioSpin MRI GmbH, Ettlingen, Germany) with a 1 H transmit resonator (inner diameter: 86 mm; vendor type-nr.: T12053V3, Bruker, Ettlingen, Germany) and a receive-only surface coil array (2×2 array rat brain coil; vendor type-nr.: T11483V3, Bruker) positioned on the head of the rats. The imaging protocol included 3D isotropic T1w FLASH imaging sequences with transversal slice orientation, 8/45 ms TE/TR, 35 mm $\times$ 35 mm $\times$ 16 mm FOV, 200 μm $\times$ 200 μm $\times$ 200 μm resolution, 175 pixel $\times$ 175 pixel $\times$ 80 pixel matrix size, 20° flip angle, 12:36 min:sec acquisition time, one average and fat suppression.

4.7. Image Analysis

For qualitative and quantitative analysis PMOD v3.7 (PMOD Technologies LLC, Zurich, Switzerland) was used. Qualitative assessment was conducted with parametric maps of the spatial BP_{nd} distribution by pixel-wised calculation (extracting signals from individual pixels). For determination of BP_{nd}, delineation of the target region (left and right striatum) as well as the reference region (cerebellum) was conducted using an implemented MRI-based rat brain atlas [73]. Therefore, PET data were first transformed to the standard matrix of the animal-specific MRI (3D isotropic T1w FLASH). Two rats did not receive a MRI scan, for transformation a representative MRI of the same experimental group was used. Two animals died during the study course and four PET/CT scanning time points were canceled. Six PET/CT scans were excluded from analysis due to incorrect injection of the radioligand and one due to development of a brain tumor (Table 1). Animal-specific MRI datasets were then transformed to Schiffer matrix and the respective transformation matrix was saved. In a final step, PET data were transformed into a Schiffer matrix using the saved transformation matrix to guarantee maximal resolution of PET data (Figure 7a,b). All transformations were performed using ridged matching method implemented in PMOD. VOIs for target and reference region were then defined with the Schiffer atlas [73] (Figure 7c,d) and time–activity curves (TAC) were extracted from dynamic PET data. For kinetic analysis the model-driven SRTM2 [54] was applied and BP_{nd} was estimated, being defined as the ratio of receptor density (B_{max}) multiplied by the radioligand affinity [74]. We assumed that receptor affinity was not changed in 6-OHDA-lesion rats compared to controls as BP of [^{18}F]fallypride is resistant to DA depletion [75–77].

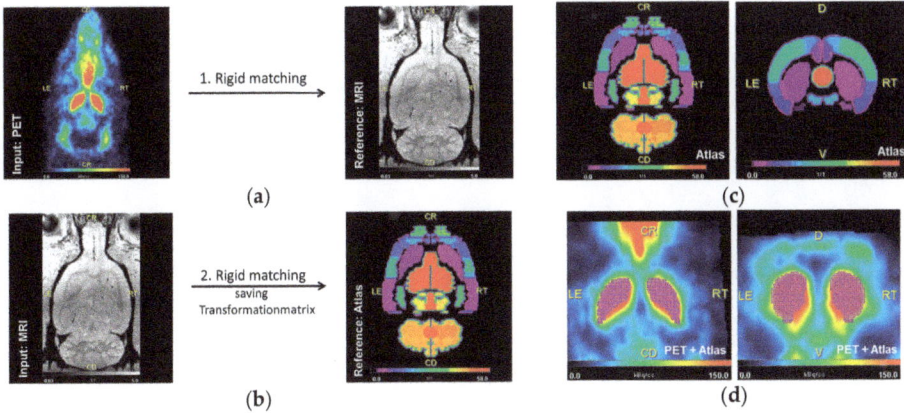

Figure 7. (**a**,**b**) Workflow of the transformation process using rigid matching. In a first step PET (input) was transformed to the matrix of the animal-specific MRI (reference) (**a**). Secondly, the animal-specific MRI (input) was transformed to the matrix of the Schiffer atlas and the transformation matrix was saved and finally applied to PET data (**b**). (**c**,**d**) Example of the delineation of the left and right striatum (purple grid) and the cerebellum (orange grid) in PET (**d**) with published Schiffer atlas (**c**) in a control animal (sham-6-OHDA + sham-BoNT-A).

4.8. Statistical Assessment

Statistical significance was examined using IBM SPSS Statistics software version 22. To test for the Gaussian distribution of all reported data, a Kolmogorov–Smirnov test was performed followed by a univariate general linear model. Between-subjects post hoc ANOVA analysis of variance was conducted with D_2/D_3R availability of the CPu (left/right) as the dependent variable and the experimental group as covariate. Following, Bonferroni correction with the factor group for each of the analyzed time points was performed (df = 5; (F = 4.598 one month post-BoNT-A or -sham-BoNT-A; F = 3.051 three months post-BoNT-A or -sham-BoNT-A, F = 2.584 six months post-BoNT-A or -sham-BoNT-A)). Statistical significance of apomorphine- and amphetamine-induced behavior was analyzed by the unpaired student's *t*-test. To analyze correlations of D_2/D_3R availability and rotational behavior in the apomorphine-induced rotation test linear regression followed by a two-sided Pearson correlation test was implemented. A *p*-value below 0.05 was considered to indicate significance.

5. Conclusions

We here provide a longitudinal study on changes of D_2/D_3R availability in the 6-OHDA-induced hemi-PD rat model. We found an increase in D_2/D_3R availability of 23% up to six months post-lesion, which was significantly reduced after striatal injection of BoNT-A. Interestingly, this decrease of pathological D_2/D_3R imbalance by intrastriatal BoNT-A injection significantly correlated with behavior in the apomorphine rotation test. Altogether, our results emphasize the therapeutical capability of BoNT-A in hemi-PD rats and provide insights in the underlying mechanisms.

Acknowledgments: The authors would like to thank Anne Möller, Joanna Förster, and Susann Lehmann for their technical assistance. We gratefully acknowledge Susann Lehmann, Iloana Klamfuß, Petra Wolff, Robin Piecha, Mathias Lietz, and Ulf Hasse for animal housing and care. We appreciate financial support by Deutsche Forschungsgemeinschaft and Universität Rostock/Universitätsmedizin Rostock within the funding program Open Access Publishing.

Author Contributions: Teresa Mann, Andreas Wree, Jens Kurth, and Tobias Lindner wrote the manuscript. Andreas Wree and Bernd Joachim Krause planned the project and Jens Kurth provided the analyzing strategy. Andreas Wree, Alexander Hawlitschka, and Teresa Mann executed stereotaxic operations. Teresa Mann analyzed data and designed figures. Teresa Mann and Alexander Hawlitschka conducted rotation tests with apomorphine and amphetamine. Alexander Hawlitschka performed TH immunostaining. Jan Stenzel planned and performed PET/CT scans. Tobias Lindner and Stefan Polei determined the MRI protocol and conducted measurements. Alexander Hohn established the synthesis protocol and synthesized the radioligand [^{18}F]fallypride. All authors contributed substantially to the project and the manuscript.

Conflicts of Interest: The authors declare no conflict of interest. The founding sponsors had no role in the design of the study; in the collection, analyses, or interpretation of data; in the writing of the manuscript, and in the decision to publish the results.

Abbreviations

BSA	Bovine serum albumin
BP$_{nd}$	non-displaceable binding potential
BoNT-A	Botulinum neurotoxin A
CPu	caudate–putamen
D$_2$/D$_3$R	D$_2$/D$_3$ receptor
DA	dopamine
hemi-PD	hemiparkinsonian
MFB	medial forebrain bundle
MRI	magnetic resonance imaging
MSN	medium spiny neuron
PD	Parkinson's disease
PET	positron emission tomography
PBS	phosphate-buffered saline
PVE	partial volume effect
SNAP25	synaptosomal-associated protein of 25-kDa
SN(pc)	substantia nigra (pars compacta)
SRTM2	simplified reference tissue model 2
TAC	time-activity curve
TH	tyrosine hydroxylase
VOI	voxels of interest
6-OHDA	6-hydroxdopamine

References

1. Vučković, M.G.; Li, Q.; Fisher, B.; Nacca, A.; Leahy, R.M.; Walsh, J.P.; Mukherjee, J.; Williams, C.; Jakowec, M.W.; Petzinger, G.M. Exercise elevates dopamine D2 receptor in a mouse model of Parkinson's disease: In vivo imaging with [^{18}F]fallypride. *Mov. Disord.* **2010**, *25*, 2777–2784. [CrossRef] [PubMed]
2. Choi, J.Y.; Kim, C.H.; Jeon, T.J.; Cho, W.G.; Lee, J.S.; Lee, S.J.; Choi, T.H.; Kim, B.S.; Yi, C.H.; Seo, Y.; et al. Evaluation of dopamine transporters and D2 receptors in hemiparkinsonian rat brains in vivo using consecutive PET scans of [^{18}F]FPCIT and [^{18}F]fallypride. *Appl. Radiat. Isot.* **2012**, *70*, 2689–2694. [CrossRef] [PubMed]
3. Wedekind, F.; Oskamp, A.; Lang, M.; Hawlitschka, A.; Zilles, K.; Wree, A.; Bauer, A. Intrastriatal administration of botulinum neurotoxin A normalizes striatal D2 R binding and reduces striatal D1 R binding in male hemiparkinsonian rats. *J. Neurosci. Res.* **2018**, *96*, 75–86. [CrossRef] [PubMed]
4. Mukherjee, J.; Christian, B.T.; Narayanan, T.K.; Shi, B.; Mantil, J. Evaluation of dopamine D-2 receptor occupancy by clozapine, risperidone, and haloperidol in vivo in the rodent and nonhuman primate brain using 18F-fallypride. *Neuropsychopharmacology* **2001**, *25*, 476–488. [CrossRef]
5. Honer, M.; Brühlmeier, M.; Missimer, J.; Schubiger, A.P.; Ametamey, S.M. Dynamic imaging of striatal D2 receptors in mice using quad-HIDAC PET. *J. Nucl. Med.* **2004**, *45*, 464–470. [PubMed]
6. Mukherjee, J.; Constantinescu, C.C.; Hoang, A.T.; Jerjian, T.; Majji, D.; Pan, M.-L. Dopamine D3 receptor binding of ^{18}F-fallypride: Evaluation using in vitro and in vivo PET imaging studies. *Synapse* **2015**, *69*, 577–591. [CrossRef] [PubMed]

7. Ding, J.; Guzman, J.N.; Tkatch, T.; Chen, S.; Goldberg, J.A.; Ebert, P.J.; Levitt, P.; Wilson, C.J.; Hamm, H.E.; Surmeier, D.J. RGS4-dependent attenuation of M4 autoreceptor function in striatal cholinergic interneurons following dopamine depletion. *Nat. Neurosci.* **2006**, *9*, 832–842. [CrossRef] [PubMed]

8. Pisani, A.; Bonsi, P.; Centonze, D.; Gubellini, P.; Bernardi, G.; Calabresi, P. Targeting striatal cholinergic interneurons in Parkinson's disease: Focus on metabotropic glutamate receptors. *Neuropharmacology* **2003**, *45*, 45–56. [CrossRef]

9. Goldberg, J.A.; Ding, J.B.; Surmeier, D.J. Muscarinic modulation of striatal function and circuitry. *Handb. Exp. Pharmacol.* **2012**, 223–241. [CrossRef]

10. Zhou, F.M.; Liang, Y.; Dani, J. A Endogenous nicotinic cholinergic activity regulates dopamine release in the striatum. *Nat. Neurosci.* **2001**, *4*, 1224–1229. [CrossRef] [PubMed]

11. Threlfell, S.; Clements, M.A.; Khodai, T.; Pienaar, I.S.; Exley, R.; Wess, J.; Cragg, S.J. Striatal muscarinic receptors promote activity dependence of dopamine transmission via distinct receptor subtypes on cholinergic interneurons in ventral versus dorsal striatum. *J. Neurosci.* **2010**, *30*, 3398–3408. [CrossRef] [PubMed]

12. Duty, S.; Jenner, P. Animal models of Parkinson's disease: A source of novel treatments and clues to the cause of the disease. *Br. J. Pharmacol.* **2011**, *164*, 1357–1391. [CrossRef] [PubMed]

13. Soto-Otero, R.; Méndez-Alvarez, E.; Hermida-Ameijeiras, A.; Muñoz-Patiño, A.M.; Labandeira-Garcia, J.L. Autoxidation and neurotoxicity of 6-hydroxydopamine in the presence of some antioxidants: Potential implication in relation to the pathogenesis of Parkinson's disease. *J. Neurochem.* **2000**, *74*, 1605–1612. [CrossRef] [PubMed]

14. Bové, J.; Perier, C. Neurotoxin-based models of Parkinson's disease. *Neuroscience* **2012**, *211*, 51–76. [CrossRef] [PubMed]

15. Carlsson, A.; Lindqyist, M.; Magnusson, T. 3,4-Dihydroxyphenylalanine and 5-hydroxytryptophan as reserpine antagonists. *Nature* **1957**, *180*, 1200. [CrossRef] [PubMed]

16. Cacabelos, R. Parkinson's Disease: From Pathogenesis to Pharmacogenomics. *Int. J. Mol. Sci.* **2017**, *18*, 551. [CrossRef] [PubMed]

17. Moritz, A.E.; Benjamin Free, R.; Sibley, D.R. Advances and challenges in the search for D2 and D3 dopamine receptor-selective compounds. *Cell. Signal.* **2018**, *41*, 75–81. [CrossRef] [PubMed]

18. Olanow, C.W.; Stocchi, F. Levodopa: A new look at an old friend. *Mov. Disord.* **2017**. [CrossRef] [PubMed]

19. Kühn, J.; Haumesser, J.K.; Beck, M.H.; Altschüler, J.; Kühn, A.A.; Nikulin, V.V.; van Riesen, C. Differential effects of levodopa and apomorphine on neuronal population oscillations in the cortico-basal ganglia loop circuit in vivo in experimental parkinsonism. *Exp. Neurol.* **2017**, *298*, 122–133. [CrossRef]

20. Horstink, M.; Tolosa, E.; Bonuccelli, U.; Deuschl, G.; Friedman, A.; Kanovsky, P.; Larsen, J.P.; Lees, A.; Oertel, W.; Poewe, W.; et al. Review of the therapeutic management of Parkinson's disease. Report of a joint task force of the European Federation of Neurological Societies and the Movement Disorder Society-European Section. Part I: Early (uncomplicated) Parkinson's disease. *Eur. J. Neurol.* **2006**, *13*, 1170–1185. [CrossRef] [PubMed]

21. Horstink, M.; Tolosa, E.; Bonuccelli, U.; Deuschl, G.; Friedman, A.; Kanovsky, P.; Larsen, J.P.; Lees, A.; Oertel, W.; Poewe, W.; et al. Review of the therapeutic management of Parkinson's disease. Report of a joint task force of the European Federation of Neurological Societies (EFNS) and the Movement Disorder Society-European Section (MDS-ES). Part II: Late (complicated) Parkinson's dise. *Eur. J. Neurol.* **2006**, *13*, 1186–1202. [CrossRef] [PubMed]

22. Fernandez, H.H. Updates in the medical management of Parkinson disease. *Clevel. Clin. J. Med.* **2012**, *79*, 28–35. [CrossRef] [PubMed]

23. Wree, A.; Mix, E.; Hawlitschka, A.; Antipova, V.; Witt, M.; Schmitt, O.; Benecke, R. Intrastriatal botulinum toxin abolishes pathologic rotational behaviour and induces axonal varicosities in the 6-OHDA rat model of Parkinson's disease. *Neurobiol. Dis.* **2011**, *41*, 291–298. [CrossRef] [PubMed]

24. Holzmann, C.; Dräger, D.; Mix, E.; Hawlitschka, A.; Antipova, V.; Benecke, R.; Wree, A. Effects of intrastriatal botulinum neurotoxin A on the behavior of Wistar rats. *Behav. Brain Res.* **2012**, *234*, 107–116. [CrossRef] [PubMed]

25. Hawlitschka, A.; Antipova, V.; Schmitt, O.; Witt, M.; Benecke, R.; Mix, E.; Wree, A. Intracerebrally applied botulinum neurotoxin in experimental neuroscience. *Curr. Pharm. Biotechnol.* **2013**, *14*, 124–130. [PubMed]

26. Antipova, V.; Hawlitschka, A.; Mix, E.; Schmitt, O.; Dräger, D.; Benecke, R.; Wree, A. Behavioral and structural effects of unilateral intrastriatal injections of botulinum neurotoxin a in the rat model of Parkinson's disease. *J. Neurosci. Res.* **2013**, *91*, 838–847. [CrossRef] [PubMed]

27. Caleo, M.; Antonucci, F.; Restani, L.; Mazzocchio, R. A reappraisal of the central effects of botulinum neurotoxin type A: By what mechanism? *J. Neurochem.* **2009**, *109*, 15–24. [CrossRef] [PubMed]

28. Binz, T.; Sikorra, S.; Mahrhold, S. Clostridial neurotoxins: Mechanism of SNARE cleavage and outlook on potential substrate specificity reengineering. *Toxins* **2010**, *2*, 665–682. [CrossRef] [PubMed]

29. Kroken, A.R.; Karalewitz, A.P.-A.; Fu, Z.; Kim, J.-J.P.; Barbieri, J.T. Novel ganglioside-mediated entry of botulinum neurotoxin serotype D into neurons. *J. Biol. Chem.* **2011**, *286*, 26828–26837. [CrossRef] [PubMed]

30. Mehlan, J.; Brosig, H.; Schmitt, O.; Mix, E.; Wree, A.; Hawlitschka, A. Intrastriatal injection of botulinum neurotoxin-A is not cytotoxic in rat brain—A histological and stereological analysis. *Brain Res.* **2016**, *1630*, 18–24. [CrossRef] [PubMed]

31. Mann, T.; Zilles, K.; Dikow, H.; Hellfritsch, A.; Cremer, M.; Piel, M.; Rösch, F.; Hawlitschka, A.; Schmitt, O.; Wree, A. Dopamine, Noradrenaline and Serotonin Receptor Densities in the Striatum of Hemiparkinsonian Rats following Botulinum Neurotoxin-A Injection. *Neuroscience* **2018**, *374*, 187–204. [CrossRef] [PubMed]

32. Grossman, R.; Paden, C.M.; Fry, P.A.; Rhodes, R.S.; Biegon, A. Persistent region-dependent neuroinflammation, NMDA receptor loss and atrophy in an animal model of penetrating brain injury. *Future Neurol.* **2012**, *7*, 329–339. [CrossRef] [PubMed]

33. Qü, M.; Buchkremer-Ratzmann, I.; Schiene, K.; Schroeter, M.; Witte, O.W.; Zilles, K. Bihemispheric reduction of GABAA receptor binding following focal cortical photothrombotic lesions in the rat brain. *Brain Res.* **1998**, *813*, 374–380. [CrossRef]

34. Jeon, B.S.; Jackson-Lewis, V.; Burke, R.E. 6-Hydroxydopamine lesion of the rat substantia nigra: Time course and morphology of cell death. *Neurodegeneration* **1995**, *4*, 131–137. [CrossRef] [PubMed]

35. Creese, I.; Burt, D.R.; Snyder, S.H. Dopamine receptor binding enhancement accompanies lesion-induced behavioral supersensitivity. *Science* **1977**, *197*, 596–598. [CrossRef] [PubMed]

36. Przedbroski, S.; Leviver, M.; Jiang, H.; Ferreira, M.; Jackson-Lewis, V.; Donaldson, D.; Togasaki, D.M. Dose-dependent lesions of the dopaminergic nigrostriatal pathway induced by instrastriatal injection of 6-hydroxydopamine. *Neuroscience* **1995**, *67*, 631–647. [CrossRef]

37. Sun, W.; Sugiyama, K.; Asakawa, T.; Yamaguchi, H.; Akamine, S.; Ouchi, Y.; Magata, Y.; Namba, H. Dynamic changes of striatal dopamine D2 receptor binding at later stages after unilateral lesions of the medial forebrain bundle in Parkinsonian rat models. *Neurosci. Lett.* **2011**, *496*, 157–162. [CrossRef] [PubMed]

38. Ungerstedt, U.; Butcher, L.L.; Butcher, S.G.; Andén, N.E.; Fuxe, K. Direct chemical stimulation of dopaminergic mechanisms in the neostriatum of the rat. *Brain Res.* **1969**, *14*, 461–471. [CrossRef]

39. Ma, Y.; Zhan, M.; OuYang, L.; Li, Y.; Chen, S.; Wu, J.; Chen, J.; Luo, C.; Lei, W. The effects of unilateral 6-OHDA lesion in medial forebrain bundle on the motor, cognitive dysfunctions and vulnerability of different striatal interneuron types in rats. *Behav. Brain Res.* **2014**, *266*, 37–45. [CrossRef] [PubMed]

40. Wang, Q.; Wei, X.; Gao, H.; Li, J.; Liao, J.; Liu, X.; Qin, B.; Yu, Y.; Deng, C.; Tang, B.; et al. Simvastatin reverses the downregulation of M1/4 receptor binding in 6-hydroxydopamine-induced parkinsonian rats: The association with improvements in long-term memory. *Neuroscience* **2014**, *267*, 57–66. [CrossRef] [PubMed]

41. Ungerstedt, U.; Arbuthnott, G.W. Quantitative recording of rotational behavior in rats after 6-hydroxy-dopamine lesions of the nigrostriatal dopamine system. *Brain Res.* **1970**, *24*, 485–493. [CrossRef]

42. Hudson, J.L.; van Horne, C.G.; Strömberg, I.; Brock, S.; Clayton, J.; Masserano, J.; Hoffer, B.J.; Gerhardt, G.A. Correlation of apomorphine- and amphetamine-induced turning with nigrostriatal dopamine content in unilateral 6-hydroxydopamine lesioned rats. *Brain Res.* **1993**, *626*, 167–174. [CrossRef]

43. Ginovart, N.; Farde, L.; Halldin, C.; Swahn, C.G. Effect of reserpine-induced depletion of synaptic dopamine on [¹¹C]raclopride binding to D2-dopamine receptors in the monkey brain. *Synapse* **1997**, *25*, 321–325. [CrossRef]

44. Seeman, P.; Guan, H.C.; Niznik, H.B. Endogenous dopamine lowers the dopamine D2 receptor density as measured by [3H]raclopride: Implications for positron emission tomography of the human brain. *Synapse* **1989**, *3*, 96–97. [CrossRef] [PubMed]

45. Young, L.T.; Wong, D.F.; Goldman, S.; Minkin, E.; Chen, C.; Matsumura, K.; Scheffel, U.; Wagner, H.N. Effects of endogenous dopamine on kinetics of [3H]N-methylspiperone and [3H]raclopride binding in the rat brain. *Synapse* **1991**, *9*, 188–194. [CrossRef] [PubMed]

46. Staunton, D.A.; Wolfe, B.B.; Groves, P.M.; Molinoff, P.B. Dopamine receptor changes following destruction of the nigrostriatal pathway: Lack of a relationship to rotational behavior. *Brain Res.* **1981**, *211*, 315–327. [CrossRef]

47. Fornaretto, M.G.; Caccia, C.; Caron, M.G.; Fariello, R.G. Dopamine Receptors Status After Unilateral Nigral 6-OHDA Lesion Hybridization Study in the Rat Brain. *Mol. Chem. Neuropathol.* **1993**, *19*, 147–162. [CrossRef] [PubMed]

48. Ryu, J.H.; Yanai, K.; Zhao, X.L.; Watanabe, T. The effect of dopamine D1 receptor stimulation on the up-regulation of histamine H3-receptors following destruction of the ascending dopaminergic neurones. *Br. J. Pharmacol.* **1996**, *118*, 585–592. [CrossRef] [PubMed]

49. Araki, T.; Tanji, H.; Kato, H.; Imai, Y.; Mizugaki, M.; Itoyama, Y. Temporal changes of dopaminergic and glutamatergic receptors in 6-hydroxydopamine-treated rat brain. *Eur. Neuropsychopharmacol.* **2000**, *10*, 365–375. [CrossRef]

50. Xu, Z.C.; Ling, G.; Sahr, R.N.; Neal-Beliveau, B.S. Asymmetrical changes of dopamine receptors in the striatum after unilateral dopamine depletion. *Brain Res.* **2005**, *1038*, 163–170. [CrossRef] [PubMed]

51. Inaji, M.; Okauchi, T.; Ando, K.; Maeda, J.; Nagai, Y.; Yoshizaki, T.; Okano, H.; Nariai, T.; Ohno, K.; Obayashi, S.; et al. Correlation between quantitative imaging and behavior in unilaterally 6-OHDA-lesioned rats. *Brain Res.* **2005**, *1064*, 136–145. [CrossRef] [PubMed]

52. Hume, S.P.; Opacka-Juffry, J.; Myers, R.; Ahier, R.G.; Ashworth, S.; Brooks, D.J.; Lammertsma, A.A. Effect of L-dopa and 6-hydroxydopamine lesioning on [¹¹C]raclopride binding in rat striatum, quantified using PET. *Synapse* **1995**, *21*, 45–53. [CrossRef] [PubMed]

53. Zhou, X.; Doorduin, J.; Elsinga, P.H.; Dierckx, R.A.J.O.; de Vries, E.F.J.; Casteels, C. Altered adenosine 2A and dopamine D2 receptor availability in the 6-hydroxydopamine-treated rats with and without levodopa-induced dyskinesia. *Neuroimage* **2017**, *157*, 209–218. [CrossRef] [PubMed]

54. Ariano, M.A.; Stromski, C.J.; Smyk-Randall, E.M.; Sibley, D.R. D2 dopamine receptor localization on striatonigral neurons. *Neurosci. Lett.* **1992**, *144*, 215–220. [CrossRef]

55. Brock, J.W.; Farooqui, S.; Ross, K.; Prasad, C. Localization of dopamine D2 receptor protein in rat brain using polyclonal antibody. *Brain Res* **1992**, *578*, 244–250. [CrossRef]

56. Yung, K.K.L.; Bolam, J.P.; Smith, A.D.; Hersch, S.M.; Ciliax, B.J.; Levey, A.I. Immunocytochemical localization of D1 and D2 dopamine receptors in the basal ganglia of the rat: Light and electron microscopy. *Neuroscience* **1995**, *65*, 709–730. [CrossRef]

57. Wu, Y.; Carson, R.E. Noise reduction in the simplified reference tissue model for neuroreceptor functional imaging. *J. CerebR. Blood Flow Metab.* **2002**, *22*, 1440–1452. [CrossRef] [PubMed]

58. Lammertsma, A.A.; Bench, C.J.; Hume, S.P.; Osman, S.; Gunn, K.; Brooks, D.J.; Frackowiak, R.S. Comparison of methods for analysis of clinical [¹¹C]raclopride studies. *J. Cerebr. Blood Flow Metab.* **1996**, *16*, 42–52. [CrossRef] [PubMed]

59. Martres, M.P.; Sales, N.; Bouthenet, M.L.; Schwartz, J.C. Localisation and pharmacological characterisation of D-2 dopamine receptors in rat cerebral neocortex and cerebellum using [¹²⁵I]iodosulpride. *Eur. J. Pharmacol.* **1985**, *118*, 211–219. [CrossRef]

60. Seneca, N.; Finnema, S.J.; Farde, L.; Gulyás, B.; Wikström, H.V.; Halldin, C.; Innis, R.B. Effect of amphetamine on dopamine D2 receptor binding in nonhuman primate brain: A comparison of the agonist radioligand [¹¹C]MNPA and antagonist [¹¹C]raclopride. *Synapse* **2006**, *59*, 260–269. [CrossRef] [PubMed]

61. Constantinescu, C.C.; Coleman, R.A.; Pan, M.-L.; Mukherjee, J. Striatal and extrastriatal microPET imaging of D2/D3 dopamine receptors in rat brain with [¹⁸F]fallypride and [¹⁸F]desmethoxyfallypride. *Synapse* **2011**, *65*, 778–787. [CrossRef] [PubMed]

62. Yoder, K.K.; Albrecht, D.S.; Kareken, D.A.; Federici, L.M.; Perry, K.M.; Patton, E.A.; Zheng, Q.-H.; Mock, B.H.; O'Connor, S.; Herring, C.M. Test-retest variability of [¹¹C]raclopride-binding potential in nontreatment-seeking alcoholics. *Synapse* **2011**, *65*, 553–561. [CrossRef] [PubMed]

63. Prieto, E.; Collantes, M.; Delgado, M.; Juri, C.; García-García, L.; Molinet, F.; Fernández-Valle, M.E.; Pozo, M.A.; Gago, B.; Martí-Climent, J.M.; et al. Statistical parametric maps of ¹⁸F-FDG PET and 3-D autoradiography in the rat brain: A cross-validation study. *Eur. J. Nucl. Med. Mol. Imaging* **2011**, *38*, 2228–2237. [CrossRef] [PubMed]

64. Hoffman, E.J.; Huang, S.C.; Phelps, M.E. Quantitation in positron emission computed tomography: 1. Effect of object size. *J. Comput. Assist. Tomogr.* **1979**, *3*, 299–308. [CrossRef] [PubMed]

Molecules **2018**, *23*, 587

65. Hoffman, E.J.; Huang, S.C.; Plummer, D.; Phelps, M.E. Quantitation in positron emission computed tomography: 6. effect of nonuniform resolution. *J. Comput. Assist. Tomogr.* **1982**, *6*, 987–999. [CrossRef] [PubMed]

66. Andersson, C.; Hamer, R.M.; Lawler, C.P.; Mailman, R.B.; Lieberman, J.A. Striatal volume changes in the rat following long-term administration of typical and atypical antipsychotic drugs. *Neuropsychopharmacology* **2002**, *27*, 143–151. [CrossRef]

67. Garcia, D.V.; Casteels, C.; Schwarz, A.J.; Dierckx, R.A.J.O.; Koole, M.; Doorduin, J. Correction: A Standardized Method for the Construction of Tracer Specific PET and SPECT Rat Brain Templates: Validation and Implementation of a Toolbox. *PLoS ONE* **2015**, *10*, e0143900. [CrossRef] [PubMed]

68. Paxinos, G.; Watson, C. *The Rat Brain in Stereotaxic Coordinates*, 6th ed.; Academic press: San Diego, CA, USA, 2007; ISBN 9780125476126.

69. Hawlitschka, A.; Holzmann, C.; Witt, S.; Spiewok, J.; Neumann, A.-M.; Schmitt, O.; Wree, A.; Antipova, V. Intrastriatally injected botulinum neurotoxin-A differently effects cholinergic and dopaminergic fibers in C57BL/6 mice. *Brain Res.* **2017**, *1676*, 46–56. [CrossRef] [PubMed]

70. Gao, M.; Wang, M.; Mock, B.H.; Glick-Wilson, B.E.; Yoder, K.K.; Hutchins, G.D.; Zheng, Q.-H. An improved synthesis of dopamine D2/D3 receptor radioligands [^{11}C]fallypride and [^{18}F]fallypride. *Appl. Radiat. Isot.* **2010**, *68*, 1079–1086. [CrossRef] [PubMed]

71. Bao, Q.; Newport, D.; Chen, M.; Stout, D.B.; Chatziioannou, A.F. Performance evaluation of the inveon dedicated PET preclinical tomograph based on the NEMA NU-4 standards. *J. Nucl. Med.* **2009**, *50*, 401–408. [CrossRef] [PubMed]

72. Kemp, B.J.; Hruska, C.B.; McFarland, A.R.; Lenox, M.W.; Lowe, V.J. NEMA NU 2-2007 performance measurements of the Siemens Inveon preclinical small animal PET system. *Phys. Med. Biol.* **2009**, *54*, 2359–2376. [CrossRef] [PubMed]

73. Schiffer, W.K.; Mirrione, M.M.; Biegon, A.; Alexoff, D.L.; Patel, V.; Dewey, S.L. Serial microPET measures of the metabolic reaction to a microdialysis probe implant. *J. Neurosci. Methods* **2006**, *155*, 272–284. [CrossRef] [PubMed]

74. Mintun, M.A.; Raichle, M.E.; Kilbourn, M.R.; Wooten, G.F.; Welch, M.J. A quantitative model for the in vivo assessment of drug binding sites with positron emission tomography. *Ann. Neurol.* **1984**, *15*, 217–227. [CrossRef] [PubMed]

75. Mukherjee, J.; Yang, Z.-Y.; Lew, R.; Brown, T.; Kronmal, S.; Cooper, M.D.; Seiden, L.S. Evaluation of d-Amphetamine Effects on the Binding of Dopamine D-2 Receptor Radioligand, 18F-Fallypride in Nonhuman Primates Using Positron Emission Tomography. *Synapse* **1997**, *27*, 1–13. [CrossRef]

76. Cropley, V.L.; Innis, R.B.; Nathan, P.J.; Brown, A.K.; Sangare, J.L.; Lerner, A.; Ryu, Y.H.; Sprague, K.E.; Pike, V.W.; Fujita, M. Small effect of dopamine release and no effect of dopamine depletion on [^{18}F]fallypride binding in healthy humans. *Synapse* **2008**, *62*, 399–408. [CrossRef] [PubMed]

77. Rominger, A.; Wagner, E.; Mille, E.; Böning, G.; Esmaeilzadeh, M.; Wängler, B.; Gildehaus, F.-J.; Nowak, S.; Bruche, A.; Tatsch, K.; et al. Endogenous competition against binding of [^{18}F]DMFP and [^{18}F]fallypride to dopamine D(2/3) receptors in brain of living mouse. *Synapse* **2010**, *64*, 313–322. [CrossRef] [PubMed]

Sample Availability: Samples of the compounds are not available from the authors.

molecules

MDPI

Article

Bridging from Brain to Tumor Imaging: (S)-(−)- and (R)-(+)-[18F]Fluspidine for Investigation of Sigma-1 Receptors in Tumor-Bearing Mice [†]

Mathias Kranz [1,2,*], Ralf Bergmann [3], Torsten Kniess [3], Christin Neuber [3], Zhengxin Cai [2], Gang Deng [4], Steffen Fischer [1], Jiangbing Zhou [4], Yiyun Huang [2], Peter Brust [1], Winnie Deuther-Conrad [1,‡] and Jens Pietzsch [3,5,‡]

[1]   Department of Neuroradiopharmaceuticals, Institute of Radiopharmaceutical Cancer Research, Helmholtz-Zentrum Dresden-Rossendorf, 04318 Leipzig, Germany; s.fischer@hzdr.de (S.F.); p.brust@hzdr.de (P.B.); w.deuther-conrad@hzdr.de (W.D.-C.)
[2]   Department of Diagnostic Radiology, PET Center, Yale University School of Medicine, New Haven, CT 06519, USA; zhengxin.cai@yale.edu (Z.C.); henry.huang@yale.edu (Y.H.)
[3]   Department of Radiopharmaceutical and Chemical Biology, Institute of Radiopharmaceutical Cancer Research, Helmholtz-Zentrum Dresden-Rossendorf, 01328 Dresden, Germany; r.bergmann@hzdr.de (R.B.); t.kniess@hzdr.de (T.K.); b.belter@hzdr.de (B.B.); c.neuber@hzdr.de (C.N.); j.pietzsch@hzdr.de (J.P.)
[4]   Department of Neurosurgery and Biomedical Engineering, Yale University School of Medicine, New Haven, CT 06519, USA; gang.deng@yale.edu (G.D.); jiangbing.zhou@yale.edu (J.Z.)
[5]   Technische Universität Dresden, School of Science, Faculty of Chemistry and Food Chemistry, 01062 Dresden, Germany
*   Correspondence: m.kranz@hzdr.de; Tel.: +49-0341-234-179-4639
†   This publication is dedicated to Professor Jörg Steinbach on the occasion of his 65th birthday.
‡   These authors contributed equally.

Received: 27 January 2018; Accepted: 18 March 2018; Published: 20 March 2018

Abstract: Sigma-1 receptors (Sig1R) are highly expressed in various human cancer cells and hence imaging of this target with positron emission tomography (PET) can contribute to a better understanding of tumor pathophysiology and support the development of antineoplastic drugs. Two Sig1R-specific radiolabeled enantiomers (S)-(−)- and (R)-(+)-[18F]fluspidine were investigated in several tumor cell lines including melanoma, squamous cell/epidermoid carcinoma, prostate carcinoma, and glioblastoma. Dynamic PET scans were performed in mice to investigate the suitability of both radiotracers for tumor imaging. The Sig1R expression in the respective tumors was confirmed by Western blot. Rather low radiotracer uptake was found in heterotopically (subcutaneously) implanted tumors. Therefore, a brain tumor model (U87-MG) with orthotopic implantation was chosen to investigate the suitability of the two Sig1R radiotracers for brain tumor imaging. High tumor uptake as well as a favorable tumor-to-background ratio was found. These results suggest that Sig1R PET imaging of brain tumors with [18F]fluspidine could be possible. Further studies with this tumor model will be performed to confirm specific binding and the integrity of the blood-brain barrier (BBB).

Keywords: [18F]fluspidine; carcinoma; glioblastoma; melanoma; sigma-1 receptor; dedicated small animal PET/CT

1. Introduction

In clinical oncology, major efforts are dedicated to the optimization of cancer treatment either with chemotherapeutics or radiation therapy [1]. For detection, diagnostic prognosis, and follow-up after treatment of tumors positron emission tomography (PET) is a widely applied tool able to image tumor-specific biochemical processes in vivo with radioactive probes [2]. One of the most widely applied radiotracers for that purpose is the glucose analog [18F]fluorodeoxyglucose ([18F]FDG) which accumulates in tissues with high metabolic activity [2,3]. However, it has shown some limitations in the detection of tumors like prostatic carcinoma DU145 [4] or PC-3 xenografts [5], A431 xenografts [6], FaDu tumors [7], A375 tumors [8], and in U87-MG glioblastoma [9,10]. Hence, not all tumor species show high glucose uptake (non-FDG-avid tumors) and consequently might be overlooked in a tumor screening study [11–14]. Furthermore, for brain tumors, the high background uptake of glucose in healthy brain makes the detection of small or low-grade tumors almost impossible [15]. Therefore, radiolabeled probes based on small molecules, amino acids, peptides, or antibodies targeting different biomarkers, e.g., tumor metabolism, cellular proliferation, or hypoxia were developed for the visualization and characterization of brain tumors with PET [16].

One promising approach for the use of PET radiotracers in oncology is the development of sigma receptor (SigR) ligands [17–19]. There is evidence that both SigR subtypes, sigma-1 receptors (Sig1R) and sigma-2 receptors (Sig2R), play important roles in cancer biology [17]. Studies with different human and rodent tumor cell lines have proven that there exists a high density of SigR binding sites in cancer [20]. While the Sig2R are regarded as potential markers for cellular proliferation [19,21,22], the stress-activated Sig1R are associated with the endoplasmic reticulum interface and are involved in the regulation of calcium signaling [23]. Furthermore, Aydar et al. has described a link between Sig1R expression and the tumor aggressiveness [24], suggesting that this receptor subtype may be a potential marker for the diagnosis and prognosis of (brain) tumors [23]. Hence, imaging of Sig1R with PET might contribute to a better understanding of the tumor physiology, the pathophysiological function of Sig1R, and the development of antineoplastic drugs [25].

Several PET radiotracers have been developed for imaging of Sig1R, such as [18F]1-3-fluoropropyl-4-((4-cyanophenoxy)-methyl)piperidine ([18F]FPS) [26], 1-(4-[18F]fluorobenzyl)-4-[(tetrahydrofuran-2-yl)methyl]piperazine [27], [11C]1-(3,4-dimethoxyphenethyl)-4-(3-phenylpropyl) piperazine ([11C]SA4503) [28], [18F]6-(3-fluoropropyl)-3-(2-(azepan-1-yl)ethyl)benzo[d]thiazol-2(3H)-one ([18F]FTC-146) [29] and the 18F-labeled 1,4-dioxa-8-azaspiro[4.5]decane derivative from our group [30]. Recently, we developed another two promising radioligands (*S*)-(−)-[18F]fluspidine and (*R*)-(+)-[18F]fluspidine [31,32], which both have been successfully used to image Sig1R in healthy mice [31] and piglets [33]. Furthermore, (*S*)-(−)-[18F]fluspidine was applied to humans in a first-in-man study [34]. The two enantiomers differ in their affinity towards Sig1R ((*R*)-(+)-fluspidine: K_i = 0.57 nM; (*S*)-(−)-fluspidine: K_i = 2.3 nM), which is suggested to influence their different pharmacokinetics [33]. The *S*-enantiomer, showing fast and reversible binding kinetics in brain, was chosen for the first-in-man evaluation [34]. Furthermore, a phase 1 clinical trial (drug occupancy study, ClinicalTrials.gov identifier: NCT03019289) is currently underway to investigate (*S*)-(−)-[18F]fluspidine for imaging and quantification of Sig1R in brain with pharmacokinetic modeling.

The current study was performed to investigate both enantiomers of [18F]fluspidine for their ability to image the Sig1R availability in tumors. Starting from the cellular level, we investigated the Sig1R synthesis with Western blot and the accumulation of (*S*)-(−)- and (*R*)-(+)-[18F]fluspidine in the human prostatic cancer cells DU145 and PC3, as well as in the cell lines A431 and FaDu (squamous carcinoma), A375 (malignant melanoma), and U87-MG (glioblastoma). Supported by the promising in vitro results, PET scans with (*S*)-(−)-[18F]fluspidine were performed in mice bearing five different heterotopically (peripheral) implanted tumors, representing both FDG-avid and non-FDG-avid cancers. Furthermore, an orthotopic brain tumor model in mice was chosen and used in PET scans to investigate the applicability of the two radiotracers with special regard to brain tumors (pilot study).

2. Results

2.1. Cell Specific Expression/Synthesis of Sig1R

The Sig1R expression was analyzed in lysates from melanoma (A375), squamous cell/epidermoid carcinoma (FaDu, A431), prostate carcinoma (DU145, PC3), glioblastoma (U87-MG), and lung carcinoma (NCI-H292) cells grown in vitro in cell culture as well as in mouse heterotopic tumor. As demonstrated in Figure 1, we detected the Sig1R protein in all cell lines as well as in the explanted tumor xenografts.

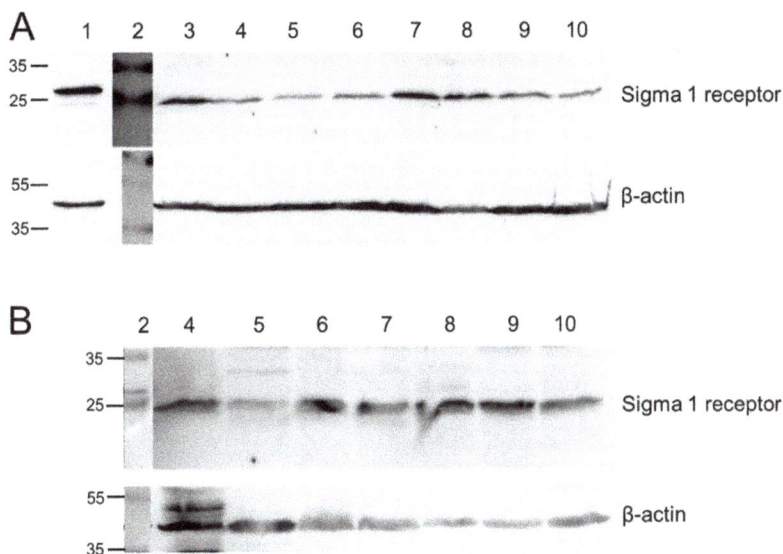

Figure 1. Detection of the Sig1R protein in (**A**) cell lysates and (**B**) tumor lysates. (1) HEK-S1R (Sig1R-overexpressing, transgenic cells, positive control), (2) protein standard, (3) HEK, (4) FaDu, (5) PC3, (6) DU145, (7) A431, (8) A375, (9) U87-MG, (10) NCI-H292. Expected band at 25 kDa. β-actin was used as loading control.

2.2. Cellular Accumulation of [^{18}F]fluspidine

Six human tumor cell lines were used to study the cellular accumulation (Figure 2) of (S)-(−)-[^{18}F]fluspidine and (R)-(+)-[^{18}F]fluspidine in vitro following a published protocol [30].

Substantial cellular accumulation of both (S)-(−)- and (R)-(+)-[^{18}F]fluspidine was observed in all investigated cancer cell lines. The extent of the cellular accumulation at 120 min differs between the cell lines and the highest cellular accumulation of (S)-(−)-[^{18}F]fluspidine was observed for A431 and A375 cells, with levels slightly higher than those of (R)-(+)-[^{18}F]fluspidine. Furthermore, high cellular accumulation of both radiotracers was observed in the two prostate carcinoma cell lines and the lowest values were obtained with FaDu and U87-MG cells.

Blocking studies were performed using 10 μM haloperidol which significantly reduced the cellular accumulation of both (S)-(−)- and (R)-(+)-[^{18}F]fluspidine in the cell lines with an inhibitory effect of 35–77%. These data demonstrate specific binding of both enantiomers of [^{18}F]fluspidine to Sig1R in all cell lines under investigation.

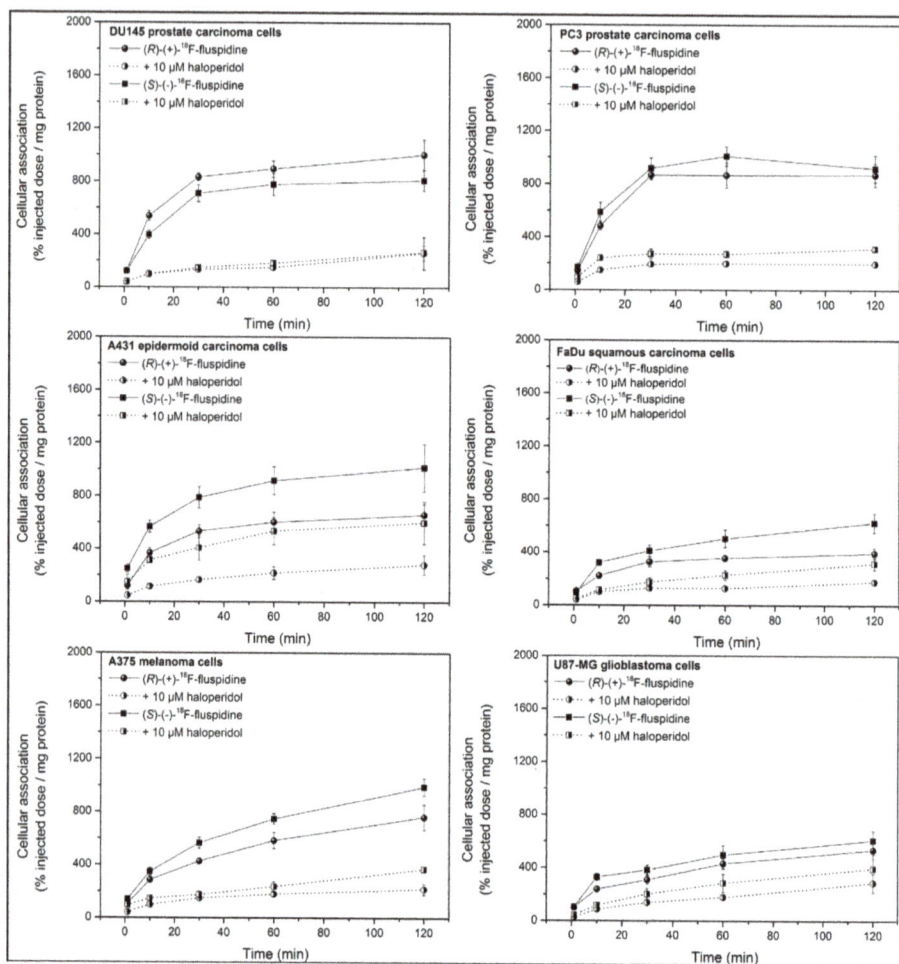

Figure 2. Cellular accumulation of (S)-(−)- and (R)-(+)-[^{18}F]fluspidine in the human tumor cells DU145, PC3, A431, FaDu, A375, and U87-MG in vitro. Blocking experiments were performed by preincubation (for 10 min) with 10 μM haloperidol. Results are given as percentage of injected dose (%ID) per mg protein (mean ± SD; $n \geq 8$ for (R)-(+)-[^{18}F]fluspidine and $n = 7$ for (S)-(−)-[^{18}F]fluspidine).

2.3. Small Animal PET Imaging

Small animal PET was performed in tumor-bearing mice after i.v. injection of (S)-(−)-[^{18}F]fluspidine. The results obtained for the five heterotopically (peripheral) implanted tumors in terms of standardized uptake values (SUV) or the tumor-to-muscle ratio are presented in Figure 3. Although there is uptake of (S)-(−)-[^{18}F]fluspidine in all investigated tumor models and the suitability of this radioligand for detection of Sig1R in tumors was indicated by in vitro autoradiography in an explanted heterotopic U87-MG tumor (Figure 3C), the extent of accumulation is low. The highest accumulation of the radiotracer was observed in FaDu (squamous cell carcinoma, $n = 2$) and PC3 (human prostate cancer, $n = 2$) tumors. Furthermore, the tumor-to-muscle ratio revealed a low signal from the surrounding tissue (Figure 3D). Administration of haloperidol did not significantly reduce the tracer uptake in the respective tumor models. Hence, no specific binding of (S)-(−)-[^{18}F]fluspidine to Sig1R was found in the heterotopically implanted tumors.

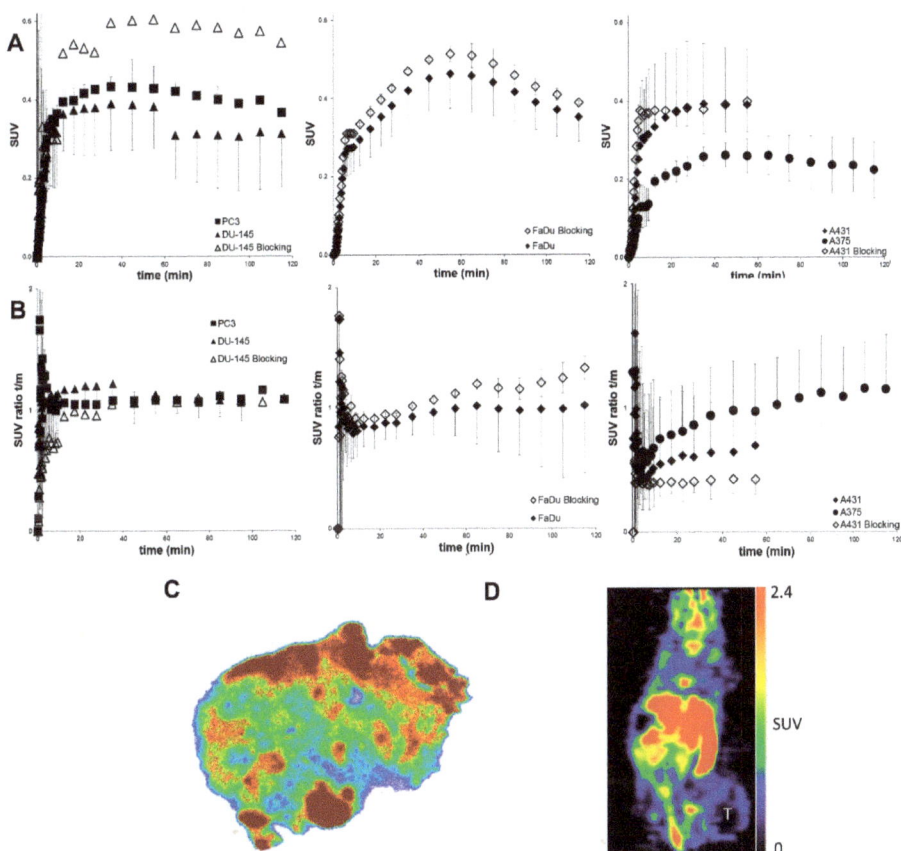

Figure 3. Small animal PET imaging in mice bearing different heterotopic tumors ($n = 2$/tumor model; $n = 1$ for DU145 blocking) after i.v. injection of (S)-(−)-[^{18}F]fluspidine. (**A**) The maximal tumor uptake corresponds to standardized uptake values (SUV) of 0.25 and 0.6 while a blocking effect is not visible (mean SUV ± SD). (**B**) Tumor-to-muscle SUV ratios of the respective animals (mean SUV ratio ± SD). (**C**) In vitro Sig1R autoradiography with (S)-(−)-[^{18}F]fluspidine of an explanted U87-MG tumor, grown heterotopically in a mouse, with heterogeneous activity distribution. (**D**) Representative coronal PET image of a FaDu tumor bearing mouse at 50–60 min p.i. (tumor highlighted, T).

Subsequently, a pilot study using small animal PET/CT after orthotopic tumor cell implantation (U87-MG) into the brain revealed high tumor uptake and a favorable tumor-to-background ratio (Figure 4A,E) following the administration of (S)-(−)-[^{18}F]fluspidine ($n = 2$) or (R)-(+)-[^{18}F]fluspidine ($n = 3$), respectively. The PET/CT images (Figure 4B–D,F–H) showed a clear separation of tumor from the brain. However, further PET/MR and ex vivo studies need to be performed to confirm specific binding of the ligands in the tumor.

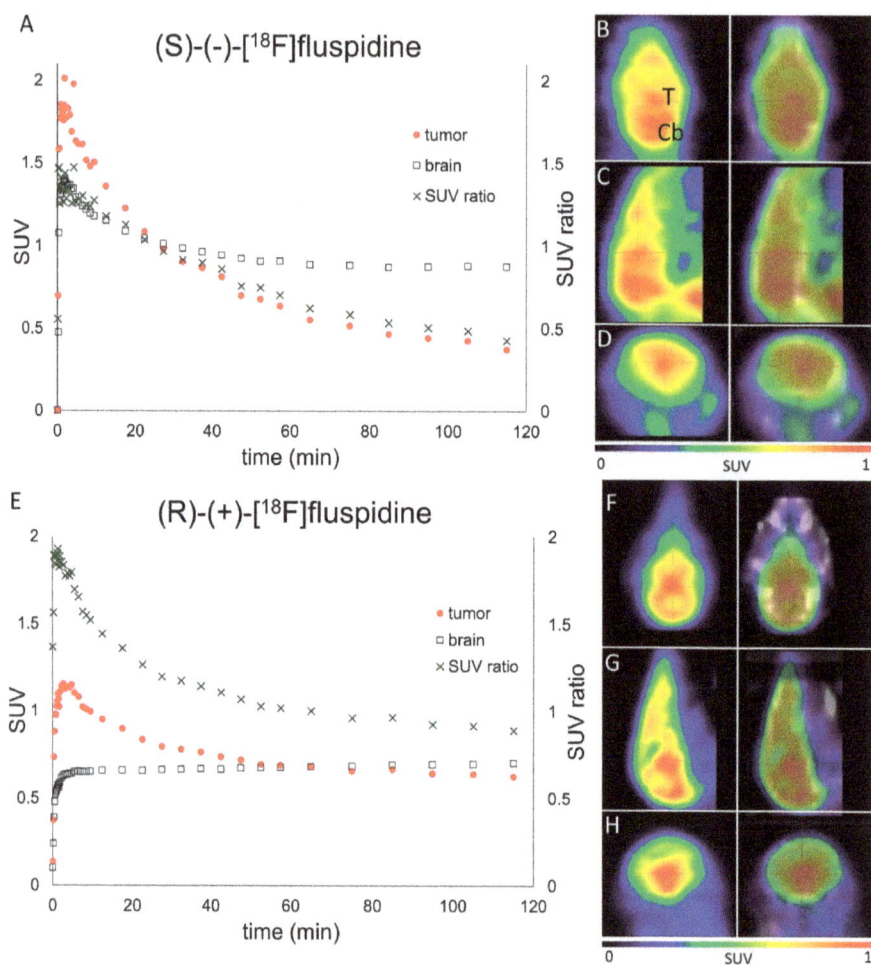

Figure 4. In vivo PET/CT imaging of mice with orthotopically implanted glioblastoma cells (U87-MG) after i.v. administration of (S)-(−)-[18F]fluspidine (n = 2) (**A–D**) or (R)-(+)-[18F]fluspidine (n = 3) (**E–H**). (**A,E**) Higher SUV of the tumor compared to the whole brain up to 25 or 50 min p.i., (S)-(−)-[18F]fluspidine or (R)-(+)-[18F]fluspidine, resulting in tumor-to-background SUV ratios >1 and hence tumor visibility at early time points. (**B–D,F–H**) Summed PET and PET/CT frames from 3–15 min p.i. in coronal (**B,F**), sagittal (**C,G**) and transaxial (**D,H**) views (T-tumor, Cb-cerebellum).

3. Discussion

Our group has recently developed (R)-(+)- and (S)-(−)-[18F]fluspidine [31], two specific radioligands for Sig1R PET with promising preclinical properties and distinctive kinetics [33,35]. In this publication, we investigated the suitability of both enantiomers of [18F]fluspidine to image Sig1R in different mouse tumor models. Current investigations [30] show strong expression of Sig1R and correlation with pathologic tumor tissue in human esophageal squamous cell carcinoma [36], prostate cancer [37,38], myeloma [39], melanoma [20,40], and glioma [41]. To assess the expression of the Sig1R protein in different human tumors, cell lines from epidermoid carcinoma, melanoma, and glioblastoma were investigated with Western blot in the current study. The results showed a positive signal for all investigated tumor cell lines, which represent both FDG-avid- and non-FDG-avid cancer entities, and confirmed substantial Sig1R expression in these tumor types.

Next, we investigated the accumulation and specific binding of (*S*)-(−)- and (*R*)-(+)-[^{18}F]fluspidine in the selected cancer cell lines in order to evaluate the potential of both radiotracers to address specific oncological questions. In general, all the cell lines showed accumulation of both enantiomers of [^{18}F]fluspidine and treatment with haloperidol significantly reduced the radiotracer binding [42,43]. Although haloperidol shows affinity to other receptors (i.e., Sig2R and dopamine D$_2$), it was used as blocking agent in the current study as the binding of [^{18}F]fluspidine is highly Sig1R selective [31]. Furthermore, 1′-benzyl-3-methoxy-3H-spiro[[2]benzofuran-1,4′-piperidine, the lead compound of the ligand development resulting in [^{18}F]fluspidine, did not show any relevant binding towards more than 60 neurotransmitter receptors, ion channels, and neurotransmitter transporters [44,45]. Hence, results from the blocking study confirm Sig1R specific binding of (*S*)-(−)- and (*R*)-(+)-[^{18}F]fluspidine in the cancer cell lines.

Given the limitations of [^{18}F]FDG for tumor imaging, we performed in vivo investigations using small animal dynamic PET to examine the tracer uptake and specific binding and therefore the suitability of (*S*)-(−)-[^{18}F]fluspidine as tumor imaging agent. Although we could show in vitro the expression of Sig1R in heterotopically grown U87-MG tumor by autoradiography with (*S*)-(−)-[^{18}F]fluspidine, the analysis of the PET scans performed in human tumor xenograft models revealed comparatively low accumulation of this radioligand in the respective tumors. It is reflected in particular by low tumor-to-muscle ratios due to low specific signal in tumor, i.e., binding displaceable by haloperidol. This is due to areas of necrosis in these tumor models where radiotracer accumulation is not expected [46]. Another possible reason is that interstitial oncotic pressure generated by proliferating cancer cells might have reduced the tumor blood flow and prohibited tracer uptake [47]. Furthermore, the lack of specific uptake for [^{18}F]fluspidine in heterotopic (peripheral) tumors [48,49] may be caused by differences in the tumor microenvironment and vascularization in these models [50,51]. Additionally, as shown by Figure 3C, the heterogeneity of the tumor tissue results in different levels of radiotracer uptake and applying ROI analysis to the whole tumor region results in low mean SUV.

In a subsequent pilot study we performed PET imaging with [^{18}F]fluspidine in orthotopically implanted brain tumors. In general, [^{18}F]fluspidine has been shown to be suitable for brain imaging [31,33,52] and thus, we hypothesize that these radioligands are suitable for imaging of Sig1R expressing brain tumors. Preliminary data from these experiments revealed a tumor-to-background SUV ratio (TBR) of >1 for early time points and thus visualization of the tumor in the mouse brain appeared to be feasible. However, at this point of investigation, without having the fraction of specific binding of both enantiomers proven with suitable experiments (e.g., pre-blocking with haloperidol or SA4503) a comparison of [^{18}F]fluspidine with established radiotracers is not possible. Further studies with this orthotopic brain tumor model are ongoing to test the BBB integrity, to confirm the specific binding and to further evaluate both enantiomers of [^{18}F]fluspidine as brain tumor imaging agents.

4. Materials and Methods

4.1. Radiochemistry

Enantiomerically pure (*S*)-(−)-[^{18}F]fluspidine and (*R*)-(+)-[^{18}F]fluspidine were prepared on a TRACERlab FX$_N$ synthesizer (GE Healthcare, Waukesha, WI, USA) as described in previous publications [33]. The radiochemical purity of (*R*)-(+)- or (*S*)-(−)-[^{18}F]fluspidine was >99%, and the molar activity at the end of the synthesis was 69.2 ± 35.8 GBq/µmol (*n* = 9) and 56.6 ± 17.3 GBq/µmol (*n* = 7), respectively.

4.2. Cell Culture

Human malignant melanoma line A375 (ATCC CRL-1619), human squamous cell/epidermoid carcinoma cell lines FaDu (ATCC HTB-43) and A431 (ATCC CRL-1555), androgen-independent human malignant prostate adenocarcinoma lines PC3 (ATCC CRL-1435) and DU145 (ATCC HTB-81), and human likely glioblastoma cell line U87-MG (ATCC HTB-14) were used. Cells were routinely cultivated in Dulbecco's modified Eagle's medium (A375, A431, PC3, U87-MG), Eagle's Minimum Essential Medium (DU145) or RPMI 1640 medium (FaDu) supplemented with 10% (v/v) heat-inactivated fetal calf serum (FCS), penicillin (100 U/mL), streptomycin (100 µg/mL), glutamine (4 mM), 1% HEPES (1 M; A431 and U87-MG cells only) at 37 °C and 5% CO_2 in a humidified incubator. Cells were passaged twice a week by mild enzymatic dissociation using 0.25% trypsin/EDTA [30,53].

4.3. Immunoblotting (Western Blot)

Tumor samples were processed as described earlier [54]. Preparation of cell lysates, SDS-PAGE and immunoblotting were performed as described elsewhere [55]. In brief, 40–80 µg protein per lane was transferred to PVDF membranes using a semi-dry transfer system (Bio-Rad Laboratories, Hercules, CA, USA). Membranes were blocked for 60–90 min with non-fat dry milk powder (5%, w/v) in Tris-buffered saline containing 0.05% (v/v) Tween 20 (TBS-T). For detection of Sig1R, membranes were incubated with primary antibodies in bovine serum albumin (BSA, 2% w/v) in TBS-T (PA5-30372, 1:500, Thermo Fisher Scientific (Waltham, MA, USA), (tumor samples), respectively, ab53852, 1:200, Abcam (Cambridge, UK) (cell lysates)), followed by incubation with peroxidase-conjugated secondary antibody (anti-rabbit IgG, A0545, 1:5000, Sigma-Aldrich, Steinheim, Germany). Proteins were visualized using Super Signal West Pico/Femto Chemiluminescent Substrate (Thermo Fisher Scientific) and a CELVIN®S Chemiluminescence Imaging system (Biostep, Burkhardtsdorf, Germany). For detection of loading control, membranes were stripped and further processed using mouse anti-β-actin antibody (A5316, 1:1000, Sigma-Aldrich) and anti-mouse IgG (A9044, 1:10,000, Sigma-Aldrich) as described elsewhere [54].

4.4. Cellular Accumulation

Radiotracer uptake studies with (S)-(−)-[^{18}F]fluspidine/(R)-(+)-[^{18}F]fluspidine (stock solution 1.50–1.75 MBq/mL; molar activity at application time: 69 GBq/µmol for (R)-(+)-[^{18}F]fluspidine and 56 GBq/µmol for (S)-(−)-[^{18}F]fluspidine) were performed in monolayer cultures. Therefore, the cells were seeded in 24-well plates at a density of 1.0×10^5 cells/mL and grown to confluence. The tracer cell uptake experiments were performed in quadruplicate in PBS at 37 °C for 1, 10, 30, 60, and 120 min using an activity of 0.3–0.5 MBq/well for each tracer in a total volume of 500 µL (independent experiments with and without blocking, 2–3 for (R)-(+)-[^{18}F]fluspidine and 2 for (S)-(−)-[^{18}F]fluspidine). For blocking experiments, the cells were pre-incubated for 10 min with 10 µM of haloperidol (100 µL). After the tracer uptake was stopped with 1 mL ice-cold PBS, the monolayer cells were washed three times with PBS and dissolved in 0.5 mL NaOH (0.1 M containing 1% sodium dodecylsulfate, w/v). The radioactivity in cell extracts was then measured with a Cobra II gamma counter (Canberra-Packard, Meriden, CT, USA) and decay-corrected. Activity measurements were corrected for nonspecific tracer binding determined in empty (cell-free) plates using the same experimental conditions. Total protein concentration in cell extracts was determined by the bicinchoninic acid assay (BCA; Pierce, Rockford, IL, USA) using bovine serum albumin as protein standard. Uptake data for all experiments are expressed as percentage of injected dose per mg protein (%ID/mg protein).

4.5. Heterotopic Tumor Model and Small Animal PET Imaging

Animal experiments were carried out according to the guidelines of the German Regulations for Animal Welfare. The protocol was approved by the local Ethical Committee for Animal Experiments (reference numbers 24D-9168.11-4/2007-2 and 24-9168.21-4/2004-1).

For the generation of subcutaneous tumors, DU145, PC3, A431, FaDu, A375, and U87-MG cells were used and cultivated as described elsewhere [30,54]. Tumor cells were harvested, washed in PBS, and transferred to 0.9% sodium chloride solution (5×10^6 cells/100 µL). Nine weeks old male (prostate cancer cells) and female (other cancer cells) NMRI Foxn1$^{nu/nu}$ mice (weight: 35.9 ± 4.6 g) were purchased from Janvier Labs or from the specific pathogen-free breeding facility of the Experimental Centre of the Faculty of Medicine Carl Gustav Carus, Technische Universität Dresden. General anesthesia of mice was induced with inhalation of desflurane 12% (v/v) (Suprane, Baxter, Unterschleißheim, Germany) in 40% oxygen/air (gas flow 0.5 L/min), and was maintained with desflurane 8% (v/v). The single-cell suspension (100 µL) was subcutaneously injected into the right hind leg of mice [56]. Tumor size was monitored trice a week by caliper measurements and tumor volume was calculated. The animals were visually inspected daily. Tumor-bearing mice entered imaging studies 18–24 days past tumor cell injection, when tumor size reached a volume of about 400 to 700 mm^3.

For PET investigations, anesthesia was performed as described above. In the PET experiments, 7.6 ± 3.0 MBq of (S)-(−)-[^{18}F]fluspidine or (R)-(+)-[^{18}F]fluspidine was administered intravenously over 1 min into a tail vein. Dynamic PET imaging was performed for up to 2 h with a dedicated small animal tomograph (microPET P4, Siemens Medical Solutions, Knoxville, TN, USA). Data acquisition was performed in 3D list mode. A transmission scan was carried out prior to the injection of the radiotracer using a ^{57}Co point source. The list mode data were sorted into sinograms using a framing scheme of 12 × 10 s, 6 × 30 s, 5 × 300 s and 9 × 600 s frames. The frames were reconstructed by Ordered Subset Expectation Maximization applied to 3D sinograms (OSEM3D) with 14 subsets, 6 OSEM3D iterations, 2 maximum a posteriori (MAP) iterations, and 0.05 beta-value for smoothing and corrected for attenuation. The pixel size was 0.8 by 0.8 by 1.2 mm, and the resolution in the center of field of view was 1.85 mm. The reconstructed data were converted into ECAT7 format and processed using the ROVER software (ABX GmbH, Radeberg, Germany). Summed frames from 30 to 60 min post injection (p.i.) were used to define the regions of interest (ROI). The ROI were located over the tumor and the muscle of the contralateral hindleg and the results expressed as SUV or SUV$_{tumor/muscle}$ ratio.

4.6. Orthotopic Brain Tumor Model: Stereotactic Intracranial Tumor Cell Inoculation and PET/CT Imaging

All procedures were approved by the Institutional Animal Care and Utilization Committee (IACUC) of Yale University. The procedures for cell culture are described in detail elsewhere [57]. Mice were purchased from Charles River Laboratories. Intracranial U87-MG-luc mouse xenografts were established in 5–6 week-old female athymic nude mice (Crl:NU(NCr)-Foxn1^{nu}). The animals were anesthetized via intraperitoneal injection of a ketamine/xylazine mixture. After positioning in a stereotactic apparatus, a skin incision was made until the bregma was visible, a 0.45 mm hole was drilled into the skull at 2 mm lateral and 0.5 mm posterior to the bregma and a 30 G needle attached to a 10 µL Hamilton syringe was inserted 3 mm deep into the brain tissue. The needle remained for 2 min at this position followed by the injection of 50,000 U87-MG cells in 5 µL of PBS (2.5 µL/min) into the right striatum with an UltraMicroPump (UMP3, World Precision Instruments, Sarasota, FL, USA). After the whole volume was injected, the needle was kept in place for another 2 min and then quickly withdrawn from the brain. The hole was closed with bone wax (Ethicon Inc., Somerville, NJ, USA) and the incision site sutured. Finally, the mice were inspected after narcosis until they were fully awake. For the three consecutive days after implantation the animals received s.c. injections of Meloxicam (5 mg/kg) for pain relief and anti-inflammatory treatment. The tumor growth was monitored weekly by bioluminescence imaging (IVIS Spectrum, PerkinElmer, Waltham, MA, USA). PET/CT experiments were performed 4 weeks after tumor cell inoculation using the Inveon PET/CT scanner (Siemens Medical Solutions, Knoxville, TN, USA). Dynamic PET scans were acquired for 2 h following i.v. injection of 3.5 MBq ± 2.6 of (S)-(−)-[^{18}F]fluspidine or (R)-(+)-[^{18}F]fluspidine. The list mode data were sorted into sinograms using a framing scheme of 12 × 10 s, 6 × 30 s, 5 × 300 s and 9 × 600 s frames. The frames were reconstructed by OSEM3D, corrected for attenuation based on the CT data resulting in a pixel size of 0.78 by 0.78 by 0.8 mm and a resolution in the center of the FOV of 1.64 mm. ROIs were defined with ROVER for whole brain and the tumor region and the results expressed as standardized uptake value.

5. Conclusions

In this study we investigated the feasibility for tumor imaging of two Sig1R specific radiolabeled enantiomers of [^{18}F]fluspidine using different mouse tumor models, with special regard to brain tumor imaging. The results support the use of the two radiotracers as imaging agents for clinical oncological applications. Both enantiomers of [^{18}F]fluspidine showed promising results in the cellular accumulation experiments and encouraged us to further investigate these radiotracers in different tumor models. Although the heterotopic models are not suitable for PET imaging with [^{18}F]fluspidine, the results obtained with an orthotopic brain tumor model support the use of (*S*)-(−)-[^{18}F]fluspidine and (*R*)-(+)-[^{18}F]fluspidine for brain tumor imaging. However, further studies with this orthotopic brain tumor model will be needed to investigate the integrity of the blood-brain barrier, to confirm the specific binding to Sig1R and to investigate the suitability of [^{18}F]fluspidine for brain tumor imaging.

Acknowledgments: The excellent technical assistance of Mareike Barth, Catharina Heinig, Regina Herrlich, Sebastian Meister, and Andrea Suhr is greatly acknowledged. This work was supported by the German Research Association (Deutsche Forschungsgemeinschaft) Germany. The authors thank the Helmholtz Association for funding a part of this work through the Helmholtz Cross-Programme Initiative "Technology and Medicine—Adaptive Systems".

Author Contributions: P.B., J.S., J.P., Y.H., B.B. and R.B. conceived and designed the experiments; M.K., R.B., T.K., B.B., C.N., G.D., S.F. and W.D.-C. performed the experiments; M.K., J.P., W.D.-C., Z.C., G.D. and C.N. analyzed the data; Z.C., G.D. and J.Z. contributed animal models and analysis tools; M.K. and W.D.-C. wrote the paper. All authors discussed the results and implications and commented on the manuscript at all stages. All authors read and approved the final manuscript.

Conflicts of Interest: The authors declare no conflict of interest.

References

1. Steel, G.G.; Peckham, M.J. Exploitable mechanisms in combined radiotherapy-chemotherapy: The concept of additivity. *Int. J. Radiat. Oncol. Biol. Phys.* **1979**, *5*, 85–91. [CrossRef]
2. Rohren, E.M.; Turkington, T.G.; Coleman, R.E. Clinical applications of PET in oncology. *Radiology* **2004**, *231*, 305–332. [CrossRef] [PubMed]
3. Gambhir, S.S.; Czernin, J.; Schwimmer, J.; Silverman, D.H.; Coleman, R.E.; Phelps, M.E. A tabulated summary of the FDG PET literature. *J. Nucl. Med.* **2001**, *42*, 1S–93S. [PubMed]
4. Kukuk, D.; Reischl, G.; Raguin, O.; Wiehr, S.; Judenhofer, M.S.; Calaminus, C.; Honndorf, V.S.; Quintanilla-Martinez, L.; Schönberger, T.; Duchamp, O. Assessment of PET tracer uptake in hormone-independent and hormone-dependent xenograft prostate cancer mouse models. *J. Nucl. Med.* **2011**, *52*, 1654–1663. [CrossRef] [PubMed]
5. Chang, E.; Liu, H.; Unterschemmann, K.; Ellinghaus, P.; Liu, S.; Gekeler, V.; Cheng, Z.; Berndorff, D.; Gambhir, S.S. ^{18}F-FAZA PET imaging response tracks the reoxygenation of tumors in mice upon treatment with the mitochondrial complex i inhibitor BAY87-2243. *Clin. Cancer Res.* **2015**, *21*, 335–346. [CrossRef] [PubMed]
6. Waldherr, C.; Mellinghoff, I.K.; Tran, C.; Halpern, B.S.; Rozengurt, N.; Safaei, A.; Weber, W.A.; Stout, D.; Satyamurthy, N.; Barrio, J.; et al. Monitoring antiproliferative responses to kinase inhibitor therapy in mice with 3'-deoxy-3'-^{18}F-fluorothymidine PET. *J. Nucl. Med.* **2005**, *46*, 114–120. [PubMed]
7. Ekshyyan, O.; Sibley, D.; Caldito, G.C.; Sunderland, J.; Vascoe, C.; Nathan, C.A.O. ^{18}F-fluorodeoxythymidine micro–positron-emission tomography versus ^{18}F-fluorodeoxyglucose micro–positron-emission tomography for in vivo minimal residual disease imaging. *Laryngoscope* **2013**, *123*, 107–111. [CrossRef] [PubMed]
8. Bruechner, K.; Bergmann, R.; Santiago, A.; Mosch, B.; Yaromina, A.; Hessel, F.; Hofheinz, F.; van den Hoff, J.; Baumann, M.; Beuthien-Baumann, B. Comparison of [^{18}F] FDG uptake and distribution with hypoxia and proliferation in fadu human squamous cell carcinoma (hscc) xenografts after single dose irradiation. *Int. J. Radiat. Biol.* **2009**, *85*, 772–780. [CrossRef] [PubMed]
9. Fu, Y.; Ong, L.-C.; Ranganath, S.H.; Zheng, L.; Kee, I.; Zhan, W.; Yu, S.; Chow, P.K.; Wang, C.-H. A dual tracer ^{18}F-fch/^{18}F-FDG PET imaging of an orthotopic brain tumor xenograft model. *PLoS ONE* **2016**, *11*, e0148123. [CrossRef] [PubMed]

10. Witney, T.H.; Pisaneschi, F.; Alam, I.S.; Trousil, S.; Kaliszczak, M.; Twyman, F.; Brickute, D.; Nguyen, Q.-D.; Schug, Z.; Gottlieb, E. Preclinical evaluation of 3-^{18}F-fluoro-2, 2-dimethylpropionic acid as an imaging agent for tumor detection. *J. Nucl. Med.* **2014**, *55*, 1506–1512. [CrossRef] [PubMed]

11. Wagner, J.D.; Schauwecker, D.S.; Davidson, D.; Wenck, S.; Jung, S.H.; Hutchins, G. FDG–PET sensitivity for melanoma lymph node metastases is dependent on tumor volume. *J. Surg. Oncol.* **2001**, *77*, 237–242. [CrossRef] [PubMed]

12. Effert, P.J.; Bares, R.; Handt, S.; Wolff, J.M.; Bull, U.; Jakse, G. Metabolic imaging of untreated prostate cancer by positron emission tomography with sup fluorine-18-labeled deoxyglucose. *J. Urol.* **1996**, *155*, 994–998. [CrossRef]

13. Tsuchida, T.; Takeuchi, H.; Okazawa, H.; Tsujikawa, T.; Fujibayashi, Y. Grading of brain glioma with 1-11 c-acetate PET: Comparison with ^{18}F-FDG PET. *Nucl. Med. Biol.* **2008**, *35*, 171–176. [CrossRef] [PubMed]

14. Seltzer, M.A.; Barbaric, Z.; Belldegrun, A.; Naitoh, J.; Dorey, F.; Phelps, M.E.; Gambhir, S.S.; Hoh, C.K. Comparison of helical computerized tomography, positron emission tomography and monoclonal antibody scans for evaluation of lymph node metastases in patients with prostate specific antigen relapse after treatment for localized prostate cancer. *J. Urol.* **1999**, *162*, 1322–1328. [CrossRef]

15. Chen, W.; Cloughesy, T.; Kamdar, N.; Satyamurthy, N.; Bergsneider, M.; Liau, L.; Mischel, P.; Czernin, J.; Phelps, M.E.; Silverman, D.H. Imaging proliferation in brain tumors with 18f-flt PET: Comparison with ^{18}F-FDG. *J. Nucl. Med.* **2005**, *46*, 945–952. [PubMed]

16. Basu, S.; Alavi, A. Molecular imaging (PET) of brain tumors. *Neuroimaging Clin. N. Am.* **2009**, *19*, 625–646. [CrossRef] [PubMed]

17. Aydar, E.; Palmer, C.P.; Djamgoz, M.B. Sigma receptors and cancer. *Cancer Res.* **2004**, *64*, 5029–5035. [CrossRef] [PubMed]

18. Mach, R.; Huang, Y.; Buchheimer, N.; Kuhner, R.; Wu, L.; Morton, T.; Wang, L.-M.; Ehrenkaufer, R.; Wallen, C.; Wheeler, K. [^{18}F] n-4'-fluorobenzyl-4-(3-bromophenyl) acetamide for imaging the sigma receptor status of tumors: Comparison with [^{18}F] FDG and [^{125}I] IUDR. *Nucl. Med. Biol.* **2001**, *28*, 451–458. [CrossRef]

19. Mach, R.H.; Smith, C.R.; Al-Nabulsi, I.; Whirrett, B.R.; Childers, S.R.; Wheeler, K.T. Σ2 receptors as potential biomarkers of proliferation in breast cancer. *Cancer Res.* **1997**, *57*, 156–161. [PubMed]

20. Vilner, B.J.; John, C.S.; Bowen, W.D. Sigma-1 and sigma-2 receptors are expressed in a wide variety of human and rodent tumor cell lines. *Cancer Res.* **1995**, *55*, 408–413. [PubMed]

21. Al-Nabulsi, I.; Mach, R.; Wang, L.; Wallen, C.; Keng, P.; Sten, K.; Childers, S.; Wheeler, K. Effect of ploidy, recruitment, environmental factors, and tamoxifen treatment on the expression of sigma-2 receptors in proliferating and quiescent tumour cells. *Br. J. Cancer* **1999**, *81*, 925. [CrossRef] [PubMed]

22. Wheeler, K.; Wang, L.; Wallen, C.; Childers, S.; Cline, J.; Keng, P.; Mach, R. Sigma-2 receptors as a biomarker of proliferation in solid tumours. *Br. J. Cancer* **2000**, *82*, 1223. [CrossRef] [PubMed]

23. Crottès, D.; Guizouarn, H.; Martin, P.; Borgese, F.; Soriani, O. The sigma-1 receptor: A regulator of cancer cell electrical plasticity? *Front. Physiol.* **2013**, *4*, 175. [CrossRef] [PubMed]

24. Aydar, E.; Onganer, P.; Perrett, R.; Djamgoz, M.B.; Palmer, C.P. The expression and functional characterization of sigma (σ) 1 receptors in breast cancer cell lines. *Cancer Lett.* **2006**, *242*, 245–257. [CrossRef] [PubMed]

25. Hashimoto, K.; Ishiwata, K. Sigma receptor ligands: Possible application as therapeutic drugs and as radiopharmaceuticals. *Curr. Pharm. Des.* **2006**, *12*, 3857–3876. [PubMed]

26. Lee Collier, T.; O'Brien, J.C.; Waterhouse, R.N. Synthesis of [18f]-1-(3-fluoropropyl)-4-(4-cyanophenoxymethyl)-piperidine: A potential sigma-1 receptor radioligand for PET. *J. Label. Compd. Radiopharm.* **1996**, *38*, 785–794. [CrossRef]

27. He, Y.; Xie, F.; Ye, J.; Deuther-Conrad, W.; Cui, B.; Wang, L.; Lu, J.; Steinbach, J.; Brust, P.; Huang, Y. 1-(4-[^{18}F]fluorobenzyl)-4-[(tetrahydrofuran-2-yl) methyl] piperazine: A novel suitable radioligand with low lipophilicity for imaging σ1 receptors in the brain. *J. Med. Chem.* **2017**, *60*, 4161–4172. [CrossRef] [PubMed]

28. Kawamura, K.; Ishiwata, K.; Tajima, H.; Ishii, S.-I.; Matsuno, K.; Homma, Y.; Senda, M. In vivo evaluation of [^{11}C]SA4503 as a PET ligand for mapping cns sigma 1 receptors. *Nucl. Med. Biol.* **2000**, *27*, 255–261. [CrossRef]

29. James, M.L.; Shen, B.; Nielsen, C.H.; Behera, D.; Buckmaster, C.L.; Mesangeau, C.; Zavaleta, C.; Vuppala, P.K.; Jamalapuram, S.; Avery, B.A. Evaluation of σ-1 receptor radioligand ^{18}F-FTC-146 in rats and squirrel monkeys using PET. *J. Nucl. Med.* **2014**, *55*, 147–153. [CrossRef] [PubMed]

30. Xie, F.; Bergmann, R.; Kniess, T.; Deuther-Conrad, W.; Mamat, C.; Neuber, C.; Liu, B.; Steinbach, J.; Brust, P.; Pietzsch, J. ^{18}F-labeled 1, 4-dioxa-8-azaspiro [4.5] decane derivative: Synthesis and biological evaluation of a σ1 receptor radioligand with low lipophilicity as potent tumor imaging agent. *J. Med. Chem.* **2015**, *58*, 5395–5407. [CrossRef] [PubMed]

31. Fischer, S.; Wiese, C.; Maestrup, E.G.; Hiller, A.; Deuther-Conrad, W.; Scheunemann, M.; Schepmann, D.; Steinbach, J.; Wünsch, B.; Brust, P. Molecular imaging of σ receptors: Synthesis and evaluation of the potent σ1 selective radioligand [^{18}F] fluspidine. *Eur. J. Nucl. Med. Mol. Imaging* **2011**, *38*, 540–551. [CrossRef] [PubMed]

32. Holl, K.; Falck, E.; Köhler, J.; Schepmann, D.; Humpf, H.U.; Brust, P.; Wünsch, B. Synthesis, characterization, and metabolism studies of fluspidine enantiomers. *Chem. Med. Chem.* **2013**, *8*, 2047–2056. [CrossRef] [PubMed]

33. Brust, P.; Deuther-Conrad, W.; Becker, G.; Patt, M.; Donat, C.K.; Stittsworth, S.; Fischer, S.; Hiller, A.; Wenzel, B.; Dukic-Stefanovic, S. Distinctive in vivo kinetics of the new σ1 receptor ligands (R)-(+)-and (S)-(–)-^{18}F-fluspidine in porcine brain. *J. Nucl. Med.* **2014**, *55*, 1730–1736. [CrossRef] [PubMed]

34. Kranz, M.; Sattler, B.; Wüst, N.; Deuther-Conrad, W.; Patt, M.; Meyer, P.M.; Fischer, S.; Donat, C.K.; Wünsch, B.; Hesse, S. Evaluation of the enantiomer specific biokinetics and radiation doses of [^{18}F]fluspidine—A new tracer in clinical translation for imaging of σ1 receptors. *Molecules* **2016**, *21*, 1164. [CrossRef] [PubMed]

35. Baum, E.; Cai, Z.; Bois, F.; Holden, D.; Lin, S.-F.; Lara-Jaime, T.; Kapinos, M.; Chen, Y.; Deuther-Conrad, W.; Fischer, S. PET imaging evaluation of four σ1 radiotracers in nonhuman primates. *J. Nucl. Med.* **2017**, *58*, 982–988. [CrossRef] [PubMed]

36. Xu, Q.-X.; Li, E.-M.; Zhang, Y.-F.; Liao, L.-D.; Xu, X.-E.; Wu, Z.-Y.; Shen, J.-H.; Xu, L.-Y. Overexpression of sigma1 receptor and its positive associations with pathologic tnm classification in esophageal squamous cell carcinoma. *J. Histochem. Cytochem.* **2012**, *60*, 457–466. [CrossRef] [PubMed]

37. Das, D.; Persaud, L.; Dejoie, J.; Happy, M.; Brannigan, O.; De Jesus, D.; Sauane, M. Tumor necrosis factor-related apoptosis-inducing ligand (trail) activates caspases in human prostate cancer cells through sigma 1 receptor. *Biochem. Biophys. Res. Commun.* **2016**, *470*, 319–323. [CrossRef] [PubMed]

38. John, C.S.; Vilner, B.J.; Geyer, B.C.; Moody, T.; Bowen, W.D. Targeting sigma receptor-binding benzamides as in vivo diagnostic and therapeutic agents for human prostate tumors. *Cancer Res.* **1999**, *59*, 4578–4583. [PubMed]

39. Brune, S.; Schepmann, D.; Lehmkuhl, K.; Frehland, B.; Wünsch, B. Characterization of ligand binding to the σ1 receptor in a human tumor cell line (rpmi 8226) and establishment of a competitive receptor binding assay. *Assay Drug Dev. Technol.* **2012**, *10*, 365–374. [CrossRef] [PubMed]

40. Rybczynska, A.A.; de Bruyn, M.; Ramakrishnan, N.K.; de Jong, J.R.; Elsinga, P.H.; Helfrich, W.; Dierckx, R.A.; van Waarde, A. In vivo responses of human A375m melanoma to a σ ligand: 18f-FDG PET imaging. *J. Nucl. Med.* **2013**, *54*, 1613–1620. [CrossRef] [PubMed]

41. Mégalizzi, V.; Decaestecker, C.; Debeir, O.; Spiegl-Kreinecker, S.; Berger, W.; Lefranc, F.; Kast, R.E.; Kiss, R. Screening of anti-glioma effects induced by sigma-1 receptor ligands: Potential new use for old anti-psychiatric medicines. *Eur. J. Cancer* **2009**, *45*, 2893–2905. [CrossRef] [PubMed]

42. Colabufo, N.A.; Berardi, F.; Contino, M.; Niso, M.; Abate, C.; Perrone, R.; Tortorella, V. Antiproliferative and cytotoxic effects of some σ2 agonists and σ1 antagonists in tumour cell lines. *Naunyn-Schmiedeberg Arch. Path.* **2004**, *370*, 106–113.

43. Moebius, F.F.; Reiter, R.J.; Bermoser, K.; Glossmann, H.; Cho, S.Y.; Paik, Y.-K. Pharmacological analysis of sterol δ8-δ7 isomerase proteins with [3h] ifenprodil. *Mol. Pharmacol.* **1998**, *54*, 591–598. [CrossRef] [PubMed]

44. Wiese, C.; Maestrup, E.G.; Schepmann, D.; Vela, J.M.; Holenz, J.; Buschmann, H.; Wünsch, B. Pharmacological and metabolic characterisation of the potent σ1 receptor ligand 1'-benzyl-3-methoxy-3h-spiro[[2]benzofuran-1,4'-piperidine]. *J. Pharm. Pharmacol.* **2009**, *61*, 631–640. [CrossRef] [PubMed]

45. Maestrup, E.G.; Wiese, C.; Schepmann, D.; Brust, P.; Wünsch, B. Synthesis, pharmacological activity and structure affinity relationships of spirocyclic σ1 receptor ligands with a (2-fluoroethyl) residue in 3-position. *Bioorg. Med. Chem.* **2011**, *19*, 393–405. [CrossRef] [PubMed]

46. McLarty, K.; Fasih, A.; Scollard, D.A.; Done, S.J.; Vines, D.C.; Green, D.E.; Costantini, D.L.; Reilly, R.M. ^{18}F-FDG small-animal PET/ct differentiates trastuzumab-responsive from unresponsive human breast cancer xenografts in athymic mice. *J. Nucl. Med.* **2009**, *50*, 1848–1856. [CrossRef] [PubMed]

47. Jain, R.K. Normalization of tumor vasculature: An emerging concept in antiangiogenic therapy. *Science* **2005**, *307*, 58–62. [CrossRef] [PubMed]

48. Cespedes, M.V.; Casanova, I.; Parreño, M.; Mangues, R. Mouse models in oncogenesis and cancer therapy. *Clin. Transl. Oncol.* **2006**, *8*, 318–329. [CrossRef] [PubMed]

49. Taillandier, L.; Antunes, L.; Angioi-Duprez, K. Models for neuro-oncological preclinical studies: Solid orthotopic and heterotopic grafts of human gliomas into nude mice. *J. Neurosci. Methods* **2003**, *125*, 147–157. [CrossRef]

50. Tagashira, H.; Bhuiyan, S.; Shioda, N.; Hasegawa, H.; Kanai, H.; Fukunaga, K. Σ 1-receptor stimulation with fluvoxamine ameliorates transverse aortic constriction-induced myocardial hypertrophy and dysfunction in mice. *Am. J. Physiol.* **2010**, *299*, H1535–H1545. [CrossRef] [PubMed]

51. Tagashira, H.; Matsumoto, T.; Taguchi, K.; Zhang, C.; Han, F.; Ishida, K.; Nemoto, S.; Kobayashi, T.; Fukunaga, K. Vascular endothelial σ1-receptor stimulation with SA4503 rescues aortic relaxation via akt/enos signaling in ovariectomized rats with aortic banding. *Circulation* **2013**, *77*, 2831–2840. [CrossRef]

52. Wiese, C.; Große Maestrup, E.; Galla, F.; Schepmann, D.; Hiller, A.; Fischer, S.; Ludwig, F.A.; Deuther-Conrad, W.; Donat, C.K.; Brust, P. Comparison of in silico, electrochemical, in vitro and in vivo metabolism of a homologous series of (radio) fluorinated σ1 receptor ligands designed for positron emission tomography. *ChemMedChem* **2016**, *11*, 2445–2458. [CrossRef] [PubMed]

53. Kniess, T.; Laube, M.; Bergmann, R.; Sehn, F.; Graf, F.; Steinbach, J.; Wuest, F.; Pietzsch, J. Radiosynthesis of a ^{18}F-labeled 2, 3-diarylsubstituted indole via mcmurry coupling for functional characterization of cyclooxygenase-2 (COX-2) in vitro and in vivo. *Bioorg. Med. Chem.* **2012**, *20*, 3410–3421. [CrossRef] [PubMed]

54. Mamat, C.; Mosch, B.; Neuber, C.; Köckerling, M.; Bergmann, R.; Pietzsch, J. Fluorine-18 radiolabeling and radiopharmacological characterization of a benzodioxolylpyrimidine-based radiotracer targeting the receptor tyrosine kinase ephb4. *ChemMedChem* **2012**, *7*, 1991–2003. [CrossRef] [PubMed]

55. Reissenweber, B.; Mosch, B.; Pietzsch, J. Experimental hypoxia does not influence gene expression and protein synthesis of eph receptors and ephrin ligands in human melanoma cells in vitro. *Melanoma Res.* **2013**, *23*, 85–95. [CrossRef] [PubMed]

56. Bergmann, R.; Ruffani, A.; Graham, B.; Spiccia, L.; Steinbach, J.; Pietzsch, J.; Stephan, H. Synthesis and radiopharmacological evaluation of ^{64}Cu-labeled bombesin analogs featuring a bis(2-pyridylmethyl)-1,4,7-triazacyclononane chelator. *Eur. J. Med. Chem.* **2013**, *70*, 434–446. [CrossRef] [PubMed]

57. Han, L.; Kong, D.K.; Zheng, M.-Q.; Murikinati, S.; Ma, C.; Yuan, P.; Li, L.; Tian, D.; Cai, Q.; Ye, C. Increased nanoparticle delivery to brain tumors by autocatalytic priming for improved treatment and imaging. *ACS Nano* **2016**, *10*, 4209–4218. [CrossRef] [PubMed]

Sample Availability: Samples of the compounds are available from the authors.

molecules

MDPI

Article

Combining Albumin-Binding Properties and Interaction with Pemetrexed to Improve the Tissue Distribution of Radiofolates

Cristina Müller [1,*], Patrycja Guzik [1], Klaudia Siwowska [1], Susan Cohrs [1], Raffaella M. Schmid [1] and Roger Schibli [1,2]

[1] Center for Radiopharmaceutical Sciences ETH-PSI-USZ, Paul Scherrer Institute, 5232 Villigen-PSI, Switzerland; patrycja.guzik@psi.ch (P.G.); klaudiasiwowska@gmail.com (K.S.); susan.cohrs@psi.ch (S.C.); raffaella.schmid@bluewin.ch (R.M.S.); roger.schibli@psi.ch (R.S.)
[2] Department of Chemistry and Applied Biosciences, ETH Zurich, 8093 Zurich, Switzerland
* Correspondence: cristina.mueller@psi.ch; Tel.: +41-56-310-44-54

Academic Editors: Peter Brust and Derek J. McPhee
Received: 24 March 2018; Accepted: 13 June 2018; Published: 16 June 2018

Abstract: Folic-acid-based radioconjugates have been developed for nuclear imaging of folate receptor (FR)-positive tumors; however, high renal uptake was unfavorable in view of a therapeutic application. Previously, it was shown that pre-injection of pemetrexed (PMX) increased the tumor-to-kidney ratio of radiofolates several-fold. In this study, PMX was combined with the currently best performing radiofolate ([^{177}Lu]cm13), which is outfitted with an albumin-binding entity. Biodistribution studies were carried out in mice bearing KB or IGROV-1 tumor xenografts, both FR-positive tumor types. SPECT/CT was performed with control mice injected with [^{177}Lu]folate only and with mice that received PMX in addition. Control mice showed high uptake of radioactivity in KB and IGROV-1 tumor xenografts, but retention in the kidneys was also high, resulting in tumor-to-kidney ratios of ~0.85 (4 h p.i.) and ~0.60 (24 h p.i.) or ~1.17 (4 h p.i.) and ~1.11 (24 h p.i.) respectively. Pre-injection of PMX improved the tumor-to-kidney ratio to values of ~1.13 (4 h p.i.) and ~0.92 (24 h p.i.) or ~1.79 (4 h p.i.) and ~1.59 (24 h p.i.), respectively, due to reduced uptake in the kidneys. It was found that a second injection of PMX—3 h or 7 h after administration of the radiofolate—improved the tumor-to-kidney ratio further to ~1.03 and ~0.99 or ~1.78 and ~1.62 at 24 h p.i. in KB and IGROV-1 tumor-bearing mice, respectively. SPECT/CT scans readily visualized the tumor xenografts, whereas accumulation of radioactivity in the kidneys was reduced in mice that received PMX. In this study, it was shown that PMX had a positive impact in terms of reducing the kidney uptake of albumin-binding radiofolates; hence, the administration of PMX resulted in ~1.3–1.7-fold higher tumor-to-kidney ratios. This is, however, a rather moderate effect in comparison to the previously shown effect of PMX on conventional radiofolates (without albumin binder), which led to 5–6-fold increased tumor-to-kidney ratios. An explanation for this result may be the different pharmacokinetic profiles of PMX and long-circulating radiofolates, respectively. Despite the promising potential of this concept, it is believed that a clinical translation would be challenging, particularly when PMX had to be injected more than once.

Keywords: pemetrexed; folic acid; radiofolate; albumin-binder; SPECT; ^{177}Lu; KB; IGROV-1

1. Introduction

Folic-acid-based radioconjugates have been developed and pre-clinically investigated over the past years for the purpose of nuclear imaging of folate receptor (FR)-positive tumors using Single Photon Emission Computed Tomography (SPECT) or Positron Emission Tomography (PET).

Only a very limited number of candidates have been translated to clinical studies, among those [^{111}In]DTPA-folate and [^{99m}Tc]EC20 (Etarfolatide[TM]) [1,2]. [^{99m}Tc]EC20 has been used for imaging of FR-positive malignancies enabling the selection of patients who could potentially profit from FR-targeted chemotherapy [3–5]. PET imaging agents using ^{18}F-labeled folate tracers are currently under investigation for the same purposes [6,7]. The application of folate radioconjugates for targeted radionuclide therapy would be extremely attractive, given the fact that a large variety of tumor types express the FR, among those several gynecological cancer types, but also other frequently occurring cancer types such as non-small cell (NSC) lung cancer [5,8]. The high accumulation of folic acid radioconjugates in the kidneys has, however, presented a drawback in this regard [9]. A tumor-to-kidney ratio of accumulated activity in the range of 0.1–0.2, as was the case for conventional folate radioconjugates (without albumin binder), prevented the realization of the therapeutic concept completely. Our group has made major efforts to develop concepts that can enable the application of therapeutic folic acid radioconjugates [10].

We were the first to demonstrate the fact that the antifolate pemetrexed (PMX, Alimta[TM], Figure 1a [11]) increases the tumor-to-background ratio of accumulated radioactivity when injected one hour prior to the radiofolate [12]. This concept was further investigated in different preclinical settings with a variety of radioligands including [^{111}In]DTPA-folate and [^{99m}Tc]EC20 that had previously been tested in patients [13–16]. The "antifolate effect" was reproducible in different animal models including xenograft and syngeneic tumor mouse models [13–15]. Hence, it could be shown repeatedly by our group and others that this concept improved the tissue distribution of any radiofolate by reducing retention in the kidneys [17]. We were also able to demonstrate, that the combination of PMX and [^{177}Lu]folate can enhance the therapeutic outcome and, in addition, reduce undesired side effects of radiofolates [18]. These findings were based on the radiosensitizing potential of PMX [19–21], and the reduction of accumulated activity in the kidneys [12]. Even though this approach revealed to be highly promising, the tumor-to-kidney ratio of accumulated activity was still <1 when using the most promising folate radioconjugates (e.g., DOTA-conjugates [22]). This fact presented a hurdle for therapeutic application of radiofolates as the risk of damage to the kidneys would be high.

More recently, we have pursued another strategy in which we modified the folate conjugate chemically by introducing an albumin-binding entity [23]. This modification was thought to enhance the blood circulation time of the radiofolate ([^{177}Lu]cm09) and, hence, improve the tissue distribution profile [23]. Indeed, this concept led to unprecedentedly high tumor-to-kidney ratios of accumulated activity (0.5–0.7 over 5 days p.i.) and had the advantage of avoiding the use of additional medication to reach the desired effect.

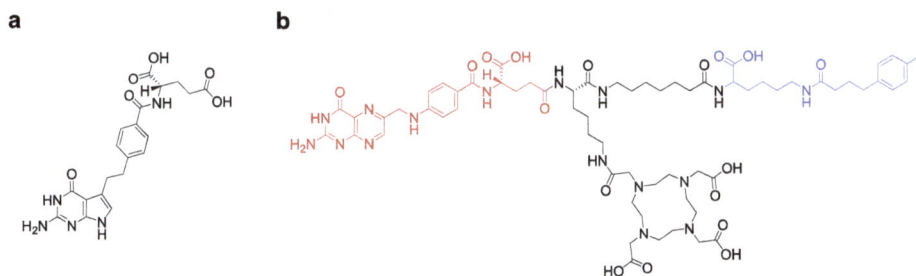

Figure 1. (**a**) Chemical structure of the antifolate pemetrexed (PMX, Alimta[TM]). (**b**) Chemical structure of the most promising albumin-binding DOTA-folate conjugate referred to as cm13 [24]; folic acid (red) serves as the targeting agent; the albumin-binding entity (blue) enables binding to serum albumin and, hence, enhanced residence time in the blood; the DOTA chelator allows stable coordination of [177]Lu.

By changing the linker entity of the radiofolate, we aimed at further improving the tissue distribution profile [24]. Integration of a short alkane spacer between folic acid and the albumin-binding entity was realized in compound cm13 [24] (Figure 1b). This modification appeared favorable based on a slightly improved tumor-to-kidney ratio of accumulated radioactivity after application of [^{177}Lu]cm13 [24].

Based on the observation that PMX improves the tissue distribution of folate radioconjugates the question arose whether the combined application of PMX and [^{177}Lu]cm13 would enable a further increase in the tumor-to-kidney ratio of accumulated radioactivity. The aim of this study was, therefore, to combine [^{177}Lu]cm13 and PMX in vivo by investigating a potentially positive effect on the tissue distribution of radioactivity. Mice with KB tumor xenografts (cervical cancer type), the most often used mouse model to test folate (radio)conjugates, were employed in the first place for this study. In addition, we investigated mice bearing IGROV-1 tumor xenografts, an ovarian cancer model, which has been used previously for the investigation of folate radioconjugates [15,25]. Biodistribution studies were performed in order to investigate [^{177}Lu]cm13 applied alone and combined with PMX injected before (and after) the radiofolate. SPECT/CT imaging was carried out to visualize the anticipated effects.

2. Results

2.1. Biodistribution Studies

2.1.1. Biodistribution in KB and IGROV-1 Tumor-Bearing Mice

Biodistribution studies were performed at 4 h and 24 h after injection of the [^{177}Lu]folate (5 MBq, 1 nmol per mouse) using nude mice bearing KB or IGROV-1 tumor xenografts (Tables 1–3). The uptake of [^{177}Lu]cm13 in KB tumors was high (22.4 ± 4.50% IA/g) at 4 h p.i. and largely retained over the time of investigation (18.6 ± 6.80% IA/g, 24 h p.i.).

Table 1. Biodistribution data obtained in KB and IGROV-1 tumor-bearing mice, 4 h after injection of [^{177}Lu]folate with and without pre-injected pemetrexed (PMX). Data are shown as % IA/g tissue, representing the average ± S.D.

	[^{177}Lu]cm13			
	-	PMX [1]	-	PMX [1]
	4 h p.i.	4 h p.i.	4 h p.i.	4 h p.i.
Tissue	KB	KB	IGROV-1	IGROV-1
	n = 5	n = 5	n = 4	n = 4
Blood	7.32 ± 0.85	9.13 ± 0.89	7.88 ± 1.70	10.8 ± 1.43
Lung	4.33 ± 0.44	4.92 ± 0.39	4.47 ± 0.97	6.14 ± 0.97
Spleen	1.51 ± 0.15	1.45 ± 0.06	1.74 ± 0.15	2.16 ± 0.45
Kidneys	26.5 ± 1.20	15.8 ± 2.60 ****	26.9 ± 2.90	16.9 ± 3.10 ***
Stomach	1.37 ± 0.35	1.41 ± 0.19	1.58 ± 0.47	1.83 ± 0.39
Intestines	1.29 ± 0.42	1.38 ± 0.32	1.10 ± 0.18	1.12 ± 0.20
Liver	3.88 ± 0.49	3.31 ± 0.43	3.27 ± 0.49	3.38 ± 0.56
Muscle	1.92 ± 0.25	1.55 ± 0.28	1.17 ± 0.43	1.28 ± 0.38
Bone	1.55 ± 0.10	1.67 ± 0.24	1.38 ± 0.30	1.57 ± 0.25
Tumor	22.4 ± 4.50	17.6 ± 0.90 ***	31.5 ± 5.60	29.2 ± 8.80
Salivary glands	6.78 ± 0.57	5.84 ± 1.25	6.17 ± 0.49	6.18 ± 0.88

[1] PMX (400 µg/mouse) was injected 1 h prior to the [^{177}Lu]folate; Statistical significance is indicated by asterisks (statistically significant difference between uptake in the tissue of control mice and PMX injected mice.) *** $p \leq 0.001$; **** $p \leq 0.0001$).

Pre-injection of PMX reduced the uptake in KB tumor (17.6 ± 0.90% IA/g, $p < 0.001$) at 4 h p.i., but had no impact on the tumor uptake at later time points (22.1 ± 3.60% IA/g; 24 h p.i.). At 4 h p.i. retention of radioactivity in the kidneys was significantly reduced when PMX was pre-injected (15.8 ± 2.60% IA/g vs. control mice: 26.5 ± 1.20% IA/g; $p < 0.0001$). The same effect was observed at 24 h p.i. (24.7 ± 5.70% IA/g vs. control mice: 30.9 ± 3.90% IA/g; $p < 0.01$). If PMX was injected for a second time, 3 h or 7 h after administration of the radiofolate, kidney uptake was even more effectively reduced at 24 h p.i. (21.8 ± 0.70% IA/g; $p < 0.0001$ and 21.0 ± 4.70% IA/g; $p < 0.0001$, respectively). In the liver, muscles and salivary glands, PMX did not affect uptake of radiofolate significantly, even though a slight increase in blood activity was seen at 4 h p.i. (~9.13% IA/g vs. control ~7.32% IA/g; $p > 0.05$), but not at later time points (Table 1).

The uptake of the radiofolate in IGROV-1 tumor xenografts was consistently higher (31.5 ± 5.60% IA/g; 4 h p.i. and 37.7 ± 5.10% IA/g, 24 h p.i.) than in KB tumor xenografts. Pre-injection of PMX did not reduce the tumor uptake significantly ($p > 0.05$). A significant difference in tumor accumulated radioactivity was determined, however, between groups of mice that received PMX according to different application schemes. It is not entirely understood why the mice that received PMX only 1 h before the application of [^{177}Lu]folate showed the highest tumor uptake (40.7 ± 9.00% IA/g), whereas mice that were injected with PMX 1 h before and 7 h after the radiofolate showed significantly reduced tumor accumulation (32.9 ± 5.30; 24 h p.i.; $p < 0.001$). As compared to control values, PMX reduced the uptake in the kidneys ($p < 0.001$) 4 h after injection of [^{177}Lu]folate and in all cases 24 h after injection of the radiofolate ($p < 0.001$). Radioactivity levels in the blood were slightly but not significantly ($p > 0.05$) increased in IGROV-1 tumor-bearing mice that received PMX before the radiofolate (~10.8% IA/g, 4 h p.i.) when compared to control mice (~7.88% IA/g, 4 h p.i.). This effect was still observable at later time points (~1.97% IA/g, 24 h p.i. vs. control mice: ~1.46% IA/g, 24 h p.i.; $p > 0.05$) (Tables 1 and 3). In the liver, muscles, and salivary glands, PMX did not have a significant effect on the retention of the radiofolate.

Table 2. Biodistribution data obtained in KB tumor-bearing mice, 24 h after injection of [^{177}Lu]folate with and without pre- and post-injected pemetrexed (PMX). Data are shown as % IA/g tissue, representing the average ± S.D.

		[^{177}Lu]cm13		
	-	PMX [1]	PMX [2]	PMX [3]
	24 h p.i.	24 h p.i.	24 h p.i.	24 h p.i.
Tissue	**KB**	**KB**	**KB**	**KB**
	$n = 4$	$n = 5$	$n = 4$	$n = 4$
Blood	1.28 ± 0.23	1.39 ± 0.14	1.37 ± 0.17	1.31 ± 0.14
Lung	1.74 ± 0.45	1.69 ± 0.22	1.66 ± 0.15	1.60 ± 0.35
Spleen	0.69 ± 0.14	0.79 ± 0.13	0.80 ± 0.13	0.72 ± 0.15
Kidneys	30.9 ± 3.90	24.7 ± 5.70 **	21.8 ± 0.70 ****	21.0 ± 4.70 ****
Stomach	0.77 ± 0.19	0.70 ± 0.20	0.80 ± 0.14	0.63 ± 0.21
Intestines	0.27 ± 0.07	0.47 ± 0.14	0.30 ± 0.07	0.34 ± 0.06
Liver	2.46 ± 0.16	1.91 ± 0.44	2.04 ± 0.57	2.05 ± 0.05
Muscle	1.56 ± 0.10	1.31 ± 0.25	1.17 ± 0.20	1.50 ± 0.49
Bone	1.09 ± 0.31	0.97 ± 0.15	0.85 ± 0.06	0.95 ± 0.16
Tumor	18.6 ± 6.80	22.1 ± 3.60	22.4 ± 3.20	20.9 ± 5.10
Salivary glands	4.18 ± 0.62	3.71 ± 0.33	3.39 ± 0.30	3.86 ± 0.44

[1] PMX (400 µg/mouse) was injected 1 h prior to the [^{177}Lu]folate; [2] PMX (twice 400 µg/mouse) was injected 1 h prior to the [^{177}Lu]folate and 3 h after [^{177}Lu]folate; [3] PMX (twice 400 µg/mouse) was injected 1 h prior to the [^{177}Lu]folate and 7 h after [^{177}Lu]folate. Statistical significance is indicated with asterisks (statistically significant difference between uptake in the tissue of control mice and PMX injected mice). ** $p \leq 0.01$; **** $p \leq 0.0001$).

Table 3. Biodistribution data obtained in IGROV-1 tumor-bearing mice, 24 h after injection of [^{177}Lu]folate with and without pre- and post-injected pemetrexed (PMX). Data are shown as % IA/g tissue, representing the average ± S.D.

	[^{177}Lu]cm13			
	-	PMX [1]	PMX [2]	PMX [3]
	24 h p.i.	24 h p.i.	24 h p.i.	24 h p.i.
Tissue	IGROV-1	IGROV-1	IGROV-1	IGROV-1
	$n = 4$	$n = 5$	$n = 5$	$n = 5$
Blood	1.46 ± 0.19	1.97 ± 0.15	2.21 ± 0.29	2.08 ± 0.31
Lung	1.88 ± 0.24	2.16 ± 0.23	2.23 ± 0.25	2.21 ± 0.33
Spleen	0.92 ± 0.17	1.19 ± 0.26	1.26 ± 0.33	1.20 ± 0.30
Kidneys	34.0 ± 2.00	26.2 ± 3.50 ***	21.8 ± 2.00 ****	20.8 ± 3.40 **** [4]
Stomach	0.69 ± 0.19	0.69 ± 0.31	0.78 ± 0.13	0.65 ± 0.17
Intestines	0.49 ± 0.12	0.44 ± 0.08	0.48 ± 0.07	0.43 ± 0.14
Liver	2.75 ± 0.57	2.47 ± 0.51	2.73 ± 0.65	2.41 ± 0.49
Muscle	1.52 ± 0.19	1.36 ± 0.37	1.28 ± 0.52	1.28 ± 0.32
Bone	0.95 ± 0.11	0.96 ± 0.12	1.02 ± 0.17	0.95 ± 0.10
Tumor	37.7 ± 5.10	40.7 ± 9.00	38.6 ± 3.50	32.9 ± 5.30 [5]
Salivary glands	4.43 ± 0.63	4.28 ± 0.43	3.95 ± 1.03	3.77 ± 0.45

[1] PMX (400 μg/mouse) was injected 1 h prior to the [^{177}Lu]folate; [2] PMX (twice 400 μg/mouse) was injected 1 h prior to the [^{177}Lu]folate and 3 h after [^{177}Lu]folate; [3] PMX (twice 400 μg/mouse) was injected 1 h prior to the [^{177}Lu]folate and 7 h after [^{177}Lu]folate. [4] This value was significantly different from the value obtained in mice that received PMX 1 h before the radiofolate; [5] This value was significantly different from the values obtained in mice that received PMX 1 h before (and 3 h after) the radiofolate. Statistical significance is indicated with asterisks (statistically significant difference between uptake in the tissue of control mice and PMX injected mice. *** $p \leq 0.001$; **** $p \leq 0.0001$).

2.1.2. Tumor-to-Background Ratios

The tumor-to-background ratios were determined at 4 h and 24 h after injection of [^{177}Lu]folate in order to better assess the effects of PMX. This appeared important since the absolute tumor uptake may have varied from mouse to mouse based on the size of the tumor xenograft. In line with the increased radioactivity in the blood 4 h after injection of the [^{177}Lu]folate in mice that received PMX, the tumor-to-blood ratios were significantly reduced in KB tumor-bearing mice (1.93 ± 0.18 vs. control mice: 3.05 ± 0.49; $p < 0.01$) as well as in IGROV-1 tumor-bearing mice (2.67 ± 0.58 vs. control mice: 4.05 ± 0.72; $p < 0.05$). At 24 h after injection of the radiofolate, tumor-to-blood ratios of KB tumor-bearing mice were in the same range among the different groups ($p > 0.05$). In IGROV-1 tumor-bearing mice, tumor-to-blood ratios were significantly decreased ($p < 0.05$) at 4 h p.i. in mice that received PMX as well as at 24 h when PMX was applied twice ($p < 0.05$).

The most important parameter to assess the effect of PMX was undoubtedly the tumor-to-kidney ratio of mice that received PMX compared to the ratio in control mice (Figure 2). In both mouse models, an increased value was observed when PMX was applied. At 4 h after injection of [^{177}Lu]folate, the tumor-to-kidney ratio was significantly increased in KB tumor-bearing mice that received PMX (1.13 ± 0.17% IA/g, 4 h p.i.; $p < 0.05$). An increased tumor-to-kidney ratio was also visible in the IGROV-1 tumor mouse model, however, in this case the effect was not significant due to the large standard deviation in the PMX-injected group.

Investigation of the 24 h-time point revealed also consistently increased tumor-to-kidney ratios when PMX was applied. The highest ratios in KB tumor-bearing mice were observed in mice that received PMX 1 h before and 3 h or 7 h after injection of the [^{177}Lu]folate (1.03 ± 0.16 and 0.99 ± 0.06, respectively). In IGROV-1 tumor-bearing mice the ratios obtained under these conditions were even higher (1.78 ± 0.18 and 1.62 ± 0.39, respectively) but only significant when PMX was injected 1 h before and 3 h after the radiofolate.

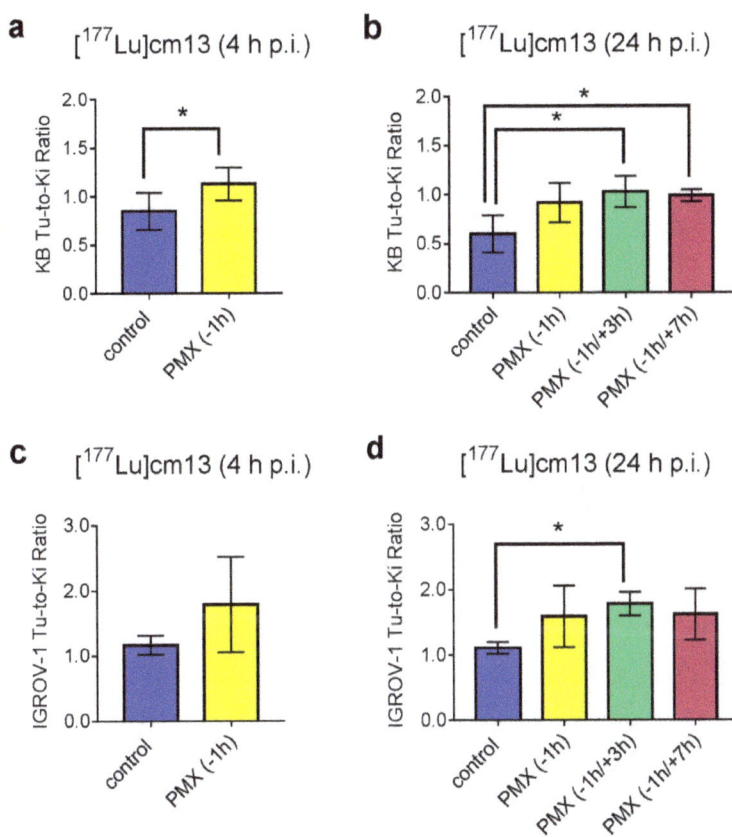

Figure 2. Tumor-to-kidney ratios of mice after injection of [^{177}Lu]folate (5 MBq, 1 nmol). (**a**) Tumor-to-kidney ratios of KB tumor-bearing mice 4 h after injection of [^{177}Lu]cm13 without pre-injected PMX (blue) or with pre-injected PMX (yellow). (**b**) Tumor-to-kidney ratios of KB tumor-bearing mice 24 h after injection of [^{177}Lu]cm13 without pre-injection of PMX (blue) or with pre-injected PMX 1 h before the radiofolate (yellow) or 1 h before and 3 h (green) or 7 h (red) after injection of the radiofolate. (**c**) Tumor-to-kidney ratios of IGROV-1 tumor-bearing mice 4 h after injection of [^{177}Lu]cm13 without pre-injected PMX (blue) or with pre-injected PMX (yellow). (**d**) Tumor-to-kidney ratios of IGROV-1 tumor-bearing mice 24 h after injection of [^{177}Lu]cm13 without pre-injection of PMX (blue) or with pre-injection of PMX 1 h before the radiofolate (yellow) or 1 h before and 3 h (green) or 7 h (red) after injection of the radiofolate. Statistically significant values are indicated with asterisks (* $p \leq 0.05$).

Tumor-to-liver ratios were in the same range for mice injected with [^{177}Lu]folate only and mice that received [^{177}Lu]folate combined with PMX, independent of which tumor xenograft (KB or IGROV-1) and time point (4 h p.i. or 24 h p.i.) was investigated and whether PMX was injected only once or twice.

2.2. In Vivo SPECT/CT Experiments

KB and IGROV-1 tumor-bearing mice were used for SPECT/CT imaging studies 4 h and 24 h after injection of the [^{177}Lu]folate only or [^{177}Lu]folate in combination with PMX, which was injected before and after the radiofolate. Measurement of the whole mice immediately before the 4 h p.i.-scan revealed that 91–95% of the injected radioactivity (non-decay corrected) was retained in the body independent on whether the mice received PMX. After 24 h, mice that received PMX showed lower radioactivity retention in the body (~62% IA retained in the body, non-decay corrected) as compared to the mice that were injected only with the [^{177}Lu]folate (~70% IA retained in the body, non-decay corrected). These data were in line with increased excretion through the kidneys in mice injected with PMX.

2.2.1. SPECT/CT Imaging of KB Tumor-Bearing Mice

SPECT/CT scans of KB tumor-bearing mice 4 h and 24 h after injection of the [^{177}Lu]folate showed high uptake of radioactivity in the tumor xenografts and accumulation of radioactivity was also observed in the kidneys (Figure 3). Based on a visual analysis, the tumor-to-kidney ratio was ~1 and did not change significantly over the time of investigation up to 24 h p.i. In mice injected with PMX 1 h before the administration of the [^{177}Lu]folate, the tumor uptake was slightly increased, while retention of radioactivity in the kidneys was reduced in comparison to the renal uptake observed in control mice. Background activity in blood circulation was reduced over time due to efficient blood clearance of the radiofolate. Other than that, the distribution profile of the radiofolate remained almost identical at 24 h p.i. in the mouse that received PMX a second time 7 h after radiofolate injection.

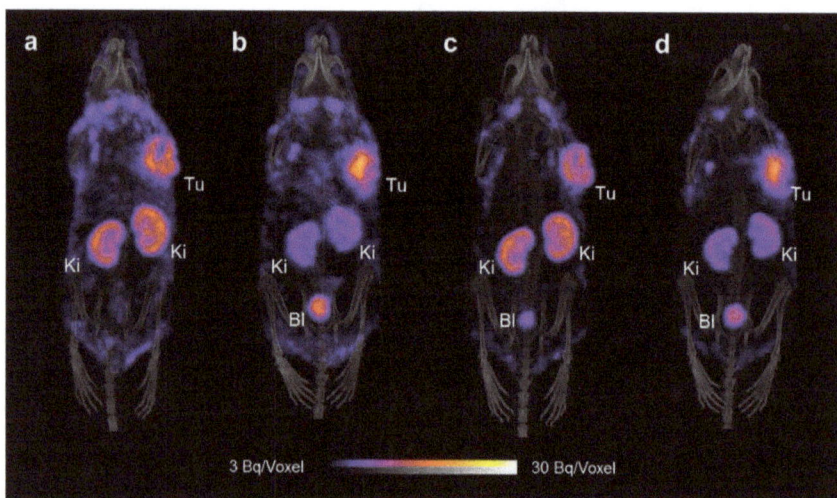

Figure 3. SPECT/CT scans of tumor-bearing mice injected with [^{177}Lu]cm13 (25 MBq; 1 nmol) shown as maximum intensity projections (MIPs). (**a**) KB tumor-bearing mouse 4 h after injection of [^{177}Lu]cm13. (**b**) KB tumor-bearing mouse 4 h after injection of [^{177}Lu]cm13 with PMX injected 1 h before the radiofolate. (**c**) KB tumor-bearing mouse 24 h after injection of [^{177}Lu]cm13. (**d**) KB tumor-bearing mouse 24 h after injection of [^{177}Lu]cm13 with PMX injected 1 h before and 7 h after the radiofolate. (Tu = KB tumor; Ki = kidney; Bl = urinary bladder).

2.2.2. SPECT/CT Imaging of IGROV-1 Tumor-Bearing Mice

SPECT/CT studies were also performed with IGROV-1 tumor-bearing mice (Figure 4). Radioactivity in the blood pool and heart (background activity) was visible at 4 h p.i. but not anymore at later time points. In line with the biodistribution data, the uptake of the radiofolate was higher in IGROV-1 tumor xenografts than in KB tumor xenografts. Based on visual analysis of the images, the tumor-to-kidney ratios were >1 at 4 h and 24 h after injection of the [^{177}Lu]folate. The favorable tissue distribution profile of the [^{177}Lu]folate observed in this model was further improved when mice received PMX before (and after) the radiofolate injection. Accumulation of radioactivity in lymph nodes (in the armpit region and next to salivary glands) was more pronounced in IGROV-1 tumor-bearing mice than in mice with KB tumor xenografts.

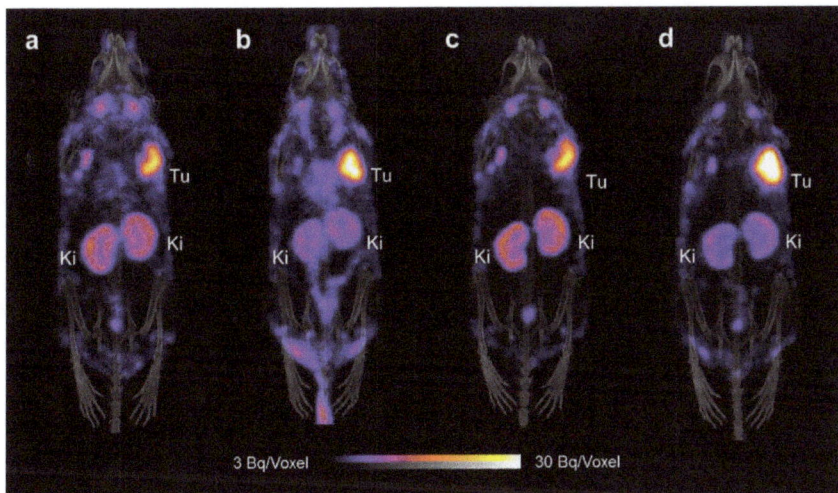

Figure 4. SPECT/CT scans of tumor-bearing mice injected with [^{177}Lu]cm13 (25 MBq; 1 nmol) shown as maximum intensity projections (MIPs). (**a**) IGROV-1 tumor-bearing mouse 4 h after injection of [^{177}Lu]cm13. (**b**) IGROV-1 tumor-bearing mouse 4 h after injection of [^{177}Lu]cm13 with PMX injected 1 h before the radiofolate. (**c**) IGROV-1 tumor-bearing mouse 24 h after injection of [^{177}Lu]cm13. (**d**) IGROV-1 tumor-bearing mouse 24 h after injection of [^{177}Lu]cm13 with PMX injected 1 h before and 7 h after the radiofolate. (Tu = IGROV-1 tumor; Ki = kidney).

3. Discussion

In this study, we aimed at combining PMX with the currently most promising albumin-binding radiofolate ([^{177}Lu]cm13) in order to optimize the tumor-to-kidney ratios further. Our current results were in agreement with those of a preliminary experiment performed with KB tumor-bearing mice and [^{177}Lu]cm09, the first DOTA-folate conjugate developed in our group, which was outfitted with an albumin-binding entity (see the Supporting Information of [23]). Although PMX increased the tumor-to-kidney ratio of conventional folate conjugates 5- to 6-fold when injected 1 h prior to the radiofolate [12,15], it was revealed that the effect was much less pronounced when PMX was combined with [^{177}Lu]cm09 [23].

In this study, it was shown that PMX was able to increase the tumor-to-kidney ratio of the albumin-binding radiofolate by a factor of ~1.3 at 4 h p.i. and by a factor of 1.5–1.7 at 24 h p.i. in the KB tumor mouse model. The situation was similar in the IGROV-1 tumor model, in which PMX increased the ratio by a factor ~1.5 at 4 h p.i. and by a factor of 1.4–1.6 at 24 h p.i. It was revealed that the distribution of the albumin-binding [^{177}Lu]folate benefited from an additional injection of PMX, 3 h or 7 h after the administration of the radiofolate, in order to further reduce the renal uptake and, therewith, increase the tumor-to-kidney ratios.

Even though the accumulation of [^{177}Lu]cm13 was clearly higher in IGROV-1 tumor xenografts than in KB tumors, the reported effect was consistent in both types of tumor mouse models. In the case of KB tumor mice the tumor-to-kidney ratios were in the range of ~1.0 when PMX was used whereas these ratios reached values of up to ~1.8 in the case of IGROV-1 tumor-bearing mice.

When performing this study, it was observed that there were inter-individual differences with regard to the tissue distribution of [^{177}Lu]folate. The absolute uptake values determined in this study were also different from previously published values; however, the tumor-to-background ratios were in the same range [24]. The effect of PMX was not exactly the same in each mouse; hence, the interaction between the two drugs appeared to be very sensitive and possibly dependent on other factors.

As already observed in previous studies, the effect of PMX is critically dependent on the time of pre-injection and injected quantity as well as on the amount of injected folate conjugate [13,23]. It is, thus, not surprising that the effect of PMX was less pronounced when combined with long-circulating radiofolates. The albumin-bound fraction of folate radioconjugates is not excreted since albumin is a large protein (>60 kDa) that cannot readily be filtered in the kidneys. Hence, only the free fraction of radiofolates (not bound to albumin) may be affected by pre-injected PMX. Most probably, PMX was excreted already when a major fraction of the radiofolate was still circulating in the blood due to albumin-binding. The fact that repeated injection of PMX was favorable to reduce renal uptake supported the hypothesis that the fast pharmacokinetics of PMX is responsible for the only moderate effect on the distribution profile of the radiofolate. A more sophisticated scheme of PMX application using repeated injections or a slow infusion over the first hours may be successful to reduce renal uptake of albumin-binding radiofolates more effectively. This would, however, be difficult to realize in a clinical setting given the fact that PMX is a chemotherapeutic agent, hence potentially toxic to the patient and not easily upscalable.

In summary, we were able to show, that a "chemical approach," which refers to the radioligand modification with an albumin-binder and a "pharmacological approach" referring to the pre-injection of PMX can be combined to further improve the tumor-to-kidney ratio of accumulated radioactivity in tumor-bearing mice. This combination led to the best tissue distribution profiles ever obtained with radiometal-based folate conjugates so far. A clinical translation of this approach would be challenging, however, particularly when PMX had to be applied at a dose that induces pharmacological/chemotherapeutic effects.

4. Materials and Methods

4.1. Preparation of [^{177}Lu]Folate

The DOTA–folate conjugate (cm13 [24], herein referred to as "folate") previously developed by our group was kindly provided by Merck & Cie (Schaffhausen, Switzerland). No-carrier added ^{177}Lu was obtained from Isotope Technologies Garching (ITG GmbH, Garching, Germany). Radiolabeling of the folate conjugate was performed in a mixture of HCl (0.05 M) and Na-acetate (0.5 M) at pH 4.5 at 95 °C and an incubation time of 10 min, as previously reported [24]. Quality control of [^{177}Lu]folate was carried out using high-performance liquid chromatography (HPLC) as previously reported [24].

4.2. Cell Culture

Human KB tumor cells (cervical carcinoma cell line, subclone of HeLa cells, ACC-136) were purchased from the German Collection of Microorganisms and Cell Cultures (DSMZ, Braunschweig, Germany). The human ovarian tumor cell line, IGROV-1, was a kind gift of Dr. Gerrit Jansen, Free University Medical Center Amsterdam, The Netherlands. Both cell lines were cultured under standard conditions (37 °C, humidified atmosphere, 5% CO_2) in folate-deficient RPMI 1640 medium (FFRPMI, Cell Culture Technologies, Gravesano, Switzerland), supplemented with 10% fetal calf serum, L-glutamine, and antibiotics.

4.3. In Vivo Studies

In vivo experiments were approved by the local veterinarian department and conducted in accordance with the Swiss law of animal protection. Athymic female nude mice (CD-1 Foxn-1/nu) were obtained from Charles River Laboratories (Sulzfeld, Germany) at the age of 5–6 weeks. All animals were fed with a folate-deficient rodent diet (ssniff Spezialdiäten GmbH, Soest, Germany). Mice were inoculated with 5×10^6 KB cells or 5×10^6 IGROV-1 cells in 100 µL phosphate buffered saline (PBS) into the subcutis of each shoulder for biodistribution studies. Additional mice were inoculated with the same number of tumor cells into the subcutis of the right shoulder for SPECT/CT

Molecules **2018**, *23*, 1465

imaging studies. For the in vivo scans, mice were anesthetized with a mixture of isoflurane (1.5–2%) and oxygen.

4.4. Biodistribution Studies

Biodistribution studies were performed with 4-5 mice per group, 12-14 days after KB cell inoculation and 14-16 days after IGROV-1 tumor cell inoculation. [^{177}Lu]folate (5 MBq, 1 nmol/mouse) was diluted in 100 μL PBS and injected into a lateral tail vein (0.05% bovine serum albumin was added to prevent adhesion to the syringe). Pemetrexed (PMX, Alimta^{TM}) was diluted in saline (4 mg/mL) and administered at defined time points (0.4 mg per injection) before and after the injection of [^{177}Lu]folate. The animals were sacrificed at 4 h and 24 h after administration of the [^{177}Lu]folate. Selected tissues and organs were collected, weighed, and radioactivity was measured using a γ-counter (PerkinElmer Wallac Wizard 1480). The results were decay corrected and presented as a percentage of the injected (radio)activity per gram of tissue mass (% IA/g).

4.5. Statistics

Biodistribution data were compared using a two-way ANOVA Sidak's multiple comparisons test (4 h p.i. time point) and a two-way ANOVA Tukey's multiple comparisons test (24 h p.i. time point; Graph Pad Prism version 7). Tumor-to-background ratios were compared using an unpaired t-test with Welch's correction (4 h p.i. time point) and ordinary one-way ANOVA Tukey's multiple comparisons test (24 h p.i. time point), respectively. Statistically significant differences were calculated using data based on the average [% IA/g]-values for the radioactivity accumulation in the blood, kidneys, liver, muscle, salivary glands and tumors. Values for tissue uptake that differed significantly from the control group are indicated with asterisks in Tables 1–3. Statistically significant differences in tissue uptake among the PMX-pre/post-treated mice were indicated as a footnote to Table 3. All statistically significant differences of tumor-to-background ratios between the groups were indicated with asterisks in Figure 2. Statistically significant values were indicated as follows: ns: $p > 0.05$; * $p \leq 0.05$; ** $p \leq 0.01$; *** $p \leq 0.001$; **** $p \leq 0.0001$.

4.6. SPECT/CT Studies

Imaging studies were performed using a small-animal SPECT/CT camera (NanoSPECT/CT^{TM}, Mediso Medical Imaging Systems, Budapest, Hungary). [^{177}Lu]folate was injected into the lateral tail vein of tumor-bearing mice (25 MBq, ~1 nmol per mouse). SPECT scans of 38 min duration were performed 4 h and 24 h after injection of the [^{177}Lu]folate after CT scans of 7.30 min duration. The images were acquired using Nucline Software (version 1.02, Mediso Ltd., Budapest, Hungary). The reconstruction was performed using HiSPECT software, version 1.4.3049 (Scivis GmbH, Göttingen, Germany). Images were analyzed using VivoQuant software (version 3.0, inviCRO Imaging Services and Software, Boston, US). Gauss post-reconstruction filter (FWHM = 1 mm) was applied twice to the SPECT images, and the scale of radioactivity was set as indicated on the images (minimum value = 3 Bq/voxel to maximum value = 30 Bq/voxel).

Author Contributions: C.M. designed and supervised the study and wrote the manuscript. P.G. and K.S. performed the experiments and analyzed the data. S.C. and R.M.S. performed biodistribution studies; R.S. gave advice and reviewed the manuscript.

Funding: The research was supported by a grant (310030_156803) of the Swiss national Science Foundation and Merck & Cie, Schaffhausen, Switzerland.

Acknowledgments: The authors would like to thank Christoph A. Umbricht for technical assistance. Konstantin Zhernosekov (Isotope Technologies Garching (ITG GmbH)) provided ^{177}Lu for this study.

References

1. Siegel, B.A.; Dehdashti, F.; Mutch, D.G.; Podoloff, D.A.; Wendt, R.; Sutton, G.P.; Burt, R.W.; Ellis, P.R.; Mathias, C.J.; Green, M.A.; et al. Evaluation of ^{111}In-DTPA-folate as a receptor-targeted diagnostic agent for ovarian cancer: Initial clinical results. *J. Nucl. Med.* **2003**, *44*, 700–707. [PubMed]

2. Fisher, R.E.; Siegel, B.A.; Edell, S.L.; Oyesiku, N.M.; Morgenstern, D.E.; Messmann, R.A.; Amato, R.J. Exploratory study of ^{99m}Tc-EC20 imaging for identifying patients with folate receptor-positive solid tumors. *J. Nucl. Med.* **2008**, *49*, 899–906. [CrossRef] [PubMed]

3. Leamon, C.P.; Reddy, J.A. Folate-targeted chemotherapy. *Adv. Drug Deliv. Rev.* **2004**, *56*, 1127–1141. [CrossRef] [PubMed]

4. Teng, L.; Xie, J.; Teng, L.; Lee, R.J. Clinical translation of folate receptor-targeted therapeutics. *Exp. Opin. Drug Deliv.* **2012**, *9*, 901–908. [CrossRef] [PubMed]

5. Assaraf, Y.G.; Leamon, C.P.; Reddy, J.A. The folate receptor as a rational therapeutic target for personalized cancer treatment. *Drug Resist. Update* **2014**, *17*, 89–95. [CrossRef] [PubMed]

6. Betzel, T.; Müller, C.; Groehn, V.; Müller, A.; Reber, J.; Fischer, C.R.; Krämer, S.D.; Schibli, R.; Ametamey, S.M. Radiosynthesis and preclinical evaluation of 3′-aza-2′-[^{18}F]fluorofolic acid: A novel PET radiotracer for folate receptor targeting. *Bioconjug. Chem.* **2013**, *24*, 205–214. [CrossRef] [PubMed]

7. Chen, Q.; Meng, X.; McQuade, P.; Rubins, D.; Lin, S.A.; Zeng, Z.; Haley, H.; Miller, P.; Gonzalez Trotter, D.; Low, P.S. Folate-PEG-NOTA-Al[18]F: A new folate based radiotracer for PET imaging of folate receptor-positive tumors. *Mol. Pharm.* **2017**, *14*, 4353–4361. [CrossRef] [PubMed]

8. Parker, N.; Turk, M.J.; Westrick, E.; Lewis, J.D.; Low, P.S.; Leamon, C.P. Folate receptor expression in carcinomas and normal tissues determined by a quantitative radioligand binding assay. *Anal. Biochem.* **2005**, *338*, 284–293. [CrossRef] [PubMed]

9. Müller, C.; Vlahov, I.R.; Santhapuram, H.K.; Leamon, C.P.; Schibli, R. Tumor targeting using ^{67}Ga-DOTA-Bz-folate—investigations of methods to improve the tissue distribution of radiofolates. *Nucl. Med. Biol.* **2011**, *38*, 715–723. [CrossRef] [PubMed]

10. Müller, C.; Schibli, R. Prospects in folate receptor-targeted radionuclide therapy. *Front. Oncol.* **2013**, *3*, 249. [CrossRef] [PubMed]

11. Curtin, N.J.; Hughes, A.N. Pemetrexed disodium, a novel antifolate with multiple targets. *Lancet Oncol* **2001**, *2*, 298–306. [CrossRef]

12. Müller, C.; Brühlmeier, M.; Schubiger, P.A.; Schibli, R. Effects of antifolate drugs on the cellular uptake of radiofolates in vitro and in vivo. *J. Nucl. Med.* **2006**, *47*, 2057–2064. [PubMed]

13. Müller, C.; Schibli, R.; Forrer, F.; Krenning, E.P.; de Jong, M. Dose-dependent effects of (anti)folate preinjection on ^{99m}Tc-radiofolate uptake in tumors and kidneys. *Nucl. Med. Biol.* **2007**, *34*, 603–608. [CrossRef] [PubMed]

14. Müller, C.; Reddy, J.A.; Leamon, C.P.; Schibli, R. Effects of the antifolates pemetrexed and CB3717 on the tissue distribution of ^{99m}Tc-EC20 in xenografted and syngeneic tumor-bearing mice. *Mol. Pharm.* **2010**, *7*, 597–604. [CrossRef] [PubMed]

15. Müller, C.; Schibli, R.; Krenning, E.P.; de Jong, M. Pemetrexed improves tumor selectivity of ^{111}In-DTPA-folate in mice with folate receptor-positive ovarian cancer. *J. Nucl. Med.* **2008**, *49*, 623–629. [CrossRef] [PubMed]

16. Reber, J.; Struthers, H.; Betzel, T.; Hohn, A.; Schibli, R.; Müller, C. Radioiodinated folic acid conjugates: Evaluation of a valuable concept to improve tumor-to-background contrast. *Mol. Pharm.* **2012**, *9*, 1213–1221. [CrossRef] [PubMed]

17. Fani, M.; Tamma, M.L.; Nicolas, G.P.; Lasri, E.; Medina, C.; Raynal, I.; Port, M.; Weber, W.A.; Maecke, H.R. In vivo imaging of folate receptor positive tumor xenografts using novel ^{68}Ga-NODAGA-folate conjugates. *Mol. Pharm.* **2012**, *9*, 1136–1145. [CrossRef] [PubMed]

18. Reber, J.; Haller, S.; Leamon, C.P.; Müller, C. ^{177}Lu-EC0800 combined with the antifolate pemetrexed: Preclinical pilot study of folate receptor targeted radionuclide tumor therapy. *Mol. Cancer Ther.* **2013**, *12*, 2436–2445. [CrossRef] [PubMed]

19. Bischof, M.; Weber, K.J.; Blatter, J.; Wannenmacher, M.; Latz, D. Interaction of pemetrexed disodium (alimta, multitargeted antifolate) and irradiation in vitro. *Int. J. Rad. Oncol. Biol. Phys.* **2002**, *52*, 1381–1388. [CrossRef]

20. Bischof, M.; Huber, P.; Stoffregen, C.; Wannenmacher, M.; Weber, K.J. Radiosensitization by pemetrexed of human colon carcinoma cells in different cell cycle phases. *Int. J. Rad. Oncol Biol. Phys.* **2003**, *57*, 289–292. [CrossRef]

21. Oleinick, N.L.; Biswas, T.; Patel, R.; Tao, M.; Patel, R.; Weeks, L.; Sharma, N.; Dowlati, A.; Gerson, S.L.; Fu, P.; et al. Radiosensitization of non-small-cell lung cancer cells and xenografts by the interactive effects of pemetrexed and methoxyamine. *Radiother. Oncol.* **2016**, *121*, 335–341. [CrossRef] [PubMed]

22. Müller, C.; Mindt, T.L.; de Jong, M.; Schibli, R. Evaluation of a novel radiofolate in tumour-bearing mice: Promising prospects for folate-based radionuclide therapy. *Eur. J. Nucl. Med. Mol. Imaging* **2009**, *36*, 938–946. [CrossRef] [PubMed]

23. Müller, C.; Struthers, H.; Winiger, C.; Zhernosekov, K.; Schibli, R. DOTA conjugate with an albumin-binding entity enables the first folic acid-targeted ^{177}Lu-radionuclide tumor therapy in mice. *J. Nucl. Med.* **2013**, *54*, 124–131. [CrossRef] [PubMed]

24. Siwowska, K.; Haller, S.; Bortoli, F.; Benešová, M.; Groehn, V.; Bernhardt, P.; Schibli, R.; Müller, C. Preclinical comparison of albumin-binding radiofolates: Impact of linker entities on the in vitro and in vivo properties. *Mol. Pharm.* **2017**, *14*, 523–532. [CrossRef] [PubMed]

25. Siwowska, K.; Schmid, R.M.; Cohrs, S.; Schibli, R.; Müller, C. Folate receptor-positive gynecological cancer cells: In vitro and in vivo characterization. *Pharmaceuticals* **2017**, *10*, 72. [CrossRef] [PubMed]

Sample Availability: Samples of the compounds are not available.

MDPI

St. Alban-Anlage 66

4052 Basel

Switzerland

Tel. +41 61 683 77 34

Fax +41 61 302 89 18

www.mdpi.com

Molecules Editorial Office

E-mail: molecules@mdpi.com

www.mdpi.com/journal/molecules

www.ingramcontent.com/pod-product-compliance
Lightning Source LLC
Chambersburg PA
CBHW051844210326
41597CB00033B/5773